Fact and Fiction in Global

FACT AND FICTION
IN GLOBAL ENERGY POLICY

Fifteen Contentious Questions

Benjamin K. Sovacool, Marilyn A. Brown,
and Scott V. Valentine

Johns Hopkins University Press

Baltimore

© 2016 Johns Hopkins University Press
All rights reserved. Published 2016
Printed in the United States of America on acid-free paper
2   4   6   8   9   7   5   3   1

Johns Hopkins University Press
2715 North Charles Street
Baltimore, Maryland 21218-4363
www.press.jhu.edu

Library of Congress Cataloging-in-Publication Data

Sovacool, Benjamin K.
Fact and fiction in global energy policy : fifteen contentious questions /
Benjamin K. Sovacool, Marilyn A. Brown, and Scott V. Valentine.
    pages cm
Includes bibliographical references and index.
ISBN 978-1-4214-1897-1 (pbk. : alk. paper) — ISBN 978-1-4214-1898-8 (electronic) —
ISBN 1-4214-1897-5 (pbk. : alk. paper) — ISBN 1-4214-1898-3 (electronic)
    1. Energy policy. 2. Energy industries. 3. Energy development.
4. Climatic changes—Environmental aspects. 5. Energy security. I. Brown, Marilyn A.
II. Valentine, Scott V. III. Title.
    HD9502.A2S67625 2016
    333.79—dc23   2015016724

A catalog record for this book is available from the British Library.

*Special discounts are available for bulk purchases of this book. For more information,
please contact Special Sales at 410-516-6936 or specialsales@press.jhu.edu.*

Johns Hopkins University Press uses environmentally friendly book materials,
including recycled text paper that is composed of at least 30 percent post-consumer
waste, whenever possible.

CONTENTS

All three authors thank the editors and reviewers at Johns Hopkins University Press for their helpful suggestions for revision and improvement. We also feel indebted to one of our earlier book projects, *Energy and American Society: Thirteen Myths*, an edited compilation published by Springer in 2007, for starting our contemplation on fact, fiction, and competing ideas in energy studies. Though this book does draw from the earlier one, we want to emphasize at least five things that set this one apart. First, whereas *Myths* was written for upper-level graduate students, our primary audience for this book is more "popular" and includes undergraduate students, lower-level graduate students wanting to learn more about energy, educated citizens, and even energy analysts. Second, *Myths* was an edited volume, so not all of the chapters were of the same quality, tone, or voice, whereas this book is a monograph with only three authors. Third, *Myths* was less comprehensive, lacking, for example, chapters on geoengineering, clean coal, shale gas, peak resources, and nuclear power— all items covered in this book. Fourth, *Myths* is already out of date. Given recent developments such as the Fukushima nuclear accident in Japan and its global ramifications, the shale gas revolution, the plummeting price of solar energy, and the wind power diffusion boom, the dynamics of the energy world have altered sufficiently to require an updated analysis. Lastly, *Myths* centered on the United States, whereas this book is broader and addresses energy concerns from a global perspective.

Benjamin K. Sovacool thanks Ethan, Cooper, Lilei, and Zachary for sacrificing countless evenings and weekends to make this book possible. Marilyn A. Brown thanks her husband and daughter, both of whom work in the energy and environmental field, for the ongoing dialogue on the myths circulating in this area. The Georgia Institute of Technology's Ivan Allen College and the Brook Byers Institute for Sustainable Systems supported Brown's time on this book. Scott V. Valentine wishes to dedicate his contributions to this book to his wife, Rebecca, and his daughter, Elle, who sacrificed family time to shed light on these contentious questions.

BCM	billion cubic meters
BTU	British Thermal Unit
CO_2e	carbon dioxide equivalent
EJ	exajoule (10^{18} joules)
Gb	gigabarrel (10^9 barrels)
Gboe	gigabarrel of oil equivalent
gC	gram of carbon
gCO_2	gram of carbon dioxide
gCO_2e	gram of carbon dioxide equivalent
GJ	gigajoule (10^9 joules)
Gt	gigaton (10^9 tons)
$GtCO_2e$	gigaton of carbon dioxide equivalent
Gtoe	gigaton of oil equivalent
GW	gigawatt (10^9 watts)
GWh	gigawatt-hour
J	joule
kW	kilowatt (10^3 watts)
kWh	kilowatt-hour
Mb	million barrels
Mcf	million cubic feet
MJ	megajoule (10^6 joules)
MMBtu	million BTUs
MT	metric ton
MTOE	metric ton of oil equivalent
MW	megawatt (10^6 watts)
MWh	megawatt-hour
PJ	petajoule (10^{15} joules)
ppm	parts per million
PW	petawatt (10^{15} watts)
PWh	petawatt-hour
tCO_2e	ton of carbon dioxide equivalent
TW	terawatt (10^{12} watts)
TWh	terawatt-hour
W	watt

Fact and Fiction in Global Energy Policy

Framing Global Energy Policy

The test of a first-rate intelligence is the ability to hold two opposing ideas in mind at the same time and still retain the ability to function.

F. Scott Fitzgerald, "The Crack-Up" (1936)

In December 2009, world leaders met in Copenhagen, Denmark, as part of the Fifteenth Conference of the Parties to the United Nations Framework Convention on Climate Change and the Fifth Meeting of the Parties to the Kyoto Protocol. They came together in an attempt to extend the life of the Kyoto Protocol by negotiating binding post-2012 targets to mitigate greenhouse gas emissions and to slow climate change. Instead, the meeting was marked by discontent. European Union leaders referred to it as "disappointing," and non-governmental organizations attacked it for being a "shameful, monumental failure."[1] One newspaper in India called it "a useless gabfest,"[2] and China's largest English newspaper proclaimed it "a dud" and a "triumph of spin over substance."[3]

Apart from the ongoing saga of industrialized and developing countries bickering over whether past, present, or future emitters should be held most responsible for climate change,[4] an unusual dynamic emerged at the conference. A new offensive was initiated by climate skeptics—some who were linked to fossil fuel special interests and some who truly believed that they had discovered new insights that the mainstream scientific community had missed. These well-organized groups were able to "hijack" a significant number of discussions, exacerbating discord.[5] One climate contrarian wore a blindfold to symbolize a belief that climate change mitigation was founded not on sound science but on the desire to profit from interventions. An environmentalist in the audience asked how the flooding of villages in India and Bangladesh caused by rising sea levels would make anybody money. Tom Zeller from the *New York Times* reported that "as debates over global warming often do, the discussion dissolved into incomprehensible shouting."[6]

What in the world happened? How could a vital part of the international process to address climate change devolve into a series of tantrums and screaming matches? Isn't scientific knowledge already reliable enough to convince people that climate change is a train wreck about to occur on a global scale unless actions are taken to address it? If so, couldn't scientific consensus on climate change mitigation help us to identify a suitable strategy for engineering our energy systems to help avert this disaster?

Major Questions and Themes

Discussions concerning energy strategy and policy are no less contentious than those that took place in Copenhagen. This book examines 15 core energy questions, demonstrating how, for each of them, no clear-cut answer exists. The questions cover the spectrum of mainstream energy issues. The first three deal with fundamental aspects of energy and society:

- Is industry the chief energy villain?
- Is energy efficiency a worthwhile investment?
- Should governments intervene in energy markets?

The next five address energy resources and new technologies:

- Do conventional energy resources have a meaningful "peak"?
- Is shale gas a bridge to a clean energy future?
- Can renewable electricity be mainstreamed?
- Is the car of the future electric?
- Can we sustainably feed and fuel the planet?

The following three discuss climate change:

- Is mitigation or adaptation the best way to address climate change?
- Should geoengineering be outlawed?
- Is clean coal an oxymoron?

And the final four analyze energy security and technological transitions:

- Is nuclear energy worth the risk?
- Is national energy independence feasible and desirable?
- Are we nearing a global energy crisis?
- Can energy transitions be expedited?

Many are questions you are likely to have seen in newspapers, on TV, or on the Internet. Some have come up in the classes we teach, others have been mentioned at professional conferences, and others arise from the peer-reviewed academic literature. In embarking on an investigation into beliefs, facts, and fictions related to these 15 questions about global energy systems, our book engages with four key themes that underpin how each chapter was written.

Dialecticism, Not Monologue

Our first theme notionally incorporates the Hegelian Dialectic developed by the philosopher Georg Wilhelm Friedrich Hegel (1770–1831). This is an approach for connecting competing plausible perspectives on an issue (thesis vs. antithesis) by reframing the conceptual lens (a synthesis). The exercise does not necessarily yield an answer, but the process of assessing issues in this way allows analysts to begin to understand the underlying tenets that support competing perspectives.[7]

Applying this to the world of energy, one begins with a thesis: a central idea (e.g., wind energy is great) that contains a degree of incompleteness that gives rise to a conflicting idea, or an antithesis (e.g., wind energy is too expensive, but maybe solar energy is great). As a result of this conflict, a new thesis is born, a synthesis, drawing from the earlier two theses (e.g., wind energy is great when used with solar energy). The synthesis is able to overcome contention by reconciling the truths contained in both the thesis and antithesis. Hegel often described this as "dialectic" because it is a continual process of intellectual development. He described it as "progressive" because each new thesis or synthesis often represents a small advancement over earlier suppositions and has progressed from potentiality and abstraction to actuality and content. Lastly, he noted that the process was intended to be "negative" as it depended on criticism and dispassionate analysis of earlier conjectures that resulted in a somewhat tedious or painful self-reconstruction of ideas.[8]

Hegel proposed that intelligence and logic should follow a "dialectic" in which it is contradiction, this entire process of trying to reconcile competing theses, that serves to advance all meaningful social change. We mimic the Hegelian Dialectic in our presentation of the 15 questions by presenting two opposing views or "sides" to a given question before we attempt to reconcile the competing perspectives in a synthesis, which we label the "common ground." Sometimes this produces a clear-cut preference for one perspective over another, while at other times it simply produces clearer insight into why the two perspectives are incommensurable. In all cases, the process serves as a lively way to learn more about how these 15 important energy questions are conceptualized. We must emphasize that the process of reaching a common ground for our questions is an analytical exercise, not an exercise in truth or personal advocacy, and that the three of us don't necessarily agree with all of the common grounds presented in the book.

Sociotechnical Systems, Not Isolated Technology

Energy systems are far more than technological packages; they are sociotechnical systems. Electricity networks, telecommunications grids, and roads and automobiles are supported by seamless webs of technical artifacts, organizations, institutional rule systems and structures, and cultural values.[9] Viewed as an individual technology, a pipeline is just a physical conduit for oil or gas, but as a sociotechnical system it includes pumping stations, operators, knowledge, financing institutions, investors, land, import and export terminals, oil refineries and natural gas sweetening facilities, energy traders, and consumers. Viewed as an individual technology, a car is just a clunky box with an engine and wheels, but as a sociotechnical system it includes roads, traffic signals, fuel stations and refineries, car manufacturers and dealerships, the maintenance industry, registration offices, insurance companies, and police and legal networks.[10] The actors involved in these sociotechnical systems have vested interests—financial and emotional commitments—that perpetuate them, and when the system is threatened, push-back occurs.

Viewing energy infrastructure as sociotechnical systems acknowledges the broad array of social, political, cultural, environmental, and technical forces that constrain and shape system evolution. Under this broader conceptualization, we come to realize that enacting a technological transition from one energy technology to another is less about facilitating technological change than about overcoming entrenched interests. Nuclear reactors, solar panels, electric vehicles, as we see repeatedly throughout this book, represent more than physical hardware occupying a fixed landscape; they are active manifestations of commercial, political, and social agendas.

Frames, Not Facts

Our 15 contentious questions reveal that conflicts in the domain of energy are not primarily due to one party lacking scientific facts or objective truths—or, at least, not always. Instead, conflicts are often caused by a clash of priorities, interests, and normative assumptions that create a number of contextual truths. At times, the context that gives rise to such truths is the result of incomplete or confusing data. In other cases, contextual truths may form from vested self-interests. In still other cases, fundamentally divergent values may be the source of contextual truths. In our coverage of these 15 questions, we seek to uncover these contextual truths and evaluate their resilience under altered contexts.

In all cases, human beings make decisions based on ideological frames. Frames underpin differing conceptions of reality, and they influence how knowledge is shaped, conditioned, and digested. French sociologist Emile Durkheim,[11] Polish physician Ludwig Fleck,[12] American science historian Thomas Kuhn,[13] American physicist Derek de Sola Price,[14] and American anthropologist Mary Douglas[15] argued that different groups of people interpret reality based on distinct "thought collectives," "paradigms," "worldviews," and "invisible colleges." University of Chicago sociologist Karin Knorr-Cetina coined the term *epistemic cultures* to describe the separate ways that scientific disciplines approach their work; high-energy physicists see the world differently than molecular biologists, for example.[16] This means that the process of cognition itself becomes socialized, which is a fancy way of saying that those bound to a particular frame are often unaware of what underpins their world views.[17]

Let's consider a few examples of how framing affects deliberations about energy. Many who believe in economic growth (which requires increased energy production) might agree with a contention that economic expansion funds improvements in average living standards, engenders upward social and economic mobility, and affords the technological solutions necessary to mitigate any environmental or social damage caused in the process. The underlying ideological frame is that money can fix anything. Many might disagree with this contention by citing examples to demonstrate that money cannot fix everything. They might highlight instances of irreparable environmental damage, extinction of flora and fauna, and threats to human health.[18] The two ideological frames—money can fix all versus money cannot fix all—can be synthesized only through contextual understanding: under certain circumstances, money cannot restore damage done.

Competing frames of this type exist in several areas of energy governance. For instance, should electricity be viewed as a commodity or as a public service? If viewed as a commodity, it might make sense for electricity companies to choose their customers carefully and focus on distributing power only where they can maximize profits, even if this means withholding services in poor and rural areas. If viewed as a public service, then electricity companies have an obligation to supply everyone regardless of cost.[19]

Competing ideological frames also underpin the debate over how energy networks should be structured—large-scale, centralized energy systems capable of serving millions of customers in a single operation, or small-scale, decentralized energy systems serving one household or neighborhood at a time. One frame values centrism, autarchy, and technocracy, whereas the other is antithetical to each.[20]

This book reveals just how ideologically fragmented energy discussions have become and how divergent a given set of frames is for each of our 15 questions. Who should be deemed responsible for the world's energy problems—the companies that supply us with energy fuels and services or the consumers of those services? What's the best way of making energy policy—letting the free market dictate supply and demand or intervening through regulations and standards? When we judge a new energy system or technology, should optimism or precaution govern how we evaluate associated risks? When we conceptualize energy security, should we prioritize national self-sufficiency (producing all of our energy ourselves) or global interconnectedness (depending on others for key energy fuels)? Should we consider environmental security as a subset of energy security or energy security as a subset of environmental security? Tap a shoulder, ask a question—you'll be sure to get a different answer each time.

Understanding, Not Agreement

A final theme that guides the structure of this book is explicit acknowledgment that universal agreement on any of the issues examined is all but impossible: disparate ideological frames preclude such consensus. As the energy historian John G. Clark noted, energy and climate policy is a domain of conflict, not cooperation, and it envelopes nexuses of differing interests.[21] Organized consumer groups contend with organized producer groups to contest energy rates and prices. Business and labor groups in the different energy sectors, such as nuclear power, oil, coal, and natural gas, compete for preferential treatment by governments. Federal, state, and local stakeholders vie over the division of authority.[22] As a result, consensus on energy and climate issues should be neither expected nor sought.

In public forums, debating the issues presented in this book can be frustrating, acerbic, and polarizing; we will see how some of this antagonism and intensity plays out through our presentation of research. However, this does not preclude analysis and discussion. By understanding the frames that undergird contentious perspectives on any energy issues, we can begin to try to identify common ground or, at least, garner reluctant acceptance of the existence of competing perspectives under different contexts. Contention in this manner is not dysfunctional. It is healthy—it keeps the players honest and promotes public education so that people can embrace energy systems that best align with their own values.

Overall, by examining competing perspectives, we hope to provide readers with the information and analytical insights needed to allow them to more effectively distinguish scientific truths from contextual truths. Our goal is to enhance understanding, not enforce agreement.

Preview of Chapters

Although the themes outlined above interconnect the chapters in various ways, the questions fall into five general categories: energy and society, energy resources and technology, climate change, energy security and energy transitions, and synthesis and conclusions.

Energy and Society

Chapter 1, which focuses on who is at fault for energy-related externalities such as air pollution, climate change, and major accidents, is about more than scapegoating. Psychological research suggests that one of the most powerful predictors of an intention to take energy problems seriously—or to change energy-related lifecycles or decisions—is the responsible party that the respondent blames for energy problems.[23] If people believe that their own consumption is wasteful and accept personal responsibility, they are likely to change their attitudes and actions. But if they are able to blame companies, politicians, foreign countries, and other consumers, they will do nothing. The nature of these responsible parties can shape investment decisions, personal behavior, and even trust (or lack of trust) in information about energy and in the institutions regulating or supplying it.[24] And so we ask: is industry really the chief energy villain?

In chapter 2, which focuses on energy efficiency, we tackle the controversial aspects of a topic many consider to be the single best option for addressing electricity and energy dilemmas. Globally, the International Energy Agency reviewed large-scale energy efficiency programs and found that they cost, on average, 3.2 cents per kilowatt-hour (kWh), well below the cost of supplying electricity from any source.[25] Others argue that low investments in energy efficiency reflect high opportunity costs—that capital may be spent more productively elsewhere.[26] How each of us views the potential of energy efficiency can therefore remarkably affect the future security and adaptability of our global energy system. Yet numerous stakeholders remain skeptical that efficiency gains can be realized or hold that interventions can backfire with a "rebound" effect. So, is energy efficiency a worthwhile investment?

The focus of chapter 3 is on markets and whether governments should intervene by setting standards or picking technology winners. Picking winners—strategically targeting technologies for government support—ranges from a light bias in favor of a given technology or group of technologies (such as production tax credits available to wind power developers in the United States prior to 2015) to a heavy bias that radically alters a national energy market (such as the German decision to phase out nuclear power plants by government-financed, expedited diffusion of renewable energy). The two strategies share the same ideological premise, one that dredges up an ongoing debate over the desirability of free markets. Yet, how governments decide to allow this drama to play out will significantly influence what our collective energy future will look like. What is the proper role of governments when it comes to energy markets?

Energy Resources and Technology

Chapter 4 assesses whether the extraction of energy resources such as oil, gas, coal, and uranium will inevitably decline despite improvements in efficiency and advancements in technology. A better understanding of whether these resources have a meaningful "peak" is a critical first step to assessing humanity's future energy options. The question may seem technical and dry, but in fact, as geographer Gavin Bridge writes, "resource depletion . . . is one of a handful of fundamental questions that societies have asked about themselves and their environment down the ages."[27] Are there meaningful peaks, will they occur soon enough to potentially be of concern, and do they matter?

In chapter 5, we investigate the topic of shale gas. Though the practice of horizontal drilling and the process of hydraulic fracturing can be applied to both oil and gas, we focus this chapter entirely on gas because it offers a potential bridge to renewable energy. Moreover, shale gas is the fastest growing source of energy in the United States and a possible boon to major energy-using entities such as China, the European Union, and Russia. Shale gas production is progressing at a dizzying rate across North America, expanding almost twentyfold from 2005 to 2013.[28] The International Energy Agency expects global production from unconventional gas resources to *double* from 2010 to 2035.[29] British Petroleum expects global shale gas production to grow sixfold from 2011 to 2030.[30] John Deutch, Massachusetts Institute of Technology chemist and former US director of Central Intelligence, wrote that the "dramatic increase in estimates of unconventional sources of natural gas" amounts to "the greatest shift in energy-reserve estimates in the last half century."[31] The usually downbeat *Economist* magazine was even pithier, calling shale gas "an unconventional bonanza" and concluding that it was "transforming the world's energy markets."[32] As the International Energy Agency put it, "natural gas is poised to enter a golden age."[33] But is shale gas a boon or bane for local communities, the climate, and global energy security?

Chapter 6 takes a close look at renewable sources of electricity such as wind, solar, geothermal, hydropower, and bioenergy. There is the conviction in all too many policy circles that conventional energy is still a necessary evil. Thomas Petersik, a former analyst for the US Energy Information Administration, expressed these concerns succinctly by stating that "by and large, renewable energy resources are too rare, too diffuse, too distant, too uncertain, and too ill-timed to provide significant supplies at the times and places of need."[34] James Hansen, the former director of NASA's Goddard Institute and staunch advocate for expedient mitigation to avert the worst perils attributed to advanced climate change, argues that nuclear power is a necessary technology in a carbon-constrained world because renewable energy is not ready for prime time.[35] Advocates of renewables retort that many renewable energy technologies are already commercially competitive with conventional energy technologies and that the attractiveness of these technologies will increase as renewable energy prices drop (thanks to enhanced economies of scale and progressive technological advances) and conventional energy prices rise (due to inflated fuel stock prices and diminishing technological progress). Which side is right: is renewable electricity destined to become a major-league player or will it be confined to serving minor-league niche markets?

Chapter 7 delves into the realm of transport and mobility. Despite the rise of the car, most of the world's population rarely travels in motorized vehicles—a small minority (10%) living in industrialized countries account for about 80% of total motorized travel.[36] As incomes rise and transportation infrastructure expands in underdeveloped countries, the demand for passenger and freight transportation is expected to grow rapidly, providing expanded access to markets, education, and health services.[37] In 2010, 94% of global transport energy demand was supported by petroleum products, reflecting the fact that internal combustion engines rule the roost.[38] Is there a better alternative—is the car of the future an electric vehicle?

In the discussion of biofuel in chapter 8, we assess perhaps the most controversial aspect of their production and use: whether they can be produced and used sustainably with minimal negative impact on the environment and food supply. Opponents point out that first-generation biofuels can replace only a small sliver of global petroleum supply, meaning that they cannot truly eliminate dependence on fossil fuels. Even then, expanding production of biofuels, usually through a few select agricultural monocultures dependent on pesticides and fertilizers, poses threats to the global climate and to land and water resources and even adversely affects food prices and global food security. Cornell scientist David Pimentel argues that "ethanol just isn't efficient and people will be left hungry if we use our croplands to grow fuel instead of food. It's just a bad idea."[39] Opinions from sponsors of biofuel are just as emotive. They argue that, if well managed, biofuel has the capacity to displace fossil fuels, or at least oil imports from unstable geopolitical regimes. Biofuel can also be produced relatively quickly compared with fossil fuels, which take millennia to form; biofuel can be obtained in a matter of days and in solid, liquid, and gaseous form.[40] In addition, biofuel industries can prop up rural economies, improve the efficiencies of agricultural development, and mitigate other forms of air pollution attributed to fossil fuel combustion, such as acid rain. Investment analyst Aaron Levitt calls ethanol "the fuel of the future,"[41] and *National Geographic* opines that "on the face of it, biofuels look like a great solution."[42] Indeed, can biomass sustainably feed and fuel the planet?

Climate Change

Chapters 9 and 10 tackle head-on the strategies for addressing climate change, a phenomenon that former World Bank chief economist Nicholas Stern labeled the greatest market failure in the history of humanity.[43] Generally, researchers have divided human responses to climate change into one of three categories:

1. Mitigation: an anthropogenic intervention to reduce the sources or enhance the sinks of greenhouse gases

2. Adaptation: adjustment in natural or human systems in response to actual or expected climatic stimuli or their effects, which moderates harm or exploits beneficial opportunities
3. Geoengineering: a broad set of methods and technologies that aim to deliberately alter the climate system to alleviate the impacts of climate change

As figure I.1 indicates, mitigation and geoengineering focus on "avoiding the unmanageable," whereas adaptation focuses on "managing the unavoidable." Chapters 9 and 10 discuss the trials and tribulations associated with these strategies. Proponents argue that mitigation not only directly addresses climate change by lowering the very pollutants that cause it but also provides other benefits such as cleaner air and improved energy security. Those subscribing to this view argue that the technological tools to allow mitigation already exist. All that is needed is the right policy prescription to thwart the attempts of special interests that are preserving a perilous status quo for self-serving reasons. Articulators in favor of adaptation point to a number of hurdles that stymie agreement on mitigation: it is prohibitively expensive, politically untenable, prone to free-riders and leakage, and unlikely to be affirmed soon by any type of international accord.[44] As one study put it succinctly, "it is useful to adapt even if nobody else does, but mitigation is meaningless unless it is as part of collective global effort."[45] A third view is that it's scientifically too late to mitigate or adapt—we've missed our opportunity, and no matter what we do, emissions already in the atmosphere will catalyze significant climate change and wreak significant social, economic, and environmental damage. Geoengineering is thus the only option that makes sense to these advocates. Are they right?

In chapter 11, we address the promise of clean coal—of systems that would capture and sequester carbon dioxide to make coal environmentally benign. This route would have us safely sink carbon back deep into the earth where it would no longer harm our atmosphere. The chapter thus cuts to the heart of our predicament about climate change: whether we can continue to support status quo practices, a highly attractive idea to stakeholders. For consumers, a future without material sacrifice is eminently attractive: a world of unfettered access to air conditioning, cold beer, private automobiles, and the Internet of Things—lifestyles where all of our needs and wants can be almost instantaneously met. Technologies such as clean coal are to energy firms what electronic nicotine-delivery systems are to cigarette manufacturers. They allow nations dependent on coal to continue to take full advantage of these resources, foster continued economic development and prosperity, and avoid negative environmental repercussions—all in one fell swoop. Opponents respond that hidden costs to clean coal make the phrase a contradiction in terms when one factors

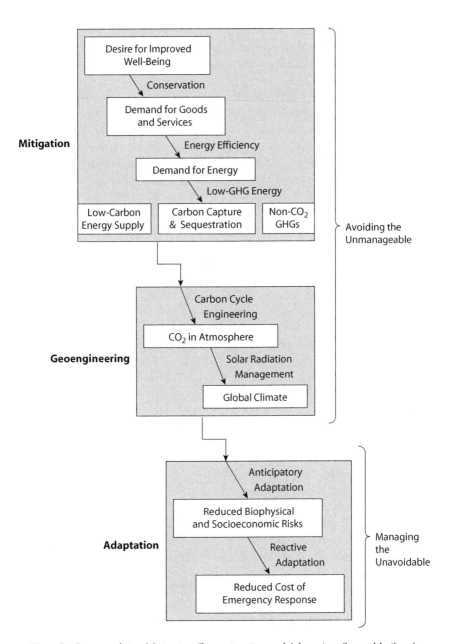

Figure I.1. Conceptualizing Mitigation, Geoengineering, and Adaptation. *Source:* Marilyn A. Brown, "The Multiple Dimensions of Carbon Management: Mitigation, Adaptation, and Geo-engineering," *Carbon Management* 1, no. 1 (2010): 27–33. *Note:* GHG = greenhouse gas.

in environmental and social damage from coal mining and processing, other pollution streams from coal generation, and the liability issues that arise from long-term storage repositories for carbon. So this chapter asks: is clean coal an oxymoron?

Energy Security and Energy Transitions

The discussion of nuclear power in chapter 12 looks at the risks involved in "Pandora's promise": nuclear energy. Nuclear reactors have become a significant source of electricity worldwide. In June 2014, 434 nuclear plants supplied 11% of the world's power, constituting 374.6 gigawatts (GW) of installed capacity generating 2,359 terawatt-hours (TWh) of electricity.[46] In the United States alone, which has 23% of the world's reactors, nuclear facilities account for 19% of national electricity generation. In France, 75% of electricity comes from nuclear sources, and nuclear energy contributes to more than 20% of national power production in Germany, South Korea, Sweden, Ukraine, and the United Kingdom. But are nuclear plants worth the risk, especially when one factors in the possibility of severe accidents, construction cost overruns, nuclear weapons proliferation, dependence on subsidies, and environmental damage at uranium mines?

We examine national energy independence in chapter 13, taking a hard look at the rationale for self-sufficiency stemming from the oil crises of the 1970s. Readers who are longer in the tooth may remember the long lines for gasoline and the damage inflicted on domestic economies as a result of OPEC (Organization of Petroleum Exporting Countries)–led supply restrictions and the bungled US policy responses that created bottlenecks. Indeed, when one reviews the history of energy-mix transitions in most countries, the 1970s catalyzed an exodus away from oil dependence that endures to this day. However, an opposing view holds that countries are energy interdependent. This interdependency arises in large part from the trading of energy fuels in the global marketplace, such as oil, gas, coal, and uranium; investment flows and consortiums for major energy projects; the institutions that govern global energy use; and shared responsibilities in attending to energy-related pollution. So, given these interdependencies, is energy independence feasible or desirable?

Chapter 14 highlights how the more dependent we become on energy systems, the more vulnerable we are to risks of technological failure and energy supply disruptions. The United States and its allies are currently mired in military activities in the Middle East that are largely deemed necessary to preserve access to oil.[47] Japanese leaders are struggling to keep their economy from imploding under a debt load that has been exacerbated by the Fukushima nuclear disaster. In Beijing, some commuters have taken to covering their heads with plastic bags to protect themselves from the particulate matter fallout associated with the city's coal-fired power plants. Meanwhile, in December 2015, delegates from all United Nations member states will converge in Paris for another round of negotiations to try to mitigate climate change. Clearly, the global need for energy has produced some vulnerabilities that lead us to ask: are we heading toward a full-blown crisis?

Chapter 15 analyzes energy transitions, a term that can refer to "a particularly significant set of changes to the patterns of energy use in a society, potentially affecting resources, carriers, converters, and services."[48] In other words, an energy transition refers to the time that elapses between the introduction of a new primary energy source, or prime mover, and its rise to claiming a substantial share of the overall market. Forces for a transition are indeed afoot. However, if we are to avoid the worst perils attributed to climate change, expedience is vital. Table I.1 provides an overview of cases (presented in chapter 15) where fast transitions have occurred. They collectively involved more than 965 million people. Although these data certainly support the contention that transitions can take place in isolated technologies or at smaller scales, could we reasonably expect a global transition and, if so, will it be quick or slow?

Table I.1. Overview of Rapid Energy Transitions

Country	Technology/fuel	Period of transition	Number of years from 1% to 25% market share	Approximate size of population affected (millions of people)
Sweden	Energy-efficient ballasts	1991–2000	7	2.3
China	Improved cookstoves	1983–1998	8	592
Indonesia	Liquefied petroleum gas stoves	2007–2010	3	216
Brazil	Flex-fuel vehicles	2004–2009	1	2
United States	Air conditioning	1947–1970	16	52.8
Kuwait	Crude oil	1946–1955	2	0.28
Netherlands	Natural gas	1959–1971	10	11.5
France	Nuclear electricity	1974–1982	11	72.8
Denmark	Combined heat and power	1976–1981	3	5.1
Canada (Ontario)*	Coal	2003–2014	11	13

*The Ontario case study is the inverse, showing how quickly a province went from 25% coal generation to zero.

Synthesis and Conclusions

The final chapter explains why and how contention manifests itself across our 15 questions. It identifies at least six likely causes behind contention. (1) Energy is big money, and strong vested interests exist in support of each major energy system. (2) Stakeholders often base their support on data and technology projections that are obsolete or out-of-date. (3) Differing interpretations of hazards and their implications can convince people to see energy systems in starkly different ways. (4) Energy systems can exclude or marginalize people from the decision-making process, causing resentment. (5) Distinct systems of values and beliefs can engender competing criteria for assessment. (6) Energy technologies can become supported dogmatically or zealously as a matter of religious or political faith—where advocates downgrade or simply ignore opposing information.

The conclusion finishes by offering six maxims that readers can follow to bring far better understanding to the analysis of energy problems. (1) To reveal competing interests, we need to understand where the power lies and how it manifests itself in energy decisions. (2) To counter the rapidity of change, we need to keep up-to-date and educate ourselves about energy technologies and issues. (3) To be more prudent about risk and uncertainty, we should attempt to make energy decisions that are more informed by an understanding of both science and society, including sound ethical principles. (4) To avoid undemocratic exclusion, we need to remember that energy decisions must meet the needs of a broad spectrum of citizens and stakeholders. (5) To reveal underlying ideologies, we need to strive to become aware of our own ideological frames that might prohibit a balanced analysis. (6) To counter energy evangelism, we need to look beyond a given energy technology and recognize that many systems can deliver the same solution.

In addressing both sides to these 15 questions and offering some conclusions, this book delves into many topics that, if not for lack of space, could have been complete questions in their own right. For instance, we can think of at least five other questions that deserve deeper treatment than the analysis presented here:

1. Are energy subsidies good for the economy?
2. Can the "smart grid" fulfill its potential?
3. Is a hydrogen economy likely?
4. Is rural electrification the best way to expand access to modern energy services?
5. Is there such a thing as a "resource curse" surrounding oil and gas?

Luckily for those interested in these questions, such topics are addressed in the book, albeit tangentially. The issue of subsidies pops up in no less than seven chapters. Arguments involving the "smart grid" and "smart" appliances or homes arise in six chapters. Hydrogen as an energy source is mentioned in four chapters. The topic of energy poverty is addressed in chapter 12 on nuclear power. Reference to the "resource curse" occurs in two chapters. The goal of choosing our 15 questions was not to be exhaustive but to select a sample of themes and technologies that enable readers to better comprehend the political, economic, technical, and social dimensions of our global energy system.

Basic Terminology

Before we further engage and explore these themes, it is of value to lay out, briefly, a few of the terms, units, concepts, and phrases employed throughout the book. (A list of abbreviations for units is included at the front of the book.)

Energy Terms and Units

When most of us discuss energy production and consumption, our statements are usually inaccurate since energy itself can be neither created nor destroyed. We do not supply or consume energy; we use different forms of it. When we discuss *energy efficiency*, we're really talking about "fuel efficiency" or "energy performance," and when we use the term *energy production*, we mean "conversion of energy carriers." Although talking about energy as if it could be produced and consumed tends to distort our understanding of what we actually "do" with it,[49] we adhere to this popular nomenclature in this book because it offers a useful shorthand widely used by analysts currently talking about energy and electricity policy. In short, it is less awkward and more colloquial.

Many common terms and units of energy measurement exist, and in this book we refer to a number of them. The terms energy "supply" and energy "demand" can encompass an array of different things. We usually use *supply* to refer to sources of energy such as power plants, refineries, and coal mines, as well as delivery mechanisms such as transmission lines or tankers. We use *demand* to refer to where energy is consumed and put to use, whether it is by households, factories, commercial enterprises, or automobiles.

Three of the most basic energy units are the joule, the British Thermal Unit (BTU), and the calorie. One joule is the energy required to light a one-watt bulb for one second. One BTU is the equivalent of about 1,055 joules. It is the amount of heat necessary to raise the temperature of one pound of water by one degree Fahrenheit (°F). A calorie is the energy necessary to raise the temperature of one gram of water by one degree Celsius (or centigrade; °C). BTUs and calories are more closely related, with 1,000 calories equivalent to 3.97 BTUs. To convert back and forth between units: one terajoule, or one trillion (10^{12}) joules, is equivalent to:

2.388×10^{11} calories

23.88 metric tons of oil equivalent (MTOE)

0.948 billion $\times 10^9$) BTUs

277,800 kilowatt-hours (kWh)

Throughout the book, we often switch between these units, though as often as possible we also try to contextualize the numbers with interesting facts or anecdotes.

Externalities

One term that surfaces frequently in the book is *externality*. In economics, an externality is a cost (or benefit) incurred as a result of an economic activity that is not incorporated into the price of the good or service.[50] Externalities can be both "positive" and "negative." A positive externality would be when an apple grower's orchards happen to be adjacent to a bee grower's farm, and an increase in activities of the apple farmer increases the amount of apple blossoms, in turn increasing the production of honey at the bee farm. The apple farmer provides the bee-keeper with additional benefits at no charge. A negative externality would be when a poor widow supporting herself by hand-laundry hangs clothes out to dry next to a factory that emits smoke that blackens it.[51] The factory pollutes the laundry without compensating our poor widow. The term "externality" highlights that it is not always those directly engaged in an economic activity that bear the costs or reap the benefits.

Consider a simple example of the importance of accounting for externalities. Imagine that you intend to travel from Seattle to Tokyo and receive a ticket quote from two airlines. You can fly either on Goldstar Airlines for $1,000 or on Misery Airlines for $900. The lower ticket quote from Misery Airlines attracts you, but the name raises some suspicions and you decide to investigate further. To your alarm, you discover that 10% of all Misery Airlines flights end in some sort of accident. Moreover, whereas Goldstar flies direct from Seattle to Tokyo taking 12 hours, the flight with Misery Airlines necessitates transfers in Anchorage, Alaska, and Vladivostok, Russia, requiring a grand total of 72 hours to get to Tokyo. These two new pieces of information represent externalities in that, for most people, the elevated safety risk and the much greater travel time (72 hours vs. 12 hours) are not adequately factored into the ticket price. For most people, making a purchase decision based only on ticket price would not yield an optimal outcome.

This phenomenon of unaccounted-for negative externalities—"failure to internalize externalities," in economic vernacular—presents grave problems when valuing energy technologies. As was the case with the airplane ticket, failure to consider all costs (and benefits) when calculating the price of electricity generation (or some other energy service) from a given technology produces a market distortion that biases the decision-making process and leads to sub-optimal decisions. We'll see this conundrum pop up in multiple chapters.

Climate Change

In this book, none of our contentious questions center on whether climate change is real. We treat that particular issue as an implacable, immutable reality. We don't want to overwhelm readers with much of the complex science, but we do believe that some background is needed to minimally understand why there is pressure to transition away from fossil fuel energy sources. In this section, we attempt to describe greenhouse gases (GHGs), radiative forcing, and atmospheric concentration. We also briefly review the consequences of GHG accumulation and where GHG emissions come from (by sector and country).

How does one define a greenhouse gas? Many chemical compounds found in the earth's atmosphere act as GHGs, including water vapor, carbon dioxide (CO_2), methane (CH_4), and nitrous oxide (N_2O). These gases play a vital role in warming our planet. Solar energy, which passes through our atmosphere, is absorbed by the earth and radiated back toward space as infrared radiation. The difference in the wavelengths of solar energy coming in (short-wave) and solar energy radiated back (long-wave) is an important factor underpinning the greenhouse effect because GHG molecules are chemically constituted to absorb long-wavelength energy—the radiated heat. When the radiated heat is absorbed, the atmosphere heats up, but this is not necessarily a bad thing: without the natural greenhouse effect, our planet's average surface temperature would be below the freezing point of water. The trouble is that human activities also emit GHGs, and the combination of natural and anthropogenic GHG emissions is causing too much heat to accumulate in the atmosphere, triggering global warming. In 1997, the Kyoto Protocol targeted six greenhouse gases associated with economic activities that need to be reduced: CO_2, CH_4, and N_2O along with three types of fluorine-containing gases (F-gases): hydrofluorocarbons (HFCs), perfluorocarbons (PFCs), and sulfur hexafluoride (SF_6). This list was expanded to include nitrogen trifluoride (NF_3) in 2012 due to its growing utilization in plasma screens and thin-film solar panels.

The potential for warming presented by a GHG depends on both its characteristics and its atmospheric concentration. On a molecular level, methane possesses greater warming potential than carbon dioxide, but carbon dioxide plays a far more influential role in warming the atmosphere because of its much higher concentration. *Radiative forcing* is a term used to quantify the effect of increased concentrations of GHGs on the climate. It is essentially a measure of how much of the incoming solar energy remains in the earth's atmosphere. This is measured in watts per square meter and is compared with

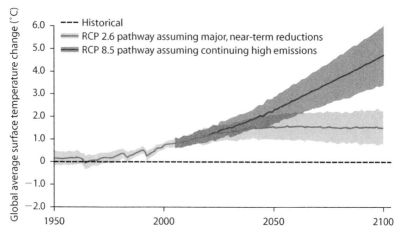

Figure I.2. Historical and Projected Global Temperature Change, 1950–2100. *Source:* Modified from Intergovernmental Panel on Climate Change (IPCC), *Mitigation of Climate Change* (Geneva: IPCC, 2014). *Note:* RCP = representative concentration pathway, based on incremental total radiative forcing in 2100 relative to 1850.

preindustrial levels to gauge global warming potential. According to the latest Intergovernmental Panel on Climate Change (IPCC) assessment, CO_2 contributes 76% to the total global stock of GHG emissions based on weightings with 100-year global warming potentials, CH_4 is second with 16%, N_2O contributes about 6%, and the combined F-gases are responsible for about 2%.

At the end of 2013, the earth's atmosphere had a CO_2 concentration of roughly 400 parts per million (ppm), and many climatologists have argued that 450 ppm is the absolute threshold that cannot be passed without serious peril. The atmospheric concentration of carbon dioxide is growing at 2.5 ppm per year, and at this rate, concentrations will surpass 750 ppm by the end of the century.

What are the consequences? If efforts to limit GHG emissions remain stalled, global mean temperatures are likely to rise by 3.5°C to 5.5°C by the end of the century relative to 1986–2005. Even if emissions are brought under control over the next few decades, as represented by the bottom trend from the IPCC in figure I.2, global mean temperatures are likely to rise by nearly 1.5°C.[52] The IPCC has defined representative concentration pathways (RCPs) to span the range of possible global warming scenarios, based on different magnitudes of incremental total radiative forcing in 2100 relative to 1850. The lowest level of global warming is represented by RCP2.6, which is the outcome of an incremental radiative forcing of 2.6 W/m² of the earth's surface; the highest is RCP8.5, which has an incremental radiative forcing of 8.5 W/m².[53] The IPCC uses these alternative pathways for evaluating climate change effects, mitigation options, and adaptation measures. Most "business as usual" scenarios in the literature are between RCP6.0 and RCP8.5.

Where do the GHG emissions, which give rise to this threat, come from? The largest sources are energy supply and use (about 50%), forestry and land-use change (18%), agriculture (13%), transportation (13%), and waste and water treatment (3%). In the energy sector, GHG emissions from electricity generation dominate global trends, tripling since 1970. Over the same period, emissions from transportation have also doubled. In terms of geographical concentration, 20 countries account for 75% of global GHG emissions, and five countries account for half: China (the largest emitter), the United States (second largest), Russia, India, and Japan.[54]

Into the Breach

With this book, readers can embark on an odyssey of inquiry into contemporary energy issues. Plausible arguments exist for supporting and opposing views for all 15 questions. This makes the global energy landscape a contested, ideological battlefield, where making almost *any* claim in support of one perspective faces criticism by somebody else. Hopefully, by the end of this book, the reader will understand that the energy world is not black and white; it is a gray zone inhabited by perspectives based on controversial analysis and competing frames. By focusing on "contending" perspectives, we highlight that energy decisions are predicated on beliefs that may or may not be supported by objective data or comprehensive analysis, constantly blurring the line between fact and fiction. Hold on to your hat. Things are about to get ugly.

NOTES

1. "Copenhagen Failure 'Disappointing,' 'Shameful,'" *EU Observer*, December 20, 2009.

2. Nitin Sethi, "After Five Days, Nothing to Show at Climate Meet," *Times of India*, December 12, 2009, 1.

3. "China Defends Bottom Line, But World Agrees Deal Is a Dud," *Sunday Morning Post* (Hong Kong), December 20, 2009, 7.

4. J. Reilly et al., "Multi-Gas Assessment of the Kyoto Protocol," *Nature* 401 (1999): 549–555; David Victor, *The Collapse of the Kyoto Protocol and the Struggle to Slow Global Warming* (Princeton: Princeton University Press, 2001); David G. Victor, "Toward Effective International Cooperation on Climate Change: Numbers, Interests, and Institutions," *Global Environmental Politics* 6, no. 3 (2006): 90–105.

5. Martin Khor, "Copenhagen Climate Summit Ends in Discord," *Third World Resurgence*, no. 233 (January 2010): 9–14.

6. Tom Zeller, "And in This Corner, Climate Contrarians," *New York Times*, December 9, 2009.

7. G. W. F. Hegel, *Science of Logic*, trans. A. V. Miller (New York: Humanities Press, 1969); G. W. F. Hegel, *Hegel's Logic* (Oxford: Oxford University Press, 1975).

8. See Quentin Lauer, *Essays in Hegelian Dialectic* (Bronx, NY: Fordham University Press, 1977); D. Berthold-Bond, *Hegel's Grand Synthesis: A Study of Being, Thought, and History* (New York: Harper, 1993).

9. Thomas P. Hughes, *Networks of Power: Electrification in Western Society, 1880–1930* (Baltimore: Johns Hopkins University Press, 1983).

10. See B. K. Sovacool, "What Are We Doing Here? Analyzing Fifteen Years of Energy Scholarship and Proposing a Social Science Research Agenda," *Energy Research and Social Science* 1, no. 1 (2014); B. K. Sovacool, "The Interpretive Flexibility of Oil and Gas Pipelines: Case Studies from Southeast Asia and the Caspian Sea," *Technological Forecasting and Social Change* 78, no. 4 (May 2011): 610–620.

11. Emile Durkheim, *Education and Sociology* (New York: Free Press, 1922).

12. Ludwig Fleck, *Genesis and Development of a Scientific Fact* (Chicago: University of Chicago Press, 1932).

13. Thomas S. Kuhn, *The Structure of Scientific Revolutions* (Chicago: University of Chicago Press, 1962); Thomas S. Kuhn, *The Essential Tension: Selected Studies in Scientific Tradition and Change* (Chicago: University of Chicago Press, 1977).

14. Derek de Sola Price and Donald Beaver, "Collaboration in an Invisible College," *American Psychologist* 21 (1966): 1011–1018.

15. Mary Douglas, *How Institutions Think* (Syracuse, NY: Syracuse University Press, 1986).

16. Karin Knorr-Cetina, *Epistemic Cultures: How the Sciences Make Knowledge* (Cambridge, MA: Harvard University Press, 1999).

17. Ronald N. Giere, "Distributed Cognition in Epistemic Cultures," *Philosophy of Science* 69, no. 4 (December 2002): 637–644.

18. Sam H. Schurr et al., *Energy in America's Future: The Choices before Us* (Baltimore: Johns Hopkins University Press, 1979).

19. Walt Patterson, *Can Public Service Survive the Market? Issues for Liberalized Electricity*, Chatham House Briefing Paper (London: Chatham House, 1999).

20. Amory Lovins, "Soft Energy Technologies," *Annual Review of Energy* 3 (1978): 508.

21. John G. Clark, *The Political Economy of World Energy: A Twentieth Century Perspective* (Charlotte: University of North Carolina Press, 1990).

22. John A. Yager, "Energy in America's Future: The Difficult Transition," in *Energy Policy in Perspective* (Washington, DC: Brookings Institution, 1981), 637–664.

23. Glenn Shippee, "Energy Consumption and Conservation Psychology: A Review and Conceptual Analysis," *Environmental Management* 4, no. 4 (1980): 297–314.

24. Michael R. Greenberg, "Energy Policy and Research: The Underappreciation of Trust," *Energy Research and Social Science* 1 (March 2014): 152–160.

25. International Energy Agency (IEA), *The Experience with Energy Efficiency Policies and Programs in IEA Countries: Learning from the Critics* (Paris: IEA, August 2005).

26. L. J. Makovich, *The Cost of Energy Efficiency Investments: The Leading Edge of Carbon Abatement* (Cambridge, MA: CERA, 2008).

27. Gavin Bridge, "Geographies of Peak Oil: The Other Carbon Problem," *Geoforum* 41 (2010): 524.

28. Jeffrey Logan et al., *Natural Gas and the Transformation of the U.S. Energy Sector: Electricity*, NREL/TP-6A50-55538 (Golden, CO: Joint Institute for Strategic Energy Analysis, November 2012).

29. IEA, *World Energy Outlook 2012* (Paris: OECD, 2012).

30. J. David Hughes, "A Reality Check on the Shale Revolution," *Nature* 494 (February 21, 2013): 308.

31. John Deutch, "The Good News about Gas: The Natural Gas Revolution and Its Consequences," *Foreign Affairs* 90 (2011): 88.

32. "Special Report: Natural Gas," *Economist*, July 14, 2012.

33. IEA, *Golden Rules for a Golden Age of Gas* (Paris: OECD, November 2012), iv.

34. Petersik quoted in B. K. Sovacool, "The Intermittency of Wind, Solar, and Renewable Electricity Generators: Technical Barrier or Rhetorical Excuse?" *Utilities Policy* 17, no. 3 (September 2009): 288.

35. P. A. Kharecha and J. E. Hansen, "Prevented Mortality and Greenhouse Gas Emissions from Historical and Projected Nuclear Power," *Environmental Science and Technology* 47, no. 9 (2013): 4889–4895.

36. Intergovernmental Panel on Climate Change (IPCC), *Mitigation of Climate Change* (Oxford: Oxford University Press, 2014).

37. Africa Union, *Transport and the Millennium Development Goals in Africa* (Addis Ababa, Ethiopia: UN Economic Commission for Africa, 2009); V. Pendakur, "Non-motorized Urban Transport as Neglected Modes," in *Urban Transport in the Developing World*, ed. H. Dimitriou and R. Gakenheimer (Cheltenham, UK: Edward Elgar, 2011).

38. IEA, *World Energy Outlook 2012* (Paris: OECD, 2012), available at www.iea.org.

39. Pimentel quoted in Sam Jaffe, "Independence Way," *Washington Monthly*, July/August 2004, 34.

40. V. Shunmugam, "Biofuel: Breaking the Myth of 'Indestructible Energy'?" *Margin: Journal of Applied Economic Research* 3 (2009): 173–189.

41. Aaron Levitt, "Cellulosic Ethanol: The Fuel of the Future?" *Investor Place*, March 29, 2012.

42. "Biofuels: The Original Car Fuel," *National Geographic*, 2014, available at http://environment.nationalgeographic.com/environment/global-warming/biofuel-profile.

43. Nicholas Stern, *The Stern Review: Report on the Economics of Climate Change* (London: Cabinet Office, Her Majesty's Treasury, 2006).

44. David Archer and Stefan Rahmstorf, *The Climate Crisis: An Introductory Guide to Climate Change* (Cambridge: Cambridge University Press, 2010).

45. A. Aaheim et al., "National Responsibilities for Adaptation Strategies: Lessons from Four Modeling Frameworks," in *Making Climate Change Work for Us: European Perspectives on Adaptation and Mitigation Strategies*, ed. Mike Hulme and Henry Neufeldt (Cambridge: Cambridge University Press, 2010), 88.

46. World Nuclear Association, "World Nuclear Power Reactors & Uranium Requirements" (June 1, 2014), available at www.world-nuclear.org/info/Facts-and-Figures/World-Nuclear-Power-Reactors-and-Uranium-Requirements.

47. K. M. Campbell and J. Price, eds., *The Global Politics of Energy* (Washington, DC: Aspen Institute, 2008).

48. Peter A. O'Connor, "Energy Transitions," *Pardee Papers*, no. 12 (November 2010): 1.

49. Walt Patterson, *Keeping the Lights On: Towards Sustainable Electricity* (London: Earthscan, 2007).

50. K. E. Case and R. C. Fair, *Principles of Economics*, 5th ed. (Upper Saddle River, NJ: Prentice Hall International, 1999).

51. US Energy Information Administration, *Electricity Generation and Environmental Externalities: Case Studies* (Washington, DC: US Department of Energy, September 1995).

52. IPCC, "Summary for Policymakers," in *Climate Change 2013: The Physical Science Basis*, Contribution of Working Group I to the Fifth Assessment Report of the Intergovernmental Panel on Climate Change, ed. T. F. Stocker and D. Qin (Geneva: IPCC, 2013).

53. IPCC, *Synthesis Report* (Geneva: IPCC, 2014).

54. IPCC, *Mitigation of Climate Change* (Geneva: IPCC, 2014).

I. ENERGY AND SOCIETY

Is Industry the Chief Energy Villain?

On April 20, 2010, an explosion occurred at British Petroleum's *Deepwater Horizon*, which was drilling exploratory wells in the Tiber Oil Field in the Gulf of Mexico. The result was a fire that burned for 36 hours before the entire rig went down like the *Titanic*. Eleven crew members died, a massive oil spill ensued until engineers sealed the well five months later, and $41 billion (at latest count) in damages, litigation, and cleanup costs were incurred.[1]

Within a few hours of the accident, critics pointed accusatory fingers at all the usual suspects: the now-disbanded Minerals Management Service, for giving BP a pass on routine inspections and lapsing into a relationship with the oil industry that US President Barack Obama denounced as "cozy"; President Obama himself, for failing to enact the reforms within the Interior Department that he had promised while campaigning for election; the oil services firm Transocean, for failing to service a faulty blowout preventer; and, of course, BP, for a "reckless" safety culture. After weeks of swirling public wrath, responsibility eventually landed on BP's shoulders; CEO Tony Hayward was eventually fired.[2]

But was BP really the king of the skunks? Another line of thought holds that while there is no denying that the public was largely aggrieved by this accident—the residents of the Gulf states being among the worst affected— the American public was also largely to blame. A chronic addiction to cheap gasoline, perceptions of energy reform as "un-American," and blind acquiescence to propaganda from the special interests that keep America locked into fossil fuel dependence have given rise to an obsession with drilling, damming, and digging the country's way out of problems. Support for enhanced fossil fuel discovery efforts (increasing access to oil and gas) obscured cheaper and possibly more beneficial actions on the demand side (such as cutting energy consumption by changing people's travel behavior or improving the fuel economy of cars and trucks). To support the illusion that lifestyles revolving around cheap oil and big cars were America's God-given right, oil companies were forced to bridge the supply-demand gap through increasingly sophisticated and risky techniques such as three-dimensional seismic imaging and tube-rotary drilling. On such a scale, accidents are par for the course. Under this rationale, if Americans want a future without other *Deepwater Horizon*–style accidents and oil-tarred wildlife, they must wean themselves off oil. The petroleum companies are merely responding to consumer demand.

These two competing narratives regarding culpability over the *Deepwater Horizon* spill highlight two different ideological frames. When most people think about energy, they conjure up images of power plants, transmission lines, and refineries and the companies that own and operate them. When they think of pollution, they think of either a factory's smoke stack belching soot or an oil spill damaging baby seals. Within this ideological frame, the energy companies' executives in their blue pin-striped suits are the villains. They profit at the expense of others, and their nefarious activities sow despair and dissent around the world.

Yet there is an alternative perspective. In the absence of consumer demand, these perceived villains would merely be businesspeople in search of jobs. It is decisions made by individuals that matter most. Individuals decide what types of automobiles to buy, and they make purchase and use decisions that dictate individual energy consumption profiles. Moreover, in most nations, individuals vote for the people who set the standards for energy provision. They collectively hold the power and make these decisions based on their own free wills. How can energy companies be responsible for generating pollution that emerges only in the process of satisfying consumer appetites?

One Side: The Energy Industry Is Responsible for Most Pollution

Proponents of the blame-the-industry thesis argue that energy suppliers and producers—stewards of the energy systems that populate the modern world—are most responsible for pollution. The ubiquity of energy supply systems is demonstrated by the sheer scale of energy infrastructure, the enormity of global energy markets, and the power of national and multinational energy companies and those that lead them. The not-so-subtle implication is that "if capital is the lifeblood of the global economy, infrastructure is its beating heart."[3] Those who control this infrastructure, the logic runs, are responsible for it and its negative (and positive) implications for society.

The Energy Industry Dominates the Globe with Its Infrastructure

During the last century, we humans were busy. From 1900 to 2000, engineers and architects built thousands and thousands of power plants, laid millions of kilometers of transmission and distribution lines for electricity, connected thousands of kilometers of pipelines for oil and natural gas, and pounded together hundreds of nuclear waste storage facilities and oil refineries.[4] The twentieth century saw the world profoundly shaped by technology such as the automobile, the airplane, and atomic energy. Millions of people dismounted from their horses, sold their oxen and ploughs, and motored into the mechanized age. Electricity, once so novel that it was prized for its "healing powers," moved from its infancy to become a major fuel for heating homes, powering industrial processes, energizing air conditioners (also invented during the century), and enabling the digital-telecommunications-media-computer-information ages. The twentieth century, in other words, was about technology.

Today, the "upstream" extractive energy industries boast hundreds of thousands of operating oil and gas wells, 760 commercial refineries, and at least 24,000 coal mines globally. The electricity sector is supported by roughly 170,000 generators producing power worldwide, spinning away in more than 75,000 power plants—about half of them coal-fired and 440 of them nuclear-powered. Given that the average human body produces between 60 and 90 W of equivalent energy per hour, it would take 13.3 billion people—almost twice as many as exist on Earth—to naturally produce as much energy as America's electricity technology does in 60 minutes.[5] As Canadian geographer Vaclav Smil put it:

> The world's fossil-fuel-based energy system . . . now has an annual throughput of more than 7 billion metric tons of hard coal and lignite, about 4 billion metric tons of crude oil, and more than 3 trillion cubic meters of natural gas. This adds up to 14 trillion watts of power. And its infrastructure—coal mines, oil and gas fields, refineries, pipelines, trains, trucks, tankers, filling stations, power plants, transformers, transmission and distribution lines, and hundreds of millions of gasoline, kerosene, diesel, and fuel oil engines—constitutes the costliest and most extensive set of installations, networks, and machines that the world has ever built, one that has taken generations and tens of trillions of dollars to put in place.[6]

Our conventional energy systems need delivery mechanisms such as pipelines, tankers, and transmission lines to function. The American Society of Mechanical Engineers estimates that the total length of pipelines around the world exceeds 3.5 million kilometers.[7] If you laid just the Canadian pipeline system end to end, it would circle the world 17 times.[8] The world's fleet of 550 oil tankers (known as VLCCs and ULCCs—very large and ultra large crude carriers) transports about 2.6 billion MTOE and refined petroleum products every year, amounting to one-third (34%) of all seaborne trade by volume, and clocking an astounding 11.7 billion nautical miles collectively traveled.[9] Indeed, the world's biggest supertanker, built in 1979 and recently decommissioned, was appropriately named the *Happy Giant*: it was so large that when its 46 tanks and 340,000 square feet of storage were filled with oil, it could not traverse the English Channel.[10] Globally, about 4 million miles of electric high-voltage transmission lines have been built, and about 150,000 miles of new transmission lines are added each year, worth $184 billion, with the fastest growth in China.[11] The highest transmission towers crisscrossing the Yangtze River in China reach 1,132 feet—that's about the height of the Empire State Building in New York. Another $300 billion or so are invested in renewable energy systems such as wind turbines, solar panels, and solar thermal water heaters each year.[12] Happy giants abound in the energy sector.

Energy Companies Are Major Polluters

Understandably, the pervasiveness of energy activities suggests that those who control these activities—those in charge of energy resources, fuels, technologies, and distribution systems—hold sway over patterns of resulting pollution. This is particularly true given the limited number of decision makers in positions of power. The number of energy "gatekeepers" in the United States, those that form public utility commissions or departments of public service responsible for licensing energy projects, is surprisingly small: only 700 state regulators will serve in office during their terms from 2010 to 2030, and each one will approve about $6.5 billion in utility capital investment.[13] Although 150 countries produce crude oil around the world,[14] the top ten national oil companies—led by giants such as Saudi Aramco or the National Iranian Oil Company—hold more than two-thirds of the world's oil reserves and half of its natural gas reserves.[15]

The numerous independent energy enterprises that exist throughout the energy supply chain are bolstered by a few mega-firms that are particularly important players in conventional energy circles. Privately owned conglomerates such as ExxonMobil, British Petroleum, Shell, and Chevron have far fewer reserves than nations enjoy but are still behemoths. ExxonMobil, the world's largest company in 2012, had $452.9 billion in revenues and $41.1 billion in profits.[16] If ExxonMobil were a country, its annual profit would exceed the gross domestic product (GDP) of nearly two-thirds of the 183 nations that comprise the World Bank's economic rankings.[17] The mining company Rio Tinto is apparently so powerful that it was largely responsible for a civil war in Papua New Guinea.[18] We should also mention the cartel-like behavior of the Organization of Petroleum Exporting Countries (OPEC), which has been consistently implicated in inflating oil and gasoline prices for the past four decades.[19]

The role that these mega energy firms play in exacerbating climate change is substantial. Climate scientist Richard Heede analyzed trends in carbon dioxide and methane emissions from 1854 to 2010 and calculated that just 90 corporations—50 leading investor-owned energy producers, 31 state-owned producers, and 9 nation-state producers of oil, natural gas, and coal—are responsible for producing nearly two-thirds of all emissions. This implies that firms such as Chevron, ExxonMobil, Saudi Aramco, the National Iranian Oil Company, Gazprom, and Peabody Energy have had the most direct control over emissions pathways. More importantly, Heede suggests that these firms possess two other important assets: production capacity and proven recoverable reserves, meaning that they possess the wherewithal to inflict far greater damage on our planet. They "hold the key to future fossil fuel production and emissions and thus, arguably, the future of the planetary climate system."[20]

Such companies also operate factories and other facilities that pollute or degrade the environment in myriad ways. Refineries for oil and gas release particulate matter that is inhaled by residents living nearby. Coal-fired power plants contribute to acid rain that bleaches streams, forests, and coral reefs. Nuclear waste repositories and abandoned uranium mines leech tritium and other contaminants into water supplies. Large hydroelectric dams contribute to deforestation through the creation of massive reservoirs and by opening up tropical forests to loggers and poachers. Even wind turbines contribute to avian mortality by knocking birds literally out of the sky. And the construction and decommissioning of solar power plants engender toxic material management challenges.

In visual media, smokestacks bellowing pollution into the sky have long become a symbolic proxy for progress and prized industrial emblems.[21] In the past, companies and manufacturers flaunted sooty plumes on advertisements, letterheads, and stock certificates, or as one author put it, "effluence signified affluence."[22] Historically, at least in industrialized nations, the pollution streams of industrial centers took a back seat to the "sublime" capacity of our economic machinery to solve social ills, lighten the toil of workers and housewives, provide faster and cleaner forms of transportation, and revolutionize food production on the farm.[23] Travel to any major city in China and you will find this mindset alive and kicking.

Today, in many nations, affluence has changed our consciousness about pollution, but that does not negate that a breathtaking amount of it still comes from the energy industry. One major international study conducted by the Economics of Ecosystems and Biodiversity found that the externalities associated with business practices can be shockingly large.[24] Using environmentally extended input-output modeling, the study projected that companies around the world produce $7.3 trillion in "unpriced" natural capital costs each year, an amount that equates to about 13% of global GDP. When broken down by category, most of these damages arise from greenhouse gas (GHG) emissions (38%), followed by water use (25%), land use (24%), air pollution (7%), land pollution (5%), and waste (1%). When broken down by industrial or regional sector, the single largest contributor to these damages is coal-fired electricity generation.

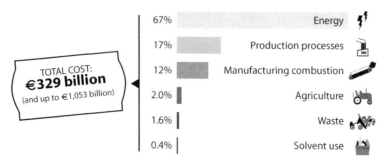

67% Energy

17% Production processes

12% Manufacturing combustion

2.0% Agriculture

1.6% Waste

0.4% Solvent use

TOTAL COST:
€329 billion
(and up to €1,053 billion)

Figure 1.1. Health and Environmental Costs of Air Pollution in Europe by Industrial Sector, 2008–2012. *Source:* European Environment Agency (EEA), *Costs of Air Pollution from European Industrial Facilities 2008–2012* (Brussels: EEA, 2014).

Even in Europe and North America, where environmental regulations are more stringent than in most other regions, energy industries are proving to be unruly neighbors. In the United States, two of the top sectors reporting releases to the Toxics Release Inventory (TRI)—a freely accessible database that provides information on environmental pollution—are the mining and electricity generation industries, together responsible for 54% of chemical pollution.[25] In the European Union, air pollution caused €329 billion to €1.05 trillion in damages from 2008 to 2012.[26] As figure 1.1 shows, the energy industry was by far the leading source of this pollution, accounting for two-thirds (67%) of all health and environmental damages. Moreover, 50% of the damage costs were caused by only 1% of the facilities assessed, implying that a select number of companies have an immense amount of control over pollution flows.

The Other Side: Consumers Matter More Than Energy Producers

A starkly different view, stated simply by Paul G. Harris, is that "the fundamental driving force behind climate change is the material consumption of people" and "the root of climate change is pollution from and caused by individuals."[27] Rather than blaming energy companies and peddlers of polluting energy technology, the articulators of this perspective point out that human beings—that is, our insatiable demand for energy, our patterns of consumption, our self-serving behavior, and so on—drive the economic activity within the global energy system. In short, we have met the enemy and she is us.

Legal scholar Michael P. Vandenbergh and his colleagues estimate that individual activities—actions under the direct, substantial control of a person but not undertaken in the scope of his or her employment—account for 30% to 40% of annual CO_2 emissions in the United States, the largest of any sector.[28] Actions such as driving down to the corner market, taking vacations, and using the hordes of electricity-sucking devices in the home accounted for 4.4 trillion pounds (2 trillion kg) of CO_2 emissions in the United States in 2000. By comparison, the entire industrial sector emitted only 3.9 trillion pounds (1.8 trillion kg) of CO_2. When put in an international context, the emissions from individual actions in the United States accounted for about 8% of the world's CO_2 emissions—larger than the total emissions from every country in the world except China.[29]

In addition to the contribution of our individual activities toward global CO_2 emissions, we are collectively responsible for a widespread siege on our environment. Individuals and households in the United States contribute nationally to:

- 94.5% of air toxics such as acetaldehyde, formaldehyde, and benzene through household disposal of trash
- 85% of petroleum pollution through individual dumping of used motor oil into municipal storm sewers and runoff from urban and suburban streets, which releases 15 times more oil to the ocean each year than the *Exxon Valdez* spill
- More than 30% of all low-level ozone precursors through auto and electricity use and the disposal of pesticides and solvents
- 14% of mercury, mostly through discarded electronics
- 11% of pesticide pollution[30]

American consumers are not alone. Individuals worldwide exert a considerable force over energy patterns through actions they make in the home, on the road, and while shopping or on vacation.

Households and Cars Have Large Energy Footprints

Consider for a moment just how hooked we are on electrical devices. According to one lightbulb manufacturing company, every year consumers in the United States spend about $1 billion to purchase 2 billion lightbulbs, which works out to 5.5 million bulbs bought per day.[31] Using the most recently available data, there are about 1.4 billion televisions throughout the world, with 400 million in China and 219 million in the United States.[32] If we assume that each one uses 200 W of electricity, powering all of those TVs for one hour requires 280 billion kWh of electricity. That's more electricity than Australia consumes over an entire year.

Televisions and lightbulbs only scratch the surface: many homes boast an inventory of microwaves, ovens, dishwashers, water heaters, refrigerators, washers, dryers, tropical fish tanks, massage chairs, ice makers, stereos, electric can openers, electric blankets, electric clocks, and the hallowed "beer fridge."[33] The think tank American Council for an Energy-Efficient Economy estimated in 2013 that the United States was home to more than 2 billion "power-hungry" devices, such as TVs and computers and specialized equipment such as elevators and icemakers, which together used more energy than scores of individual countries did and cost $70 billion in energy expenses annually.[34] These devices collectively consume 7.8 quadrillion BTUs each year—an amount greater than the primary energy use of a hundred other countries combined.

We can use every single one of these devices efficiently or inefficiently, but all too often we opt for the latter.[35] In the United Kingdom, consumers spend £1.2 billion a year on the electricity needed to fuel so-called vampires such as stereos, dust-busters, and wireless routers that consumers needlessly leave on when out of the house.[36] In the United States, standby power accounts for more than 100 billion kWh of annual electricity consumption and more than $10 billion in annual energy costs (or $100 per household).[37] Put another way, about 20 huge power plants operate to energize appliances and equipment that are turned off, just to keep them in standby mode.[38]

What's more, we love our cars. Worldwide, we own approximately one billion personal automobiles. Each of these cars and motorbikes needs material inputs such as steel, plastic, and glass that must be manufactured and assembled in energy-intensive processes before they are sold to consumers, who then fuel them up with a petroleum-based energy source that causes urban smog and contributes to global warming. Environmental law professor Jason J. Czarnezki estimates that American drivers have direct control of or responsibility for 112 billion gallons of fuel consumed and more than 2 trillion pounds of CO_2 emitted each year.[39] Moreover, our automobiles necessitate and rely on 11.1 million miles of paved roads—enough to pave a path to the moon and back 46 times.[40]

The damage caused by our reliance on automobiles is exacerbated by epic-scale indifference. The National Academies of Science has estimated that if Americans kept the tires properly inflated on all of their vehicles, they would save 548,000 barrels of oil per day, more than the amount contained in many large domestic oil fields.[41] Similarly, the commonplace practice of vehicle idling (keeping motors on at stop lights or while temporarily parked) accounts for at least 40 billion liters (10.6 billion gallons) of wasted gasoline a year, or 1.6% of all US CO_2 emissions.[42] The idling of long- and short-haul trucks is another major emitter because drivers often operate their engines or auxiliary power systems for climate comfort while they rest or wait to load or unload.[43] The International Energy Agency has calculated that relatively simple changes in driving behavior—such as optimizing the shifting of gears, stopping idling, avoiding rapid acceleration, driving at efficient speeds, and reducing weight (by removing unnecessary items from the trunk) and wind resistance (by removing luggage or ski racks when not in use)—could save "up to 20% of the fuel used by some drivers" and "as much as 10% on average across all drivers on a lasting basis."[44]

Cooking Practices and Diet Matter

Another key source of consumers' energy-related pollution comes from how we cook and what we choose to eat. In the developing economies of Asia and Africa, most households still rely on wood, dung, or charcoal to cook food or heat homes. However, this results in a sobering amount of indoor pollution, producing conditions akin to smoking hundreds of cigarettes a day in a closely confined space. Indoor air pollution ranks fourth among global disease risk factors at almost 5%, coming after only high blood pressure (almost 8%), tobacco smoking and second-hand smoke (about 7%), and alcohol use (about 6%).[45] This places it well ahead of physical inactivity and obesity, drug use, and unsafe sex as a leading cause of death. In India and all of South Asia, smoke from household cookstoves is the highest health risk factor, topping both smoking and high blood pressure. It is the second-highest health risk factor in sub-Saharan Africa, third in Southeast Asia, and fifth in East Asia. Air pollution from conventional cookstoves is responsible for 4 million deaths each year, 3.5 million direct premature annual deaths and 500,000 deaths from secondhand cook-fire smoke.[46] The annual cost of indoor air pollution—a cost not reflected in the price of energy—has been estimated at between $212 billion and $1.1 trillion.[47] Almost all of these deaths occur in developing countries, and more than half of those who die are children.[48]

In terms of what we're cooking and eating, land use and dietary patterns have changed everywhere. Our crops depend increasingly on fossil fuel–based fertilizers to produce the food and more fossil fuels to get it to market. Indeed, the average morsel most people consume traveled 1,500 to 2,500 miles (2,400 to 4,020 km) to reach their mouths. Even locally grown food is often shipped from a nearby farm to be washed and packaged elsewhere, then transported back home. One study looking at cartons of strawberry yoghurt produced in Germany found that the average carton traversed 5,000 miles (8,000 km) of roads during production and distribution.[49]

Unsurprisingly, the biggest agriculture-related energy offenses are found in industrialized nations. In less-developed societies in Africa, every calorie expended by a farmer to plant and tend crops can produce up to 60 calories of food. However, the ratio is almost exactly the reverse in modern animal husbandry, where 65 calories of energy are expended to produce a single calorie of meat.[50] In modern societies, it takes about 5 gallons (19 liters) of fresh water to process a bushel of corn, and every calorie of processed food requires 10 calories of fossil fuel energy.

The sources of food we crave today are more energy- and carbon-intensive than they used to be. Producing foods that many of us eat every week—shrimp, lamb, beef, and pork—causes extensive CO_2 emissions.[51] One study recently calculated that an individual who switches to a plant-based diet would do more to stop global warming than switching from a sports utility vehicle to a Toyota Camry.[52] Yet, an opposite trend is evident. More people are transitioning to a meat-centered diet, with global meat consumption increasing five-fold in the past five decades.[53] Industries are not leading this change; individuals are, every time they fire up the barbecue.

Conspicuous Consumption Is Conspicuously Damaging

The acquisition of material goods for social reasons—what the nineteenth-century economist Thorstein Veblen termed "conspicuous consumption"[54]—also inflates energy consumption, exacerbating pollution. Some affluent families heat their outdoor swimming pools to bath temperature all year round or purchase yachts that consume 31 liters of fuel per kilometer traveled.[55] British columnist George Monbiot mused that "as the owner of one of these yachts I'll do more damage to the biosphere in 10 minutes than most Africans inflict in a lifetime."[56] In Japan, some households have toilets that can warm and wash one's bottom, "whisk away odors" with built-in fans, produce water noises to drown out undesirable sounds, and play relaxing music. The problem is that such toilets use more electricity than dishwashers or clothes dryers and account for about 4% of household energy consumption nationwide.[57] Four percent of annual household consumption in Japan equates to the need to run three of its nuclear reactors all year round.[58]

Conspicuous consumption may explain why 90% of personal consumption globally occurs in industrialized countries, home to only 20% of the world's population. People in these affluent countries use, on average, 32 times as many resources as people living in developing economies do. Over just three days, for instance, GHG emissions generated by a typical British family will exceed annual emissions from a comparable household in Tanzania.[59]

The impact of individual energy consumption is increasingly evident outside industrialized countries, as well. In India, the billionaire Mukesh Ambani has built the world's "biggest and most expensive home" for his wife and three children. The 27-story "Antilia" house, which cost $1 billion to construct, boasts 400,000 square feet of living space replete with a dance studio, a cinema that seats 50 people, three helicopter pads, and a car park for 160 vehicles.[60] That same $1 billion would have been sufficient to provide 57 million Indian households at the "bottom of the pyramid" with solar lanterns, cookstoves, and small hydropower systems.[61]

Philip Cafaro has determined that individuals can mitigate immense amounts of carbon—as much as 15 billion tons (gigatons) per year by 2060—simply by deciding where they take their vacations and by forgoing international air travel.[62] Similarly, we could eliminate 2.38 billion tons of GHG emissions by adopting practices such as eating one less meat dish per week, removing subsidies for cattle production, and banning confined animal feedlot operations. Cafaro estimated that similar large potential reductions could come from targeting "luxury" and "population" activities—vacationing, dining out, purchasing diamond rings, choosing to have more children—all things under our own direct control. More importantly, these are actions that we can begin tomorrow, meaning they will have a more immediate impact on climate change abatement than waiting 10 years (or more) to build a nuclear power plant or waiting for our current automobile to break down so we can replace it with an electric vehicle. Given all this evidence, how can individuals not be culpable for the energy predicament we are in?

Common Ground: Production and Consumption Mutually Interact

As one might surmise, these two sides need not be viewed as competing perspectives. As box 1.1 suggests, energy production and consumption, or supply and demand, are mutually interactive or constitutive. They create a potent tag team causing pollution and emissions. As is the case with illegal narcotics, both the cartels selling the product and the addicts using it share a degree of responsibility.

Without question, energy infrastructure enables human development. Over the past century, the global energy sector has represented one of the largest, perhaps even *the* largest, sector for infrastructure investment. The tentacles of the conventional energy network entwine hundreds of millions of workers and contribute significantly to global GDP. That said, the global energy system is not masterfully controlled only by a handful of corporate elites; there is no evil wizard behind the curtain, pulling the strings. These energy moguls succeed by responding to our demands, needs, desires, and wants. Ordinary consumers— those of us addicted to electronic gadgets, living in energy-intensive homes, driving gas-guzzling cars, and jetting around the world—exert substantial, direct control over patterns of energy use through our lifestyles and values.[63] It is all about choice; how we choose to live defines consumption patterns and the scope and severity of the energy problems that such patterns engender.

The United Nations forcefully emphasized this point about choice in a study that investigated the difference between GHG emissions generated by a person conscious of a low-carbon lifestyle and a person living life according

Energy Production, Consumption, and the Common Ground

Is industry the chief energy villain?

ONE SIDE: Energy supply systems dominate the globe, and the energy companies behind the systems are to blame for most energy-related pollution.

OTHER SIDE: Individual consumers are largely to blame for pollution, given their decisions about housing, electric appliances, cars, and food.

COMMON GROUND: Energy producers and consumers are mutually responsible for energy pollution.

to global trends.[64] Although both arguably attained the same level of comfort over the course of their day, the carbon footprint of the person with the low-carbon lifestyle amounted to 14 kg, while that of the person living up to current standards amounted to 38 kg. Table 1.1 summarizes this study and shows how choosing more "climate aware" practices can cut emissions by almost two-thirds without any major sacrifices in quality of life. Other studies done in North America have produced similar results.[65]

Though many of us may feel better blaming others—politicians, companies, cartels—for our collective energy woes, the true culprits are, by and large, ourselves. Though our global energy networks can retain a degree of path dependency or lock-in, they exist, ultimately, to serve people. As the environmentalist Wendell Berry put it, although "it is understandable that much of our attention, anxiety, and energy is focused on exceptional cases, the outrages and extreme abuses of the industrial economy," in truth "the root of the problem is always to be found in private life."[66]

This perspective—that we are each complicit in society's energy problems—has profound implications for how each of us thinks about energy. Our homes, cars, and factories no longer become sources of pure enjoyment and utility; they are also dumping grounds and extremely potent sources of pollution. Goods we acquire and the services we purchase are conduits for pollution. Watching a television that draws electricity from a power grid dominated by coal-fired power is not much different than throwing soot over your neighbor's fence. The actions of family, friends, and neighbors challenge the image of belching factories as the offending parties. In addition, this finding alters our notion of a "victim," for it demonstrates that we can be both victims and abusers. Promisingly though, this view also suggests a degree of control over our collective energy future. If individuals can influence how energy is produced and

Table 1.1. Carbon Dioxide Emissions by Lifestyle for Two German Consumers

Appliance/activity	"Climate-aware" lifestyle	"Normal" lifestyle
Alarm clock	Hand-wound	Electric
Shower	With water-saving head	Without water-saving head
Drying hair	Letting it dry	Electric dryer
Breakfast	Bread in the toaster	Bread in the oven
Boiling water	Kettle	Electric stove
Toothbrush	Regular	Electric
DVD player/TV/DSL modem/wireless router	Turns off when not using	Keeps on standby
Heating	Turns down thermostat one degree (during the winter) when out of house	Leaves heat at the same temperature
Light	Energy-saving bulbs	Regular bulbs
Refrigerator	Class A++ labeling (most efficient)	Class A labeling (somewhat efficient)
Commuting	Mass transit	Private automobile
Apple snack	From Bavaria by truck	From New Zealand by air
Office devices	Turns off when at lunch	Leaves on when at lunch
Lunch	Chicken	Beef
Afternoon snack	Strawberries from Italy by truck	Strawberries from South Africa by air
Laundry	Washed at 60°C	Washed at 90°C
Laundry drying	Clothesline	Dryer
Dinner	Fresh vegetables	Frozen vegetables
Dishwasher	Label A (most efficient)	Label D (less efficient)
Sport/exercise	Jogging	Running machine
Total emissions/day	14 kilograms	38 kilograms

Source: Modified from United Nations Environment Programme (UNEP), *Kick the Habit: A UN Guide to Climate Neutrality* (Paris: UNEP, 2008).

consumed, then we can catalyze change. We can demand less energy-intensive goods and services, drive more efficient cars, purchase better electric appliances, eat less meat, and conserve water. We no longer become victims loosely connected to climate change and global energy insecurity but, instead, become active participants whose lifestyles play a starring role.

NOTES

1. B. K. Sovacool, "Notable Energy Accidents and Disasters," in *Handbook of Energy*, vol. 2, ed. J. Cutler and J. Cleveland (London: Elsevier Science and Technology, 2013), 772–773.

2. A. L. D'Agostino and B. K. Sovacool, "An American Oil Spill," *Project Syndicate*, July 1, 2010, 1.

3. K&L Gates, *Energy, Infrastructure, and Resources* (Boston: K&L Gates, 2013), 3.

4. B. K. Sovacool, "Energy Policy and Climate Change," in *The Handbook of Global Climate and Environment Policy*, ed. Robert Falkner (New York: John Wiley & Sons, 2013), 446–467.

5. B. K. Sovacool, *The Dirty Energy Dilemma* (Westport, CT: Praeger; 2008).

6. Vaclav Smil, "A Skeptic Looks at Alternative Energy," *IEEE Spectrum*, June 28, 2012.

7. Phil Hopkins, *Oil and Gas Pipelines: Yesterday and Today* (New York: Pipeline Systems Division International Petroleum Technology Institute, American Society of Mechanical Engineers, 2007).

8. Liz Kilmas, "Obama during Debate: There's Enough 'Pipeline to Wrap around the Earth Once,' " *The Blaze*, October 16, 2012.

9. Erik Ranham, "Just How Many VLCCs Do We Need?" *Intertanko*, March 2012.

10. William B. Hayler and John M. Keever, *American Merchant Seaman's Manual* (Cambridge, MD: Cornell Maritime Press, 2003). See also Mark Huber, *Tanker Operations: A Handbook for the Person-in-Charge (PIC)* (Cambridge, MD: Cornell Maritime Press, 2001).

11. ABS Energy Research, *Global Transmission & Distribution Report* (Houston, TX: ABS Energy Research, 2010).

12. Renewable Energy Policy Network for the 21st Century (REN21), *Global Status Report* (Paris: REN 21, 2013).

13. Ron Binz et al., *Practicing Risk-Aware Electricity Regulation* (Montpelier, VT: CERES and RAP, 2012).

14. Eni, *The World Oil and Gas Review* (Rome, Italy: Eni, 2012).

15. British Petroleum, *Statistical Review of World Energy 2012* (London: BP, 2012); Eni, *World Oil and Gas*.

16. "Fortune 500: Our Annual Ranking of America's Largest Corporations," *CNN Money*, December 1, 2012, available at http://money.cnn.com/magazines/fortune/fortune 500/2012/snapshots/387.html.

17. Marianne Lavelle, "Exxon's Profits: Measuring a Record Windfall," *US News and World Report*, February 1, 2008.

18. Brian Thomson, "Rio Tinto Caused War: Somare," *The Age* (Australia), June 26, 2011.

19. See "Elon Poll Reveals NC Residents Blame Oil Companies, OPEC for Gas Prices," *Citizen Wells* (blog), April 4, 2012; Steven Lacey, "Polls: Americans Support 60 mpg Fuel Economy Standard; Blame Oil Companies, OPEC for Price Hikes," *Climate Progress*, May 17, 2011; Russell Belk, John Painter, and Richard Semenik, "Preferred Solutions to the Energy Crisis as a Function of Causal Attributions," *Journal of Consumer Research* 8 (December 1981): 306–312; Carl F. Hummel, Lynn Levitt, and Ross J. Loomis, "Perceptions of the Energy Crisis: Who Is Blamed and How Do Citizens React to Environment-Lifestyle Trade-offs?" *Environment and Behavior* 10 (1978): 37–88.

20. Richard Heede, "Tracing Anthropogenic Carbon Dioxide and Methane Emissions to Fossil Fuel and Cement Producers, 1854–2010," *Climatic Change* 122, no. 1–2 (January 2014): 231.

21. David E. Nye, *Consuming Power: A Social History of American Energies* (Cambridge, MA: MIT Press, 1999).

22. Edward Tenner, *Why Things Bite Back: Technology and the Revenge of Unintended Consequences* (New York: Alfred A. Knopf, 1997), 131.

23. David E. Nye, *American Technological Sublime* (Cambridge, MA: MIT Press, 1994).

24. The Economics of Ecosystems and Biodiversity (TEEB), *Natural Capital at Risk: The Top 100 Externalities of Business* (Brussels: TEEB, April 2013).

25. US Environmental Protection Agency (EPA), *TRI Releases* (Washington, DC: EPA, September 2009).

26. European Environment Agency (EEA), *Costs of Air Pollution from European Industrial Facilities 2008–2012* (Brussels: EEA, 2014).

27. Paul G. Harris, *What's Wrong with Climate Politics and How to Fix It* (New York: Polity Press, 2013), 93–94.

28. Michael P. Vandenbergh and Anne C. Steinemann, "The Carbon-Neutral Individual," *New York University Law Review* 82 (2007): 1673–1745. See also Michael P. Vandenbergh et al., "Implementing the Behavioral Wedge: Designing and Adopting Effective Carbon Emissions Reduction Programs," *Environmental Law Reporter* 40 (2010): 10,547–10,554.

29. Vandenbergh and Steinemann, "Carbon-Neutral Individual."

30. Michael P. Vandenbergh, "From Smokestack to SUV: The Individual as Regulated Entity in the New Era of Environmental Law," *Vanderbilt Law Review* 57 (2004): 517–628.

31. Sylvania, "Lighting: About Us" (press release, August 4, 2011).

32. Nation Master Statistics, "Televisions (Most Recent) by Country" (November 2011), available at www.nationmaster.com/graph/med_tel-media-televisions.

33. Readers interested in the history of these other types of appliances should sample Ruth Schwartz Cohen, *More Work for Mother: The Ironies of Household Technology from Open Hearth to the Microwave* (New York: Basic Books, 1983), and Elizabeth Shove, *Comfort, Cleanliness, and Convenience: The Social Organization of Normality* (Oxford: Berg Press, 2003).

34. American Council for an Energy-Efficient Economy (ACEEE), *Miscellaneous Energy Loads in Buildings* (Washington, DC: ACEEE, 2013). See also ACEEE, "Power-Hungry Devices Use $70 Billion of Energy Annually" (press release, June 26, 2013).

35. Kathryn B. Janda, "Buildings Don't Use Energy: People Do," *Architectural Science Review* 54, no. 1 (2011): 15–22.

36. David J. C. MacKay, in *Sustainable Energy—Without the Hot Air* (Cambridge: UIT Cambridge, 2008), argues that each household spends £50 per year on vampires. We multiply this amount by the United Kingdom's 24.3 million homes to reach £1.2 billion a year.

37. US Environmental Protection Agency and Department of Energy, "Standby Power and Energy Vampires" (December 2013), available at www.energystar.gov/index.cfm?c=about.vampires.

38. Amory B. Lovins, "Energy Myth Nine—Energy Efficiency Improvements Have Already Reached Their Potential," in *Energy and American Society—Thirteen Myths*, ed. B. K. Sovacool and M. A. Brown (New York: Springer Publishing, 2007).

39. Jason J. Czarnezki, *Everyday Environmentalism: Law, Nature & Individual Behavior* (Washington, DC: Environmental Law Institute, 2011).

40. Numbers from chapter 2, "The Global Energy System," in Benjamin K. Sovacool and Michael Dworkin, *Global Energy Justice: Problems, Principles, and Practices* (Cambridge: Cambridge University Press, 2014).

41. B. K. Sovacool, "Solving the Oil Independence Problem: Is It Possible?" *Energy Policy* 35, no. 11 (November 2007): 5505–5514.

42. Jack N. Barkenbus, "Eco-Driving: An Overlooked Climate Change Initiative," *Energy Policy* 38, no. 2 (February 2010): 762–769; Amanda R. Carrico et al., "Costly Myths: An Analysis of Idling Beliefs and Behavior in Personal Motor Vehicles," *Energy Policy* 37, no. 8 (2009): 2881–2888.

43. L. L. Gaines and C.-J. Brodrick Hartman, *Energy Use and Emissions Comparison of Idling Reduction Options for Heavy-Duty Diesel Trucks*, Argonne National Laboratory Paper No. 09-3395 (Argonne, IL: Argonne National Laboratory, 2009).

44. International Energy Agency (IEA), *Transport Energy and CO₂: Moving Towards Sustainability* (Paris: OECD, 2009), 7–8.

45. S. S. Lim et al., "A Comparative Risk Assessment of Burden of Disease and Injury Attributable to 67 Risk Factors and Risk Factor Clusters in 21 Regions, 1990–2010: A Systematic Analysis for the Global Burden of Disease Study 2010," *Lancet* 380 (2012): 2224–2260.

46. "Secondhand cook-fire smoke" refers to smoke leaving chimneys and accumulating outdoors in populated areas.

47. United Nations Environment Programme (UNEP), *Natural Selection: Evolving Choices for Renewable Energy Technology and Policy* (New York: United Nations, 2000). Figures are updated to 2010 dollars.

48. Gwénaëlle Legros, *The Energy Access Situation in Developing Countries: A Review Focusing on the Least Developed Countries and Sub-Saharan Africa* (New York: World Health Organization and United Nations Development Programme, 2009).

49. Herbert Girardet and Miguel Mendonca, *A Renewable World: Energy, Ecology, Equality* (London: Green Books, 2009), 187.

50. Richard B. Wilk, "Culture and Energy Consumption," in *Energy: Science, Policy, and the Pursuit of Sustainability*, ed. Robert Bent, Lloyd Orr, and Randall Baker (Washington, DC: Island Press, 2002), 112.

51. UNEP, *Kick the Habit: A UN Guide to Climate Neutrality* (New York: UNEP, 2008).

52. Claudia H. Deutsch, "Trying to Connect the Dinner Plate to Climate Change," *New York Times*, August 29, 2007.

53. See Noam Mohr, *A New Global Warming Strategy: How Environmentalists Are Overlooking Vegetarianism as the Most Effective Tool against Climate Change in Our Lifetimes* (Washington, DC: Earthsave International, August 2005). See also Anthony J. McMichael et al., "Food, Livestock Production, Energy, Climate Change, and Health," *Lancet* 370, no. 9594 (2007): 1253–1263.

54. Thorstein Veblen, *The Theory of the Leisure Class* (New York: Modern Library, 2001 [1899]).

55. George Monbiot, "Stop Blaming the Poor: It's the Wally Yachters Who Are Burning the Planet," *Guardian* (London), September 28, 2009.

56. Ibid.

57. Blaine Harden, "In Energy-Stingy Japan, an Extravagant Indulgence: Posh Privies," *Washington Post*, June 25, 2008.

58. According to the IEA, the residential sector in Japan consumed 286,016 GWh of electricity in 2011; 4% of this is 11,441 GWh. A 700 MW reactor at 70% capacity (the number the World Nuclear Association uses in its "average" calculations for Japan) will generate 4,292 GWh each year, which means, roughly, that the equivalent of three nuclear reactors are needed to power those toilets. Statistics from IEA, *Energy Statistics of OECD Countries* (Paris: OECD, 2011), and World Nuclear Association, "Nuclear Power in Japan" (May 2013), available at www.world-nuclear.org/info/Country-Profiles/Countries-G-N/Japan.

59. Harris, *What's Wrong with Climate Politics*, 101.

60. Diane Pham, "World's Largest and Most Expensive Family Home Completed," *Inhabitat: Architecture*, October 14, 2010.

61. For this claim we have taken the figure of $2 billion for 114 million households and cut it in half. See Sreyamsa Bairiganjan et al., *Power to the People: Investing in Clean Energy for the Base of the Pyramid in India* (Washington, DC: World Resources Institute, 2010).

62. Philip Cafaro, "Beyond Business as Usual: Alternative Wedges to Avoid Catastrophic Climate Change and Create Sustainable Societies," in *The Ethics of Global Climate Change*, ed. Denis Arnold (Cambridge: Cambridge University Press, 2011), 192–215

63. Fereidoon P. Sioshansi, *Generating Electricity in a Carbon-Constrained World* (Amsterdam: Elsevier, 2010), 565–566

64. UNEP, *Kick the Habit*.

65. See John A. "Skip" Laitner, Karen Ehrhardt-Martinez, and Vanessa McKinney, "Examining the Scale of the Behavior Energy Efficiency Conundrum" (paper presented at the 2009 ACEEE Summer Study, Washington, DC, August 2009); Gerald T. Gardner and Paul C. Stern, "The Short List: The Most Effective Actions U.S. Households Can Take to Curb Climate Change," *Environment*, September/October 2008.

66. Wendell Berry, "Conservation Is Good Work," in *Wild Earth: Wild Ideas for a World Out of Balance*, ed. Tom Butler (New York: Milkweed Editions, 2002), 153–154.

Is Energy Efficiency
a Worthwhile Investment?

Imagine an energy resource so omnipotent that it could improve energy security, save money, bolster employment, and protect the environment in one fell swoop. This resource is universally abundant and commercially available, ready to be implemented without the need for further development. It could provide millions of high-paying jobs and does not need to be drilled, dug, or syphoned out of the earth. It would not instigate global panic over radiation, spill oil into Prince William Sound or the Gulf of Mexico, spit toxic sludge into Asian rivers, seep contaminants into Africa's water supply, destroy Amazonian rainforests, or release greenhouse gases (GHGs) into the atmosphere. It would operate automatically and remain on and ready to be dispatched without delay or intervention by energy providers. Moreover, it has existed for years in various time-tested, empirically proven, reliable manifestations. Oh, and one other detail we forgot to mention: it's invisible.[1]

What is this wondrous resource? Nuclear fusion? Solar radiation? Wind energy? None of the above. This miracle technology is none other than good, old-fashioned enhancements to managing energy demand: using energy more efficiently. Energy efficiency does not mean doing less or suffering without; it means doing more with less through smarter use of technology: lightbulbs that use less power, thermostats that anticipate occupants' needs, enhanced insulation, hybrid electric vehicles, variable-speed industrial motors with smart controls, and walkable cities.

At least, so goes one side of the argument. Advocates of energy efficiency extol a variety of cost-effective energy-efficient technologies that are currently available for deployment, technologies that benefit consumers, the economy, and the climate. As one of our colleagues joked, in the energy field, "the answer is almost always energy efficiency—regardless of the question." However, due to a variety of disincentives and implementation barriers, the market currently turns a blind eye to these efficiency benefits. Efficiency advocates argue that planners and policymakers need to utilize a variety of tools and policies to make market players sit up and take notice.[2]

A contrarian view holds that energy efficiency initiatives are largely couched in pretense—smoke and mirrors that will not keep the devices that power the global economy chugging along. These contrarians argue that priority must be given to expanding power supply, initiatives certain to provide needed energy at a given point in time, 24 hours a day. To many in this camp, energy efficiency gains are an illusion, attributable to other factors, and the benefits are far less than claimed. Advocates of this view argue that, at best, energy efficiency improvements have largely been exploited already. At worst, energy efficiency efforts can do more harm than good by encouraging a "rebound effect" that increases overall energy usage.

One Side: Managing Energy Demand Is Essential

The first view is that improving energy efficiency and managing demand are akin to stooping down to pick money off the sidewalk: they require little effort that is more than matched by the payoff. Advocates argue that energy efficiency initiatives can save millions of dollars in wasted energy consumption while improving quality of life and environmental conditions. Energy efficiency can be considered a low- to no-risk energy strategy and is sufficiently reliable, predictable, and enforceable to become a surrogate alternative source of energy that merits prioritization. Why build a new power plant to accommodate growth in energy demand when the increase can be serviced simply by using energy more efficiently in other areas? Extracting more kilometers per liter of gasoline, generating more BTUs per ton of coal, losing fewer kilowatt-hours per kilometer of transmission line, and washing more pairs of socks per kilowatt-hour are the kind of initiatives that should be prioritized before breaking ground for a new power plant.

Three arguments are often advanced to support this thesis: (1) managing demand has been successful in the past, (2) improving energy efficiency delivers social and environmental benefits, and (3) efficiency improvements are replicable and verifiable.

Energy Efficiency Has a Tremendously Successful Track Record

Perhaps the strongest argument in favor of efficiency is the historical record, which shows how improved energy efficiency has enabled economies to grow while they use less energy. These types of efficiency activities have been so successful in the United States that total primary energy use per capita in 2000 was almost identical to energy use per capita in 1973: over this 27-year period, economic output (measured as gross domestic product [GDP] per capita) increased 74%, yet national energy intensity fell 42%.[3] If the United States had not dramatically reduced its energy intensity over the years, energy consumption would have risen 225% from 1973 to 2005 instead of increasing only 30%, and consumers would have spent at least $700 billion more on energy purchases in 2012 (or more than $1.9 billion every day).[4] The amount of energy saved—about 75 EJ (exajoules; an exajoule is one quintillion joules)—is the equivalent of 18 million railcars filled with coal, a line of railcars that could wrap around the earth seven times.[5]

Just how successful have energy efficiency initiatives been? The International Energy Agency (IEA) reports that for 11 highly industrialized countries, energy efficiency initiatives have saved more energy (almost 1.5 billion MTOE) than any single supply-side energy resource, as shown in figure 2.1. Between 2005 and 2010, these efficiency measures saved the equivalent of $420 billion worth of oil. If not for energy efficiency, consumers in these countries would be using (and paying for) two-thirds more energy than in current practice.

Examples of successful energy efficiency policies and programs abound. Many focus, understandably, on influencing technological development. Japan's Top Runner program shows how adept and nimble government interventions can be in disseminating best practice by closely monitoring technological advances.[6] Japan established the Top Runner program in 1999 as a means of setting energy efficiency standards for 19 different products, ranging from automobiles and refrigerators to computers and DVD players. Within a prescribed time period, the sales-weighted average efficiency of appliances sold by each manufacturer and distributor must meet or exceed the new minimum energy standard. The resulting energy improvements have reached as high as 67.8% for air conditioners, 78% for fluorescent lights, and 99.1% for computers. The program's goals are strengthened by a policy of "naming and shaming" companies that fail to meet standards. Overall, Top Runner is expected to deliver $3 billion in net benefits in markets for lighting, vehicles, and appliances.[7]

In Portland, Oregon, a different type of energy efficiency strategy, focusing on altering household behavior, demonstrated how a little adds up to a lot. With the help of a carbon dioxide emissions calculator available on the Internet and a guidebook entitled *Low Carbon Diet: A 30-Day Program to Lose 5,000 Pounds*, participating residents shortened their showers, reset their water heaters, wore extra sweaters, and turned down their thermostats. Some installed energy-efficient appliances or insulated their attics. Motorists inflated car tires, tuned engines, traded in gas-guzzlers for fuel-efficient cars, or left their vehicles at home. By adopting these and other practices, participating households were able to cut average household CO_2 emissions by 22% in one month, a reduction equivalent to 6,700 pounds of CO_2 emissions per household.[8]

Some strategies aim to single out particularly irksome, inefficient-energy offenders. Singapore limits vehicle ownership through licensing, deters the use of private vehicles through road pricing schemes, prohibits engine idling, and encourages the use of public mass transit through subsidies and enhanced public transit connectivity.[9] The Singaporean Ministry of Transport estimates that almost 5 million daily trips (about 60% of the total) involved mass rapid

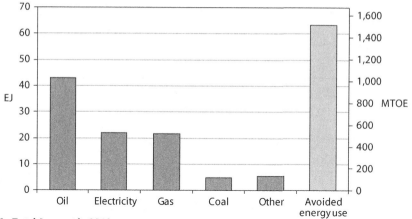

A. Total Amount in 2010

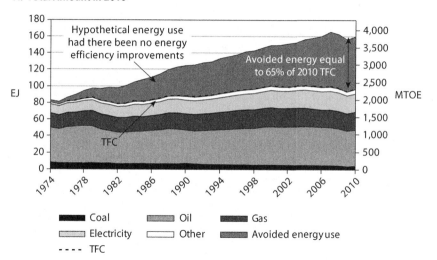

Legend:
- Coal
- Oil
- Gas
- Electricity
- Other
- Avoided energy use
- - - - - TFC

B. Amount over Time

Figure 2.1. Contribution of Energy Efficiency Compared with Other Energy Resources for 11 Countries, 1974–2010. *Source:* International Energy Agency, "The 'First Fuel' Contribution of Energy Efficiency Compared to Other Energy Resources Consumed in 2010 in 11 IEA Member Countries" (December 2013). *Note:* The countries are Australia, Denmark, Finland, France, Germany, Italy, Japan, the Netherlands, Sweden, the United Kingdom, and the United States. "Avoided energy use" represents the difference between total final consumption (TFC) in 2010 and the volume of energy that would have been consumed had there been no improvement in energy efficiency since 1974, based on a long-term IEA decomposition analysis. EJ = exajoule; MTOE = metric ton of oil equivalent.

transit, light rail transit, and buses—impressive figures given that the country has a population of less than 6 million. In short, the menu of possible approaches to promoting and improving energy efficiency is long, reflecting the diverse array of energy markets, resources, and infrastructures.[10]

Improvement in Efficiency Delivers Economic, Social, and Environmental Benefits

Another argument in favor of energy efficiency projects is that they are cost-effective and have short payback periods. As one study surmised, "the cheapest megawatt hour of electricity is the one that is not produced."[11] The National Research Council reviewed 17 major energy efficiency programs funded by the US Department of Energy from 1978 to 2000, covering residential consumption efficiency such as the development of energy-efficient refrigerators or compact fluorescent lights, commercial improvements such as electronic ballasts, and industrial improvements such as oxygen-fueled glass furnaces and lost-foam casting for steel making.[12] It estimated that the total net realized economic savings from these programs amounted to $30 billion (in 1999 dollars), while the total cost of the programs was only $7 billion. An assessment of the Warm Front Program in the United Kingdom, which provided energy audits and grants for low-income housing investments in energy efficiency, concluded that over its lifetime, the scheme cost £2.4 billion but would yield £87.2 billion in savings.[13]

The common finding in all these studies is that energy efficiency initiatives often save money for the diligent souls who are willing to explore options. When compared with other supply-side system options, efficiency options cost far less than conventional and renewable sources—even in the United States, where energy is cheaper than in many other countries.[14] In other words, if energy efficiency were an energy-generating technology, it would be the cheapest in town. Or as physicist Amory Lovins mused, its positive savings mean that energy efficiency isn't just a free lunch; it's a free lunch you get paid to eat.[15]

Efficiency has social and environmental benefits, too. Any sizable energy-generation system will exert some sort of environmental and social impact. This is what makes energy efficiency initiatives so attractive: they are benign ways to solve energy dilemmas; that is, they do not solve problems by creating new problems. As the National Academy of Sciences noted, "Energy efficiency requires none of the environmental disruption seen in extracting coal, petroleum, natural gas, or uranium; depends on no wind turbines or hydroelectric dams or thermal power plants; emits no greenhouse gases or other pollutants; and can mitigate energy security risks associated with imported oil."[16]

For instance, the capacity of energy efficiency to reduce GHG emissions has resulted in efficiency initiatives taking center stage in many national climate change platforms. As justification for including energy efficiency improvements in state plans for curbing CO_2 emissions from existing US fossil power plants, the US Environmental Protection Agency found that significant improvements in end-use energy efficiency can be realized at less cost than the savings from avoided power system costs.[17]

There is substantial evidence that energy efficiency is the best way to mitigate GHG emissions while also meeting the growing requirements for energy services that accompany expanding economic growth.[18] A comprehensive report from the independent consulting firm McKinsey & Company concluded that a host of residential and industrial energy efficiency options were far more cost-effective than building power plants, even those running on natural gas or renewable fuels.[19] For the United States, McKinsey calculated that "a holistic approach would yield gross energy savings worth more than $1.2 trillion, well above the $520 billion needed through 2020 for upfront investment in efficiency measures (not including program costs). Such a program is estimated to reduce end-use energy consumption in 2020 by 9.1 quadrillion BTUs, roughly 23 percent of projected demand, potentially abating up to 1.1 gigatons of greenhouse gases annually."[20] Similarly, a peer-reviewed evaluation of the Energy Star program determined that from 1992 to 2006, the program saved 4.8 EJ of primary energy, circumvented 82 million tons of carbon-equivalent emissions, and would prevent a further 203 million tons of emissions from 2007 to 2015.[21] A separate study that examined the cumulative primary energy savings from the Energy Star program determined that it would save $70 billion worth of cumulative energy over the first decade of this century and that managing the program would cost only a small fraction of that amount.[22] A final study tells us that in the electricity sector of the United States, energy efficiency efforts can cost-effectively save 123 million tons of carbon per year, or one-fifth of all direct US household emissions.[23]

Proponents argue that energy efficiency initiatives also deliver a notable social benefit: job creation. Evidence indicates that investments in efficiency initiatives tend to generate job opportunities in industries that are comparatively labor-intensive, meaning that these initiatives produce more net jobs per dollar invested.[24] For example, California's energy efficiency initiatives have generated an estimated $56 billion in net economic benefits since 1972, yielding an employment dividend of 1.5 million jobs.[25] An IEA study also found that consumers and businesses redirect their energy bill savings from efficiency improvements to areas of the economy that are more labor-intensive and productive than energy purchases. The study concluded that reducing energy consumption by 15% during 1995 to 2010 resulted in 770,000 additional jobs, equivalent to a 0.44% increase in overall employment rate for nations of the Organization for Economic Cooperation and Development (OECD), generating $14 billion in additional income per year.[26]

Improvements in technology can also magnify future social and environmental benefits by acting as springboards to innovations. Think about the transformational nature of LED technology in comparison to older luminescent lightbulbs. These new lighting sources offer the same light quality at a small fraction of electricity consumed. Not only do LED devices save energy and reduce landfill costs, but the technologies that underpin them give rise to new product possibilities. Such advances help insulate societies from potential energy shocks and give them much-needed flexibility in addressing future threats. In support of this claim, complex modeling undertaken for the Global Energy Assessment considered multiple scenarios—including no nuclear power or expanded nuclear power, limited renewables or expanded renewables, and fossil fuels with or without carbon capture—and found that only one, an energy efficiency scenario, gave policymakers the adaptability to respond to likely or even unlikely changes in demand or supply choices.[27] This finding was replicated in an analysis of simulated disruptions to the US electricity system—such as those threatened by increasing climate variability. Energy efficiency played a key role in the least-cost approach to grid adaptation and recovery.[28]

Efficiency Improvements Are Replicable, Verifiable, and Firm

The notion that there are still millions of dollars in savings to be gleaned through energy efficiency programs serves as a provocation to free-market economists who find it hard to believe that competitive market conditions have not already resulted in optimal efficiency. Consequently, energy efficiency programs have been subjected to a very high level of scrutiny. Program evaluations and reviews of efficiency initiatives are routinely required, producing a wealth of information on which practices are most effective and which ones do not work.

Today we have an entire energy efficiency evaluation industry dedicated to enhancing methods, standards, and implementation of energy efficiency programs. Harmonized procedures now exist for the measurement and verification of energy savings. Energy service companies have been in the vanguard to verify energy savings, particularly for ratepayer-funded programs. One of the most important tools for measuring and verifying energy savings from energy efficiency projects is the International Performance Measurement and Verification Protocol (IMP).[29] The IMP has been translated into many languages and is used in many countries, including China, India, Brazil, Ukraine, the United Kingdom, and Australia. Well-developed measurement and verification protocols are also available for other types of efficiency programs, such as bidding into wholesale markets, third-party financing, and white certificates, which certify that 1 MWh of electricity savings or 1 Mcf of natural gas savings have been achieved. The French government manages such protocols in support of its Certificates d'économie d'énergie program, which has saved 255 TWh of electricity by energy suppliers over three years.[30]

A final important point in favor of energy efficiency is that it saves you more than you think and can be as reliable as—or more reliable than—conventional energy sources. One unit of energy saved can be more valuable than one unit generated or supplied. Let's examine why. Richard Cowart, director of the think tank Regulatory Assistance Project, estimated that every dollar invested in energy efficiency mitigated uncertainty associated with reduced load, wear, and maintenance needs for the entire fossil fuel chain, even in hours when reliability problems were not anticipated by system managers. This is because efficiency gains depressed the costs of locally used fuels such as oil, coal, and natural gas and reduced demand in peak hours, the most expensive times to produce power. Efficiency gains also lessened costly pollutants and emissions from generators, improved the reliability of existing generators, and moderated transmission congestion problems. Furthermore, these initiatives, once put in place, were always at hand—available without delay or needed intervention by system operators.[31] The New York Independent Systems Operator, for example, sets a reserve criterion of 18% during times of peak demand to ensure overall system reliability. Accordingly, each megawatt-hour of peak demand that customers avoid through energy efficiency means that utilities can subtract 1.18 MWh of total capacity needed. Quite literally, every 1 kWh avoided through energy efficiency equates to 1.18 kWh of avoided supply.[32] This could be why some advocates have suggested treating efficiency as equivalent to "firm power," a source of energy just as reliable as base-load power from nuclear power plants or dispatchable units running on natural gas.[33] It could also explain why the IEA referred to energy efficiency as an "invisible powerhouse" worth at least $310 billion per year.[34]

The Other Side: Managing Energy Demand Is Unnecessary and Ineffective

Those with an alternative view propose that, for a variety of reasons, energy efficiency efforts are unnecessary and managing demand is an ineffective way of responding to energy problems. Proponents of this view advance a number of arguments that challenge the effectiveness of energy efficiency initiatives: (1) energy markets already adequately incentivize profitable energy investments, (2) energy efficiency initiatives can curtail economic development and erode revenues, and (3) rebounds negate any discernible gains made through energy efficiency initiatives.

Functioning Energy Markets Are Already Efficient

Free-market economists are apt to argue that one will rarely find a $20 bill lying on a busy sidewalk because, inevitably, some passerby will have seen the value of the bill as sufficient to justify the time and effort spent in stooping down to pick it up. Applying this worldview to energy, many analysts assume that markets work so well that most cost-effective efficiency investments have already been made, and thus claims that efficiency initiatives can lead to sizable savings are invalid.[35] In other words, further energy efficiency initiatives come at a net detriment to society: "the marketplace determines a level of efficiency, and altering this energy efficiency level comes at a cost."[36] There is no need to promote energy efficiency because it will occur naturally as markets operate freely. Indeed, chapter 3 (on government intervention) is dedicated entirely to this topic.

Economists subscribing to this view use the term *Pareto optimality* to explain their perspectives. In a Pareto-efficient market, no trades remain that can make any individual better off without making another individual worse off. In such an environment, prices are assumed to be reasonably accurate reflections of total costs, and remonstrations over imperfect prices are greatly overstated.[37] As summarized by one group of economists, "markets should be considered innocent until proven guilty."[38] Implicit in this worldview is the acceptance of rational actor theory: consumers have access to complete information and base their decisions on optimizing personal utility.

In a world of utility-maximizing consumers and perfectly operating markets, government intervention produces suboptimal outcomes, even if such intervention is intended to improve efficiency. One study noted that "an efficiency investment with a positive return that does not make the cut in a marketplace has a net positive cost" because "it requires giving up something that consumers have revealed they value more."[39]

Inherent to this worldview is the belief that markets will logically work things out on their own: "the market does not fail to deliver energy supply, energy efficiency, or energy security. Private markets automatically perform cost/benefit analyses and ensure that long-run benefits to consumers are maximized."[40] Or, as another economist put it, "few additional incentives are needed to sell energy-efficient appliances or automobiles because the rewards are real and automatic."[41] In many states, industry is allowed to "opt out" of the energy efficiency programs run by utility companies because of a pervasive belief that "companies have already realized all the cost-effective industrial energy efficiency opportunities that exist."[42] According to Massachusetts Institute of Technology economist Paul Joskow, "estimates of un-tapped economical energy-efficiency opportunities are nothing more than fantasy."[43] National Bureau of

Economic Research economists Hunt Alcott and Michael Greenstone, in turn, note that "the empirical magnitudes of the investment inefficiencies [causing the energy efficiency gap] appear to be smaller, indeed substantially smaller, than the massive potential savings calculated in engineering analyses."[44] The libertarian think tank Cato Institute concludes that "the invisible hand of the marketplace is far superior in providing for efficient energy use and conservation than is the dead hand of government planners."[45]

Efficiency Raises Prices and Hampers Economic Development

More sobering is the criticism that energy efficiency initiatives can inflate costs, which undermines industrial competitiveness. The most direct link between energy efficiency and economic stagnation refers to the risk of "market destruction" and "revenue erosion." Such concerns were purported to be the foundation for the US decision not to ratify the Kyoto Protocol.[46] Similar sentiments have been expressed by former Chinese Premier Wen Jiabao with regard to China's climate change strategy.[47]

A core premise underpinning such fears stems from the belief that most profitable energy efficiency initiatives have already been tapped by corporations. Therefore, the initiatives that remain must inflate costs. For energy-intensive firms, these additional costs can be seen as eroding competitiveness. For power utilities, energy efficiency can be viewed as a threat to business growth. And for municipal gas utilities in the United States, conservation is often opposed because it is perceived to "destroy markets" for gas by simultaneously lowering both demand and price.[48] Others note that energy efficiency has eroded electricity and gas utility revenues as customers consume fewer units of energy in their homes and factories.[49]

Many utility planners and analysts argue that energy efficiency should be seen as a customer service, not as a utility resource that can be mapped onto a utility supply curve alongside nuclear energy, coal, natural gas, and renewables.[50] Efficiency practices, at best, represent only "a very oblique approach that seems to owe more to the current tide of green favor than to sober consideration of the facts."[51]

Efficiency Can Lead to Dangerous Rebounds

Opponents point to evidence that enhanced energy efficiency can actually lead to increased energy consumption, as demand "rebounds" in response to cheaper energy services. This rebound or "take-back" effect is manifested in two ways: directly, through amplified usage, and indirectly, through additional consumption that may come from efficiency savings.[52] As economist Daniel J. Khazzoom explains, if a car's efficiency triples, the gasoline required to meet old travel requirements drops by two-thirds, and the driver can now travel three miles for the price of one. This is the economic equivalent of saying that the price of gasoline has dropped significantly from what it used to be, incentivizing further travel.[53]

A variant of this thinking is known as the "Jevons Paradox," which argues that energy efficiency yields savings that liberate resources for people to employ elsewhere, often in activities that consume more net energy.[54] There is an intuitive logic to this. If you decide to sell your car and use public transportation instead, some of the operational savings go toward public transportation and some might wind up channeled into new purchases of household appliances, computers, smartphones, and so on, or perhaps a vacation to the Maldives.

There is some empirical evidence to suggest that the rebound effect is real. As table 2.1 illustrates, the magnitude of the reduction in energy savings caused by the rebound effect appears to be wide-ranging and associated with a broad array of efficiency initiatives, ranging from food and heating to electricity and transport.[55] In most cases the rebound effect seems to be far less than the original amount of energy savings, but in some cases the rebound effect can negate the initial savings. Indeed, energy analyst Jaume Freire-González contends that energy efficiency programs can amplify consumption to exceed the energy saved by the efficiency measures.[56] A European Commission report echoes this concern in the case of solid state lighting (SSL) and LEDs: "In the future, SSL applications may be widely deployed beyond the mere replacement of existing lighting systems, such as integration into furniture or buildings. In the long run, this could reduce the expected energy savings."[57]

The true impact of the rebound effect is hotly debated, perhaps because the causal links are yet to be fully validated. Some scholars argue that energy rebounds may be real but are small relative to the typical energy savings. For OECD countries, long-run rebound effects for household heating and cooling have been estimated at less than 30%.[58] Energy efficiency initiatives centering on lighting and appliances have smaller rebound effects, estimated to be around 0% and 5% to 12%, respectively.[59] Another meta-survey found that direct rebound effects ranged from 1% to 15% for space heating and electrical appliance usage and from 3% to 22 % for transportation—not nearly enough to offset most of the energy efficiency gains achieved.[60] One IEA study concluded that the rebound effect was very small over the lifetime of a given energy efficiency initiative: less than 10% for residential appliances, residential lighting, and commercial lighting; less than 20% for industrial process improvements and innovations in automobile efficiency.[61] As four economists recently noted in a meta-survey of the topic in *Nature*, "a vast academic literature shows that rebounds are too small to derail energy-efficiency policies."[62]

There is some empirical support for such a contention. Globally over the past 40 years, for instance, the energy-to-GDP ratio has declined steadily, suggesting

Table 2.1. Estimates of Direct and Indirect Household Rebound Effects

Number of commodity groups studied	Abatement action	Area	Measure	Effects captured	Energy/emissions	Estimated rebound effect (%)
150	Efficiency and sufficiency	Food; heating	Greenhouse gases	Income	Direct and indirect	45–123
300	Sufficiency	Food; travel; utilities	Carbon	Income	Direct and indirect	7–300
13	Efficiency	Transport; utilities	Carbon	Income and substitution	Direct and indirect	120–175
13	Efficiency	Transport; utilities	Energy	Income and substitution	Direct and indirect	12–38
6	Efficiency	Transport; heating; electricity	Energy	Income and substitution	Direct only	37–86
16	Sufficiency	Transport; heating; food	Greenhouse gases	Income	Direct and indirect	7–51
74	Efficiency	Transport; electricity	Greenhouse gases	Income	Direct and indirect	7–25
36	Efficiency and sufficiency	Transport; lighting	Greenhouse gases	Income	Direct and indirect	5–40

Source: Modified from Mona Chitnis et al., "Turning Lights into Flights: Estimating Direct and Indirect Rebound Effects for UK Households," Energy Policy 55 (April 2013): 234–250.

that economic growth can occur without a corresponding increase in energy consumption.[63] Nevertheless, there is a clear upward trend in demand for electronic devices, and whether that is due to a rebound effect or to increases in economic affluence, it cannot be addressed through energy efficiency initiatives alone.

Common Ground: Efficiency Should Be Balanced with Investments in Supply

At first blush, the two views appear utterly incompatible. One camp argues that the potential for energy efficiency initiatives is limited because markets have already exploited the cost-effective opportunities and contends that, because of the rebound effect, policies to promote energy efficiency have been far less successful than they appear on the surface. The opposing camp's position can be encapsulated in this analogy: Two economists are walking down the street and come across a $20 bill lying on the sidewalk. "Look!" exclaims one of the economists. "Isn't that a $20 bill?" "Of course not," replies the companion. "If it were a $20 bill, someone would have picked it up." With that, the two economists step over the $20 bill and resume their walk.

There is a sufficient body of evidence to indicate that there are indeed numerous $20 bills lying around for innovative minds to pick up. The rebound effect may be real, but it does not negate the value of energy efficiency investments. Simply put, using energy (or any other resource) more efficiently improves economic well-being. To say that it is not a good idea to improve energy efficiency because it leads to further consumption is like saying it is not a good idea to graduate from elementary school because it will simply lead to further education at the secondary level.

When energy efficiency skeptics prevail, expanding supply becomes the only way to meet increasing energy demand. Capital is diverted to large-scale projects employing technologies that evolve slowly because there is little incentive to innovate. When the energy efficiency advocates prevail, there is an infusion of new jobs in labor-intensive industries employing building appliance and equipment installers, motor specialists, lighting contractors, energy management professionals, and retrofitters. In such a world, the revenues of energy suppliers erode unless they join the supply chain of energy efficiency services and are appropriately rewarded.

In reality, both points of view hold some virtue. One area of common ground is shown in box 2.1. Properly designed energy efficiency programs may be justified when markets fail, when barriers prevail, and when the rebound effect is less than the efficiency savings. For instance, one relatively simple regulatory reform can simultaneously promote efficiency and protect energy companies from revenue erosion: decoupling the profits of electricity and natural

BOX 2.1

Energy Efficiency and the Common Ground

Is energy efficiency a worthwhile investment?

ONE SIDE: Energy efficiency interventions represent a least-cost, environmentally friendly, socially productive, verifiable way to save large amounts of energy across a variety of sectors.

OTHER SIDE: Energy efficiency programs represent an unnecessary and potentially dangerous intervention in energy markets, leading to the erosion of revenues and substantial rebound effects.

COMMON GROUND: We need to balance investments in efficiency with "supply-side" options and ensure best practices are met.

gas utilities from their sales volumes. Some US states are adopting this policy, which allows electricity utilities to keep a small share of the savings they achieve for their customers. In other words, the utilities are "rewarded for cutting your bill, not for selling you more energy."[64] After decoupling, Idaho Power's investments in demand-side management programs tripled between 2006 and 2009, and energy savings increased 220% (to 148 GWh per year); California utilities saw a fivefold increase in investments in energy efficiency from 1998 and 2008; and Utah saw about $50 million a year in savings of natural gas.[65]

The economist Harry Saunders adds that if forecasts and projections of efficiency improvements explicitly accounted for rebound effects, then even though rebounds could never be eliminated, they could be planned for and thus minimized.[66] The Intergovernmental Panel on Climate Change (IPCC) surveyed hundreds of studies on rebound effects across numerous sectors around the world and noted that these effects could be diminished by strong policy intervention and/or increases in energy prices.[67] If energy efficiency savings are used for expanded consumption and this is deemed to be a socially or environmentally undesirable outcome, negative impacts can be managed with strategic policies.

A final argument synthesizing both views is that energy efficiency, by itself, is necessary but insufficient for addressing all of our energy problems. This is partly because new sources of energy supply are needed to expand access to modern energy services for the billions of people that do not have it, and partly because increases in per capita energy demand—driven by increased living standards, larger houses, more cars, and more electronic devices—continue to partially offset improvements in efficiency. In other words, both energy efficiency and smart, strategic investments in clean supply are needed to meet future energy demand in a manner that will not exacerbate climate change. A more sustainable future requires a balance of investment in energy supply options to meet the growing needs of society and in demand-side options to reduce energy waste.

NOTES

1. These two paragraphs draw from M. A. Brown and B. K. Sovacool, "A Source of Energy Hiding in Plain Sight," *Yale Global Online*, February 18, 2009.

2. E. Vine, M. Kushler, and D. York, "Energy Myth Ten—Energy Efficiency Measures Are Unreliable, Unpredictable, and Unenforceable," in *Energy and American Society—Thirteen Myths*, ed. B. K. Sovacool and M. A. Brown (New York: Springer Publishing, 2007), 265–288.

3. Energy intensity refers to the energy consumed per dollar of GDP. The original figure of $438 billion in 2000 is adjusted here to 2007 dollars.

4. Peter A. Seligmann and Michael Totten, "Pursuing Sustainable Planetary Prosperity" (2013), 18, available at www.chinausfocus.com/wp-content/uploads/2013/05/Chapter-18.pdf.

5. Ibid.

6. P. J. S. Siderius and H. Nakagami, "A MEPS Is a MEPS Is a MEPS: Comparing Ecodesign and Top Runner Schemes for Setting Product Efficiency Standards," *Energy Efficiency* 6, no. 1 (2012): 1–19. See also K. Osamu, "The Role of Standards: The Japanese Top Runner Program for End-Use Efficiency: Historical Case Studies of Energy Technology Innovation," in *The Global Energy Assessment*, ed. A. Grubler et al. (Cambridge: Cambridge University Press, 2012), chap. 24; Osamu Kimura, *Japanese Top Runner Approach for Energy Efficiency Standards* (Tokyo: Central Research Institute of the Electric Power Industry, 2012).

7. International Energy Agency (IEA), *Energy Efficiency Market Report 2013* (Paris: OECD, 2013).

8. Sarah Rabkin and David Gershon, "Changing the World One Household at a Time: Portland's 30-Day Program to Lose 5,000 pounds," in *Creating a Climate for Change: Communicating Climate Change and Facilitating Social Change*, ed. Susanne C. Moser and Lisa Dilling (Cambridge: Cambridge University Press, 2007), 292–302.

9. Georgina Santos, Wai Wing Li, and Winston T. H. Koh, "Transport Policies in Singapore," in *Road Pricing: Theory and Evidence*, ed. Georgina Santo (Oxford: Elsevier, 2004), 209–235.

10. Marilyn A. Brown, "Innovative Energy-Efficiency Policies: An International Review," *Wiley Interdisciplinary Reviews (WIREs): Energy and Environment* 4, no. 1 (2015): 1–24.

11. M. Croucher, "Potential Problems and Limitations of Energy Conservation and Energy Efficiency," *Energy Policy* 39, no. 10 (2011): 5796.

12. Committee on Benefits of DOE R&D on Energy Efficiency and Fossil Energy, Board on Energy and Environmental Systems, Division on Engineering and Physical Sciences, *Energy Research at DOE: Was It Worth It? Energy Efficiency and Fossil Energy Research 1978 to 2000* (Washington, DC: National Research Council, 2001).

13. B. K. Sovacool, "Affordability and Fuel Poverty in England," in *Energy and Ethics: Justice and the Global Energy Challenge* (New York: Palgrave MacMillan, 2013), 43–65.

14. Yu Wang and Marilyn A. Brown, "Policy Drivers for Improving Electricity End-Use Efficiency in the USA: An Economic-Engineering Analysis," *Energy Efficiency* 7 (2014): 517–546.

15. Loins quoted in Robert Bradley Jr., *Capitalism at Work: Business, Government, and Energy* (Salem, MA: M&M Scrivener Press, 2009), 251.

16. America's Energy Future Energy Efficiency Technologies Subcommittee, *Real Prospects for Energy Efficiency in the United States* (Washington, DC: National Academy of Sciences, National Academy of Engineering, and National Research Council, 2010), ix–x.

17. Electric Power Research Institute, *U.S. Energy Efficiency Potential through 2035: Final Report* (Palo Alto, CA: Electric Power Research Institute, April 2014); Wang and Brown, "Policy Drivers."

18. M. A. Brown and B. K. Sovacool, *Climate Change and Global Energy Security: Technology and Policy Options* (Cambridge, MA: MIT Press, 2011).

19. McKinsey & Company, *Impact of the Financial Crisis on Carbon Economics: Version 2.1 of the Global Greenhouse Gas Abatement Cost Curve, 2010* (Boston: McKinsey & Company, 2010), available at www.mckinsey.com/client_service/sustainability/latest_thinking/greenhouse_gas_abatement_cost_curves.

20. McKinsey & Company, *Unlocking Energy Efficiency in the U.S. Economy* (Boston: McKinsey & Company, 2009), iii.

21. Marla C. Sanchez et al., "Savings Estimates for the United States Environmental Protection Agency's Energy Star Voluntary Product Labeling Program," *Energy Policy* 36, no. 6 (June 2008): 2098–2108.

22. R. Brown, C. Webber, and J. G. Koomey, "Status and Future Directions of the Energy Star Program," *Energy* 27, no. 5 (May 2002): 505–520.

23. Omar I. Asensio and Magali A. Delmas, "Nonprice Incentives and Energy Conservation," *Proceedings of the National Academy of Sciences USA* (in press, 2015).

24. American Council for an Energy-Efficient Economy (ACEEE), *Fact Sheet: Energy Efficiency and Job Creation* (Washington, DC: ACEEE, 2011); Paul Baer, Marilyn A. Brown, and Gyungwon Kim, "The Job Generation Impacts of Expanding Industrial Cogeneration," *Ecological Economics* 110 (2015): 141–153.

25. Ralph Cavanagh, "Graphs, Words, and Deeds: Reflections on Commissioner Rosenfeld and California's Energy Efficiency Leadership," *Innovations* 4, no. 4 (Fall 2009): 81–89.

26. Howard Geller and Sophie Attali, *The Experience with Energy Efficiency Policies and Programs in IEA Countries* (Paris: IEA, August 2005).

27. K. Riahi et al., "Energy Pathways for Sustainable Development," in *Global Energy Assessment—Toward a Sustainable Future* (Cambridge: Cambridge University Press; Laxenburg, Austria: International Institute for Applied Systems Analysis, 2012), 1203–1306.

28. Alexander Smith and Marilyn A. Brown, "Policy Considerations for Adapting Power Systems to Climate Change," *Electricity Journal* 27, no. 9 (2014): 112–125.

29. US Department of Energy (DOE), *International Performance Measurement and Verification Protocol* (Washington, DC: DOE, 2002), doi:DOE/GO-102002-1554.

30. IEA, *Energy Efficiency Market Report 2013*.

31. Richard Cowart, *Efficient Reliability: The Critical Role of Demand-Side Resources in Power Systems and Markets* (Washington, DC: National Association of Regulatory Utility Commissioners, June 2001).

32. Charles Komanoff, "Securing Power through Energy Conservation and Efficiency in New York: Profiting from California's Experience" (report for the Pace Law School Energy Project and the Natural Resources Defense Council, Pace Law School, May 2002), 1–22.

33. Marilyn A. Brown and Benjamin K. Sovacool, "Promoting a Level Playing Field for Energy Options: Electricity Alternatives and the Case of the Indian Point Energy Center," *Energy Efficiency* 1, no. 1 (2008): 35–48.

34. IEA, *Energy Efficiency Market Report 2014* (Paris: OECD, 2014).

35. J. Taylor and P. Van Doren, "Energy Myth Five—Price Signals Are Insufficient to Induce Efficient Energy Investments," in Sovacool and Brown, *Energy and American Society*, 125–144.

36. L. J. Makovich, *The Cost of Energy Efficiency Investments* (Cambridge, MA: CERA, 2008), 2.

37. Taylor and Van Doren, "Energy Myth Five."

38. Adam B. Jaffe and Robert N. Stavins, "The Energy-Efficiency Gap," *Energy Policy* 22 (1994): 805.

39. Makovich, *Cost of Energy Efficiency Investments*, 15.

40. R. J. Sutherland and J. Taylor, "Time to Overhaul Federal Energy R&D," *Policy Analysis* 424 (2002): 2.

41. F. P. Sioshansi, "Restraining Energy Demand," *Energy Policy* 22, no. 5 (1994): 380.

42. A. M. Shipley and R. N. Elliot, *Ripe for the Picking: Have We Exhausted the Low-Hanging Fruit in the Industrial Sector?* (Washington, DC: American Council for an Energy-Efficient Economy, 2006), 3.

43. Paul Joskow, "Utility-Subsidized Energy-Efficiency Programs," *Annual Review of Energy and the Environment* 20 (1995): 529

44. Hunt Allcott and Michael Greenstone, "Is There an Energy Efficiency Gap?" *Journal of Economic Perspectives* 26, no. 1 (2012): 3.

45. J. Taylor, "Energy Conservation and Efficiency: The Case against Coercion," *Policy Analysis* 189 (1993): 1.

46. "Bush: Kyoto Treaty Would Have Hurt US Economy," *NBCNews.com*, June 30, 2005, available at www.nbcnews.com/id/8422343/ns/politics/t/bush-kyoto-treaty-would -have-hurt-economy/#.U1326FWSx8E.

47. P. Christoff, "Cold Climate in Copenhagen: China and the United States at COP15," *Environmental Politics* 19, no. 1 (2010): 637–656.

48. "Municipal Gas Utilities Focus on Price Impacts and Marketing Strategy," *Pipeline and Gas Journal*, January 1, 2007, 29.

49. Ken Costello, *The Challenges of Ratemaking for State Utility Commissions* (National Regulatory Research Institute, Silver Spring, MD: NARUC Subcommittee on Gas, February 9, 2014).

50. P. Joskow and D. Marron, "What Does a Megawatt Really Cost? Further Thoughts and Evidence," *Electricity Journal* 6, no. 6 (1993): 14–26.

51. L. Brookes, "The Greenhouse Effect: The Fallacies in the Energy Efficiency Solution," *Energy Policy* 18 (1990): 199.

52. D. Owen, "The Efficiency Dilemma," *New Yorker*, December 10, 2010, 78–85. See also L. A. Greening, D. L. Greene, and C. Difiglio, "Energy Efficiency and Consumption—The Rebound Effect: A Survey," *Energy Policy* 28, no. 6–7 (2000): 389–401.

53. Daniel J. Khazzoom, "Economic Implications of Mandated Efficiency Standards for Household Appliances," *Energy Journal* 1, no. 4 (1980): 21–39.

54. Blake Alcott, "Jevons' Paradox," *Ecological Economics* 54, no. 1 (July 1, 2005): 9–21.

55. Mona Chitnis et al., "Turning Lights into Flights: Estimating Direct and Indirect Rebound Effects for UK Households," *Energy Policy* 55 (April 2013): 234–250.

56. Jaume Freire-González, "Empirical Evidence of Direct Rebound Effect in Catalonia," *Energy Policy* 38, no. 5 (2010): 2309–2314.

57. European Commission, *Lighting the Future: Accelerating the Deployment of Innovative Lighting Technologies*, Green Paper (Brussels: European Commission, 2011), 3.

58. S. Sorrell, J. Dimitropoulos, and M. Sommerville, "Empirical Estimates of the Direct Rebound Effect: A Review," *Energy Policy* 37 (2009): 1356–1371.

59. Greening et al., "Energy Efficiency and Consumption."

60. Figures based on table 1 in Ines M. L. Azevedo, "Consumer End-Use Energy Efficiency and Rebound Effects," *Annual Review of Environment and Resources* 39 (2014): 393–418.

61. Howard Geller and Sophie Attali, *The Experience with Energy Efficiency Policies and Programs in IEA Countries* (Paris: IEA, August 2005).

62. Kenneth Gillingham et al., "The Rebound Effect Is Overplayed," *Nature* 493 (January 2013): 475–476.

63. D. Cullenward et al., "Psychohistory Revisited: Fundamental Issues in Forecasting Climate Futures," *Climatic Change* 104, no. 3–4 (February 2011): 457–472.

64. Matt J. Hirschland, Jeremy M. Oppenheim, and Allan P. Webb, "Using Energy More Efficiently: An Interview with the Rocky Mountain Institute's Amory Lovins," *McKinsey Quarterly* (online), July 2008.

65. Dylan Sullivan, Devra Wang, and Drew Bennett, "Essential to Energy Efficiency, but Easy to Explain: Frequently Asked Questions about Decoupling," *Electricity Journal* 24, no. 8 (October 2011): 56–70.

66. Harry D. Saunders, "Historical Evidence for Energy Efficiency Rebound in 30 US Sectors and a Toolkit for Rebound Analysts," *Technological Forecasting and Social Change* 80, no. 7 (September 2013): 1317–1330.

67. Ottmar Edenhofer et al., *Technical Summary of Working Group III: Mitigation of Climate Change* (Geneva: IPCC, 2014).

Should Governments Intervene in Energy Markets?

The economist Milton Friedman once opined that "if you put the federal government in charge of the Sahara Desert, in five years there'd be a shortage of sand."[1] This jaded perspective on government efficacy underpins the creed of free-market proponents: when given a choice between market-led or government-led development initiatives, the market will always produce a more efficient outcome. But it does give rise to some interesting questions. When should markets be regulated or incentivized? When is a particular bundle of goods and services best provided by a government?

This chapter presents two competing theses relating to government intervention ("picking winners") in energy markets. The first is that energy technologies, like most other goods, ought to be decided by market forces alone. Unlike the slow, heavy, rigid hand of the state, free markets optimize the allocation of resources. Free-market advocates argue that the problem of managing the negative energy externalities that litter the chapters of this book can be accommodated by the market; this merely requires that externalities be priced and enfolded into energy tariffs. Picking technology winners, as some governments have done, only distorts the efficiency of the market. If the government must intervene, it should utilize market instruments such as taxes and tradable permits rather than command-and-control regulations, which suffer from the same fetters of government inefficiency.

The competing thesis responds that energy systems ought to be decided by governments. The idea of perfect energy markets, according to proponents of this thesis, is a myth, and pricing externalities is more difficult than it seems. Markets are prone to the inescapable perils of power and strategic gaming, and what's more, "picking winners" in the energy sector is a strategy that has worked before.

One Side: The Energy Mix Should Be Decided by Free-Market Forces

The free-market position that government should cede to free-market forces the decision over which energy technologies to adopt is predicated on at least four tenets: (1) free markets allocate resources in a manner that optimizes social utility, (2) the externalities that distort decision making in free-market environments can be rectified by internalizing them into prices, (3) picking winners unnecessarily and artificially limits options, and (4) market tools also have a rich and successful history of addressing environmental problems.

Free Markets Allocate Resources Efficiently

The argument that markets usually work better than governments at producing efficient outcomes begins by pointing to the benefits stemming from a free-market ideology.[2] In theory, the market economy is a powerful force for making our lives better, and the only way that firms can make profits is by delivering goods that we want to buy. The free-market system uses demand-driven prices rather than government ideology or company strategy to allocate scarce resources. Since there is a finite amount of everything worth having, the most basic function of any economic system is to decide who gets what. By relying on prices to allocate goods, markets are seen as self-correcting. In perfect markets, every market transaction maximizes the net benefits that consumers and producers receive from participating in markets. Firms are acting in their own best interests, and so are consumers. In the short term, there might be some negative aspects of this system, but over the longer run the market does indeed respond to the needs of consumers and producers. Ultimately, economist Charles Wheelan concludes, "a market economy is to economics what democracy is to government: a decent, if flawed, choice among many bad alternatives."[3]

Those seeking examples of free-market effectiveness in the energy sector often point to the demerits of a government-run electricity-supply industry. Until the 1980s, most electric utilities remained state-owned, vertically integrated monopolies.[4] Many of these electricity suppliers and government-linked companies were overstaffed and opaque and denied ratepayers the ability to choose service providers.[5] They were, in short, inefficient. However, starting in the 1980s, many electricity markets were reformed through initiatives such as commercializing electricity enterprises and requiring that they pay taxes and market-based interest rates, unbundling utility services, and developing market regulations to be overseen by an agency insulated from political and commercial influence.[6] In many nations, when the electricity sectors were opened up to market forces, decisions and resources became more efficient and profitable. In developing countries, where rates of corruption and regulatory capture are high, evidence indicates that market reforms improve transparency and accountability.[7] Liberalized or restructured electricity sectors also tend to exhibit improvements in system-wide efficiency, lower tariffs, and a greater degree of desirable consumer choice.[8]

Modern-day proponents extend the arguments in favor of market-led development to the energy sector as a whole. As one US senator campaigning for reelection in 2014 put it:

> Unfortunately, government regulations and restrictions have for many years enacted policies that favor some forms of energy production over others. Governments shouldn't be picking winners and losers; they should allow all forms of energy production without giving special benefits or incentives to a select few . . . The protection of our environment will never be accomplished by central planners. Only individuals and communities voluntarily working together with private, free-market incentives will be able to achieve that goal. The right policy is energy freedom. As with every other example, the free market allows businesses and ideas to compete freely to produce the most efficient forms of energy at the lowest possible cost to consumers.[9]

A study in 2012 declared that "a free market . . . is the fastest and most efficient way to find the energy balance of the future."[10]

But what should we make of the negative externalities associated with free markets—the social and environmental losses for the sake of corporate gains?

Externalities Can Be Internalized

Those in favor of market forces acknowledge the existence of various externalities associated with the energy sector (perhaps the biggest of them being climate change). However, they contend that such externalities can be quantified and then incorporated into energy prices—sometimes with minimal government involvement, other times without any government involvement at all and led by the private sector. Neoclassical economists argue that restoring economic order to markets that are affected by externalities is straightforward: internalize them. Once this is done, consumers can proceed with the free-market perk of choice: deciding how many kilowatt-hours of nuclear power and how many kilowatt-hours of wind power they would prefer to consume. Properly addressing these externalities, the thinking goes, is simply a matter of "getting the price right."[11]

Pricing energy in this way, moreover, brings numerous advantages. It can convince firms to quit polluting.[12] Pricing externalities can bring issues of utility and justice into the marketplace, front and center. As they stand now, externalities are undemocratic and exclusionary. The most affected parties— often the poor or disenfranchised—are underrepresented in the marketplace and have external costs imposed upon them,[13] particularly air pollution and toxics.[14] Pricing externalities properly aligns price signals. It's not that we don't pay for energy externalities as a society; we do so through hospital admissions, lost worker days, higher taxes, blighted landscapes, death, pain, and suffering.

Society and nature "pay" eventually; it's just that those costs don't show up on our energy and gasoline bills. Put simply: the price of energy generated by a given technology rarely matches its cost.[15] Including externalities in a comprehensive manner rectifies this.

Picking Winners Limits Choices

A third reason that market forces work best, proponents suggest, is that they avoid artificially limiting choices through command-and-control-style decision making and regulation. Command-and-control policies explicitly set fixed targets to be achieved and dictate the methods to achieve them. The expectation is that such efforts enable planners or regulators to directly, appropriately, feasibly, and effectively address problems by mandating a desired solution. While command-and-control policies may sound good in principle, ecologist C. S. Holling and zoologist G. K. Meffe hold that "the command-and-control approach implicitly assumes that the problem is well-bounded, clearly defined, relatively simple, and generally linear with respect to cause and effect. But when these same methods of control are applied to a complex, non-linear, and poorly understood natural world, and when the same predictable outcomes are expected but rarely obtained, severe ecological, social, and economic repercussions result."[16] A long list of esteemed scholars have therefore attacked command-and-control measures for producing suboptimal outcomes and compromising efficiency. Moreover, critics contend that these measures notoriously suffer from sluggish bureaucratic procedures,[17] and they cite numerous cases where planners pick clearly inferior technical solutions to appease particular industries—think about nuclear reactors on the Japanese coastline, corn-based ethanol distilleries in the United States, or coal-fired power plants in Indonesia and China.[18]

Picking winners might not only entrench obsolete or undesirable systems; in a world of limited resources, it also risks missing the mark. For instance, suppose one wanted to invest $1 trillion in cleaner transport infrastructure. In the United States, the budget could be allocated in the following manner for natural gas: $210 billion invested in upstream infrastructure, $290 billion in long-distance transmission pipelines, and $500 billion in fueling stations and distribution networks.[19] Alternatively, the budget could be allocated to electric vehicles: $550 billion in commercial charging stations and $450 billion for enhanced battery research.[20] Or, the funds allocated might be just about enough to build long-distance pipelines and filling stations necessary to fuel 100 million hydrogen vehicles.[21] Picking any one of these three "winners" would exhaust resources to explore or develop the other two; therefore, picking one technology is mutually exclusive with picking the others: it can sideline other technologies, stopping them dead in their tracks.

Market Tools Work

Finally, advocates contend that if pesky difficulties still bedevil the free and full functioning of energy markets, one can and should implement market-based policy tools to solve them. Indeed, the past few decades have seen the advent of auctions, pollution and effluent charges, tradable permits, and a variety of other tools that incentivize market players to resolve energy and environmental problems. Economists and policy analysts have called on such tools to improve the efficiency of food production and agriculture,[22] regulate logging and forestry,[23] phase out leaded gasoline and chlorofluorocarbons,[24] reduce acid rain, meet air quality standards,[25] distribute taxicab licenses, set quotas on tobacco and dairy products,[26] restore ecosystems, enhance water quality,[27] encourage sustainable fisheries,[28] limit greenhouse gas emissions,[29] create common property rights to protect elephant herds,[30] and even control population growth.[31] One survey found nine applications of tradable permits to improve ambient air quality, 75 applications in fisheries, three for managing water resources, five for controlling water pollution, and five for regulating land use.[32]

The driving force behind these market-oriented policy mechanisms is the belief that a system of transferable property rights enables planners to choose an "optimal" level of pollution by setting a fixed number of pollution licenses or allowances that are then either sold or assigned to polluters. The issuance of tradable licenses engenders an open market that in turn establishes an explicit price for the right to pollute. In theory, the market automatically ensures that the reduction in pollution occurs at the least possible cost, but over time, as the price of pollution rises (through limiting or contracting availability of the pollution permits), it becomes economically disadvantageous for firms to continue to pollute and technological change is incentivized. Economists Maureen Cropper and Wallace Oates tell us that by giving participants direct control over compliance strategies, tradable permits and quotas often improve efficiency.[33] Jan-Peter Voss writes that tradable credits have become their own regime on the international stage and "something of a global standard in environmental governance."[34] Law professor Jonathan Adler writes that "the empirical evidence shows quite clearly that ecological concerns are better cared for when incorporated into market institutions."[35]

Such licenses or credits can take a variety of forms, from the regional cap-and-trade systems in the United States and Europe to the baseline credit markets seen as part of the Kyoto Protocol's Clean Development Mechanism. Some of the approaches do indeed have a record of success. Following the 1990 Clean Air Act Amendments—which created an Acid Rain Program that permitted firms to trade credits for sulfur dioxide—wet sulfate deposition, the major component of acid rain, declined by 25% to 40% from 1990 to 2006 across most areas of the northwestern and midwestern United States. Annual ecological and health benefits from the Acid Rain Program's emission reductions have been estimated at $142 billion and include averting nearly 19,000 premature deaths between 1990 and 2010. Meanwhile, annual compliance costs associated with the program have amounted to only $3.5 billion.[36] A more recent study that uses updated health impact models estimated the annual cost of the Clean Air Act at around $3 billion but total monetized benefits (mostly from reduced particulate matter pollution) of $100 billion.[37] Clearly, market-based tools seem to operate more efficiently and cheaply than government intervention.

The Other Side: The Energy Mix Should Be Decided by Governments

Advocates for government control over the energy mix base their claims on four streams of logic: (1) energy markets are rigged games, they never work as planned; (2) opponents contest the ability of economists to assign value to externalities in a manner that is devoid of political interference and contentious value judgments; (3) debate over the ability to internalize external costs is moot because wealthy stakeholders with vested interests in preserving the status quo would block attempts to effectively internalize external costs or would "game" market instruments and carbon trading schemes; and (4) many successful energy transitions have been predicated on strong intervention—that is, picking winners. In sum, these streams support the contention that government can be "a leading agent in achieving the type of innovative breakthroughs that allow companies, and economies, to grow."[38]

Perfect Energy Markets Are a Fiction

If we all lived within the pages of an economics textbook, it would be easy to support the proposition that public entities have track records of ineffective management. In the abstract world of economic theory, unfettered competition maximizes production efficiencies and personal utility. Everyone enjoys equal access to market information, labor and capital are immobile, and fishing for salmon or climbing trees for coconuts (or some other nonsubstitutive product) would typically represent the only type of economic activity. In such a world, the "invisible hand" of the market ensures that consumer preferences for coconuts or salmon reach a state of maximum consumer utility. Any temporary deviations from this blissful state of utility maximization are naturally rectified without the need for salmon-bone spears and coconut rakes. Clearly, government intervention is superfluous in this abstract world; but then again, so are economists.

Fortunately for those of us who aspire to a diet that extends beyond salmon and coconuts, we do not live within the pages of an economics textbook;

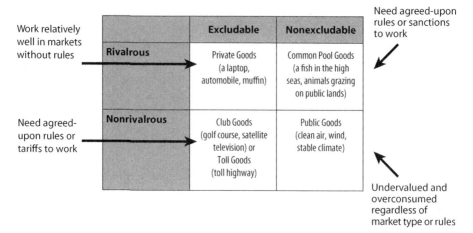

Figure 3.1. Typology of Public and Private Goods and Market Efficiency.

lamentably, the real world poses some practical problems that compound the task of applying economic theory to energy planning. One fundamental problem is that markets "work" only at distributing certain types of goods. They tend to be efficient at distributing private goods such as bicycles or hamburgers— where property rights can be completely defined and protected, owners can exclude others from access, and property rights can be transferred or sold— but less effective for common pool goods or club goods, which need agreed-upon rules or sanctions (figure 3.1). Unfettered economic markets are almost completely ineffective at distributing public goods such as clean air or improved energy security.

In addition, critics of free markets would argue that in energy markets, competition is not unfettered. Powerful firms and stakeholders leverage financial might to preserve self-interests. Market information is not symmetric: those who enjoy wealth and power also enjoy privileged access to information, thereby gaining a market advantage. Labor and capital are mobile, and this means that the Holy Grail of economic theory—comparative advantage—turns out to be a sham. Capital flows from high-cost markets to low-cost markets. Jobs in established markets are lost, and new empires arise in developing nations.[39] Some of these nouveau riche are actually vieux riche—"old wealth" that has been savvy enough to follow the money—while other members of the nouveau riche are truly new players on the global economic stage.

"Now wait a minute!" a thoughtful reader might remark. "Economists are very bright people. I know this because they wear really nice suits, travel in chauffeured vehicles, and often receive prestigious awards. Clearly they must be aware of these real-world deviations from economic theory!" Well, most economists are aware of these deviations, and there is a solution; or so they contend. They argue that markets require at least six interrelated components—information, minimal transaction costs, rational consumers, competition, internalization, and excludability—to function properly. The solution to energy problems, then, is simple: ensure that the six criteria in table 3.1 for efficient markets are met and, voila, everything works out.

Critics argue that the six criteria for efficient markets can never be met. As table 3.1 indicates, each of the six components suffers from serious, real-world challenges. The invalidation of the six criteria constitutes a market failure or market imperfection, a circumstance where the allocation of resources and prices will not yield efficient outcomes without government intervention. As the Nobel Prize–winning economist Joseph Stiglitz pointed out, there is only one way that markets can be perfect but a virtually infinite number of ways they can be imperfect, so we should expect the latter, not the former.[40] This "market-failure model" of public policy, the argument runs, justifies strong-handed government regulation or interference to fix broken markets, a theme that crops up for many of the topics in this book, including energy efficiency (chapter 2), renewable electricity (chapter 6), electric vehicles (chapter 7), clean coal (chapter 11), and nuclear power (chapter 12).

Valuing Externalities Is Impossible

If the free market is to be responsible for guiding development of the energy sector, all of the costs need to be assigned to the relevant energy technologies in order to compare costs equitably; but this never happens. The rise to prominence of nuclear power in Japan illustrates how failing to price externalities distorts market appeal. In 2007, Japan's Ministry of Economy, Trade, and Industry estimated the cost of nuclear power generation to be in the range of ¥4.8 to ¥6.2 per kilowatt-hour. This makes nuclear power the cheapest source of energy in Japan—even cheaper than coal-fired power, which was estimated in the range of ¥5.0 to ¥6.5 per kilowatt-hour.[41] In reality, the ministry's estimate failed to account for long-term nuclear waste disposal and the cost of major accidents such as Fukushima. It also excluded near-term expenses and funding for spent fuel management that the government has poured into nuclear power research—estimated at over $2 billion annually between 1981 and 2001 and over $2.5 billion since 2001.[42] Adding more than $100 billion in government subsidies would significantly alter the cost profile of nuclear power, and under free-market conditions, nuclear power would be as attractive as a lion at a convention of mice.

Admittedly, internalizing some of the externalities described earlier can be a straightforward process involving measurement devices and spreadsheets.

Table 3.1. Contrasting Theoretical and Real Energy Markets

Criterion	Theory	Reality
Information	All participants in the market must be fully informed about the quantitative and qualitative characteristics of goods and services (and substitutes for them) and the terms of exchange among them.	No one has perfect information, and asymmetric information often leads to the purchase of "bad" products and services, as demonstrated by sales of "lemons" in the automobile market.* Selection biases, misinformation, bounded rationality, and advertising further shade reception of information.
Transaction costs	Market exchanges must be instantaneous and costless.	Markets are not frictionless. It takes financial and human resources to participate in markets (exacerbated by access to capital, knowledge, education).†
Rational consumers	Consumers must maximize utility, and producers must maximize profits; economic actors must be able to collect and process all relevant information, hold rational expectations about prices and products, and make decisions that always promote their self-interest.	Individuals and firms are limited in their ability to use, store, retrieve, and analyze information—they exhibit bounded rationality.‡ People strongly prefer avoiding losses to acquiring gains, leading to high discount rates and inertia.§
Competition	No specific firm or individual can influence any market price by decreasing or increasing supply of goods and services; there must be many buyers and sellers; they must act without collusion; firms cannot use their market power to influence the market themselves; predatory practices by incumbent firms against insurgent firms must be restricted; there must be no barriers to entry and exit.	Monopolies and anticompetitive practices occur all the time (cartels such as OPEC; allocation of resources to the rich/OECD; patent blocking and other exclusionary uses of intellectual property rights).**

Internalization	All costs associated with exchanges must be borne solely by the participants of the transaction or internalized in prices so that all assets in the economic system are adequately priced and there are no positive and negative externalities.	Significant unpriced benefits and costs (i.e., positive and negative externalities) are prevalent in economic systems.[††]
Excludability	Those involved in the exchange must be able to exclude others from benefiting.	Public goods and the tragedy of the commons plague market transactions (e.g., energy security, climate change mitigation, lowered wholesale prices).[‡‡]

[*] G. A. Akerlof, "The Market for 'Lemons': Quality Uncertainty and the Market Mechanism," *Quarterly Journal of Economics* 84, no. 3 (August 1970): 488–500.

[†] O. E. Williamson, "Transaction-Cost Economics: The Governance of Contractual Relations," *Journal of Law and Economics* 22 (1979): 233–261.

[‡] H. A. Simon, "A Behavioral Model of Rational Choice," *Quarterly Journal of Economics* 69, no. 1 (February 1955): 99–118; H. A. Simon, "Rational Decision Making in Business Organizations," *American Economic Review* 69, no. 4 (September 1979): 493–513.

[§] D. Kahneman, J. L. Knetsch, and R. H. Thaler, "Anomalies: The Endowment Effect, Loss Aversion, and Status Quo Bias," *Journal of Economic Perspectives* 5, no. 1 (1991): 193–206.

[**] B. K. Sovacool, "Placing a Glove on the Invisible Hand: How Intellectual Property Rights May Impede Innovation in Energy Research and Development (R&D)," *Albany Law Journal of Science and Technology* 18, no. 2 (Fall 2008): 381–440.

[††] D. L. Weimer and A. R. Vining, *Policy Analysis: Concepts and Practice*, 5th ed. (Upper Saddle River, NJ: Prentice Hall, 2011); M. A. Brown and B. K. Sovacool, *Climate Change and Global Energy Security: Technology and Policy Options* (Cambridge, MA: MIT Press, 2011).

[‡‡] G. Hardin, "The Tragedy of the Commons," *Science* 162 (1968): 1243–1248.

Unfortunately, many externalities are not so easily quantified. Damages that have no agreed-upon market price (e.g., premature deaths, species extinctions, and degradation of community aesthetics, air quality, and water quality) provoke debate over valuation methods. Economists have designed approaches to estimate these costs, such as contingent valuation (by asking consumers to state their willingness to pay, contingent on a specific hypothetical scenario or situation), conjoint analysis (by asking consumers about the utility of different bundles of attributes that have been manipulated to isolate the influence of an individual variable), and hedonic pricing (by inferring willingness to pay through capitalized property values or other indirect means, where the environmental good is bundled into a marketed good such as a house).[43] Each of these methods attempts to place an objective value on a subjective cost, and herein lies the rub.

How much is a life worth? To answer that, some highly contentious questions must be answered. Is a life in China worth the same as a life in the United States? Is the life of an 80-year-old grandmother worth the same as the life of a 10-year-old boy? Is the life of your neighbor worth the same as the life of one of your family members? This type of valuation challenge is not easily resolved by a single methodology because people will disagree over how the valuation should be carried out. As Frank Ackerman and Lisa Heinzerling muse, we are left "knowing the price of everything and the value of nothing."[44] In sum, articulators of this view argue that pricing externalities offers us no answers to pressing questions of energy security or sustainability; it simply leads to more questions.

The Perils of Power Are Unavoidable

Even if the valuation challenges affiliated with externalities could be resolved, attempts to internalize these externalities into the cost of the relevant energy technologies would face intractable opposition. The energy sector is big business. In total, global energy revenues for 2014 are estimated at $11.25 trillion (about 15% of global gross domestic product).[45]

In other words, massive fortunes come under attack amidst an energy transition. Consequently, the fossil fuel industry has exerted considerable influence to oppose change. How substantial? James Hansen, former head of NASA's Goddard Institute, likens the misinformation perpetuated by the fossil fuel industry special interest groups to the studies commissioned by tobacco companies that attempted to cast doubt on the smoking-cancer link. He contends that CEOs of fossil fuel companies "know what they are doing and are aware of long-term consequences of continued business as usual." He further charges that these CEOs should be tried for "high crimes against humanity and nature."[46]

Stakeholders that benefit from continued support for conventional energy are not limited to individuals, firms, or industry advocacy groups: some nations also stand to gain from perpetuating the status quo. Energy superpowers such as the Organization of Petroleum Exporting Countries (OPEC) nations (oil), Canada (coal and oil), Australia (coal and uranium), and Russia (natural gas) stand to lose billions if an energy transition were to cause such resources to be left in the ground. For many nations with endowments of fossil fuel resources, these resources represent a value-added way to improve competitive advantage in global markets. Accordingly, some nations are far less inclined to support expedient transitions.

Advocates of government-led energy investment strategies would further argue that even when policy instruments are employed to try to correct for market externalities, opposition from vested interests and policy design flaws impair efficacy. There is strong evidence that large swing states in the 1988 US presidential election and those states either holding competitive gubernatorial campaigns in 1990 or being represented on the House Energy and Commerce Committee tended to receive more exemptions and favorable treatment in the Clean Air Act Amendments of 1990.[47] There is further evidence that compliance with the 1990 Amendments may have cost $1.2 billion more than it should have because state legislatures continued to protect coalmining jobs and electric utilities with targeted subsidies that offset the new incentive structure of the Clean Air Act—meaning they were "less than diligent" in finding least-cost compliance mechanisms.[48] Lastly, the 1990 Act "grandfathered in" old coal-fired power plants, which meant that even in 2014, 80% of coal units in the United States were not expected to meet minimum pollution requirements for nitrogen oxide and/or sulfur dioxide.[49]

Indeed, examples of market-leveling policy instruments being gamed are far from unsubstantiated rumor. One review of tradable permit markets around the world concluded that all of them were prone to unavoidable problems related to compromises in program design and high transaction costs, as well as price volatility and "leakage."[50] Table 3.2 illustrates these issues with a summary of the performance of eight tradable permit markets. The point is that not only are free markets unfair markets, but attempts to level competition are rarely successful.

Picking Winners Has Worked Before

A final proposition supporting targeted government action is that it has worked before in energy governance and transitional initiatives. Heavy interventions of this type frequently employ command-and-control regulations or pollution taxes.

For instance, in 1917, during World War I, President Woodrow Wilson created the United States Fuel Administration (USFA) to give the government

Table 3.2. Summary of Challenges Facing Tradable Permit Markets

Mechanism	Sector	Location/scheme	Scale	Summary of problems
Fox River water permits	Paper mills and wastewater treatment plants	Wisconsin	Local	Regulators set excessive restrictions on trades to the degree that only one trade ever occurred.
US leaded gasoline phase-out	Leaded gasoline / refineries	United States	National	Transaction costs diluted the efficacy of optimal trading levels.
US Clean Air Act Amendments of 1990	Ambient air pollution (sulfur dioxide and acid rain)	United States	National	Exemptions were given to favor upcoming elections, and polluting coal-fired power plants were grandfathered. Transaction costs accounted for 7% to 25% of the value of traded credits. Credit prices ranged from $14 to $319 per ton of sulfur dioxide. Wrong-way trades may have contributed to acid rain formation in the eastern United States.
Regional clean air incentives market (RECLAIM)	Ambient air pollution (nitrous oxide and sulfur dioxide)	Southern California	Local	Regulators gave allowances in excess of program caps and allowed firms to select their own base year. Transaction costs sometimes exceeded the value of credits being traded. Prices of credits varied by a factor of 10 for some years, based on weather and strategic trading.
Renewable energy credits (RECs)	Renewable electricity generation	United States	Regional	Political concessions have allowed clean coal and natural gas fuel cells to count as "renewable." Transaction costs have accounted for about 10% of the value of credits traded. Credit prices have varied by a factor of five from 2002 to 2005. Trading has locked in dirtier sources of electricity generation in areas dependent on credits.

Type	Sector	Example	Scale	Observations
Individual transferrable quotas (ITQs)	Fisheries	New Zealand	National	Designers distributed quotas based on historical best catch, not sustainable yield. Transaction costs account for 1% to 3% of the value of all trades. Market gluts and boom-and-bust cycles have accelerated industry consolidation and driven small firms out of the market. ITQs have incentivized fishers to high-grade fish (catch high-value stocks and discard low-value stocks) and fish down the food chain.
Carbon credits	Industry	European Union–Emissions Trading Scheme (EU-ETS)	International	Largest emitters were granted more credits than they needed. Prices of credits have varied 100-fold from €0.10 to €30. Electric utilities have used credit revenue to invest in fossil fuels, and chemical and lime industries have shifted production facilities out of Europe.
Carbon offsets	Forestry, transport, energy and electricity	Clean Development Mechanism of the Kyoto Protocol	Global	Some types of offsets have degraded land and destroyed ecosystems. Transaction costs have accounted for about 5% of the value of credits. Leakage has locked in asymmetries between carbon-intensive and carbon-efficient areas, and some credit revenues have been invested back into fossil fuels.

Source: B. K. Sovacool, "The Policy Challenges of Tradable Credits: A Critical Review of Eight Markets," *Energy Policy* 39, no. 2 (February 2011): 575–585.

complete control over the distribution of oil and coal.[51] The USFA was assigned this role only after the nationalization of railroads because price fixing
and other strategies failed to engender networks that could move fuels expeditiously from the trans-Mississippi West to the industrial areas of the Atlantic
East. An arm of the USFA, the Federal Fuel Distributor, continued to successfully manage the distribution of fuels until 1923.[52] As another example, in response to OPEC's oil embargo of April 1973, President Richard Nixon created
the Federal Energy Office to strictly ration energy fuels. The agency was given
broad powers to implement a mandatory fuel rationing program to reduce
consumption of oil by 3.5 million barrels per day. The program worked by
mandating sharp cutbacks in the quantities of oil products, especially gasoline,
albeit inducing notoriously long lines at gasoline stations.[53]

Globally, there are numerous examples of successful government-led energy
transitions. European countries hit hard by the oil shocks of the 1970s took
drastic actions such as, in Denmark, banning the driving of personal automobiles on Sundays and switching off every other street light in the country. By
the late 1980s, Denmark had virtually weaned itself entirely from petroleum as
a fuel stock for electricity generation. France, in responding to the same challenge, exhibited a far heavier hand, utilizing fines and energy "police" to enforce
energy efficiency measures. Indeed, the rise of nuclear power in France is an
excellent example of how nation-led initiatives can alter energy-mix profiles.

The past two decades have also seen governments intervene in energy markets to incentivize cleaner technologies and penalize dirty ones. To the best of
our knowledge, 15 countries around the world have implemented some type of
carbon tax, some going back more than two decades.[54] Mandatory regulations
or policies such as renewable portfolio standards or feed-in tariffs have been
adopted by 102 countries or provinces, a jump from only 45 in 2005. Brazil is
mandating 75% renewable electricity by 2030; India, 20,000 MW of solar
by 2022; and Kenya, 4,000 MW of geothermal by 2030. Biofuel blending
mandates existed in 41 states and provinces and 25 countries. In 2008, Spain
became the first country to mandate solar water heating nationwide. And in
Jiangsu, one of the most populous provinces in China, all new residential
buildings with up to 12 stories must use solar water heating.[55] Chapter 6, which
takes a hard look at fossil fuels and renewables in the electricity sector, also
documents how dozens of governments vigorously interfere with energy markets through direct and indirect subsidies.

These examples demonstrate that a coordinated approach by a central government agency, congress, parliament, state, or local legislature endowed with
sufficient authority can work when the market fails to catalyze desired change
at a desired pace. The successes of these measures also suggest that many people
will readily accept government intervention when the private sector proves in-

capable of meeting needs equitably and have been willing to accept extensive regulatory controls for long periods of time. Other examples—cookstoves in China, ethanol in Brazil, combined heat and power in Denmark, renewables in Germany—illustrate situations where strong government intervention was seen as a necessary and essential part of catalyzing the energy transitions discussed in chapter 15.

Common Ground: Governments Should Banish Losers and Nurture Potential

Free-market advocates contend that picking winners—defined as providing government support to one technology over another—should be avoided because it distorts markets and leads to suboptimal outcomes. Many advocates further contend that environmental economists have developed a number of valuation techniques and tools to properly internalize these externalities within a free-market environment. However, in the real world, political biases, unequitable market conditions, and externalities create market distortions that require government intervention. Many believe that these externalities can never be sufficiently internalized to justify ceding the decision to natural market forces. Even if this were feasible, powerful stakeholders that benefit from dominant technologies would never allow a free-market solution that reflects true energy costs.

This gives rise to a conundrum. If it is true that the free market cannot yield an optimal energy mix, how should energy policy be managed? Should the task be ceded to governments—entities that free-market economists condemn as being inefficient and ineffective? Should the government pick winners?

We argue that evidence supports a decisive case for yes . . . and no. Governments should not "pick winners," but they should temper their commitment to inferior technologies that involve large negative externalities, should provide support for currently superior technologies, and should nurture technological competition to ensure that firms from all technological platforms continue to press for innovation. In short, a two-stage strategy shown in box 3.1 is warranted.

The first stage in any energy strategy should be to banish the losers and withdraw support for inferior technologies that are maintaining supremacy through accumulated wealth. Put in more concrete terms, one can loosely quantify the negative externalities associated with different energy systems; doing so suggests that oil, coal, and nuclear power have immense social costs. This implies that governments should ensure that these systems either operate in less damaging ways or be forced to account for their externalities.

The second stage is to nurture potential through market instruments, and if possible design them to minimize gaming and the obfuscations of power. In other words, completely unregulated open markets and total national control

BOX 3.1

Picking Technology Winners and the Common Ground

Should governments intervene in energy markets?

ONE SIDE: Free markets offer the best way to manage the energy sector, externalities can be valued and priced when needed, and market tools such as tradable credits have an excellent track record.

OTHER SIDE: Perfect energy markets will never exist, market failures demand government intervention, valuing externalities is difficult if not impossible, and market tools can be manipulated and opposed.

COMMON GROUND: Governments need to banish losers and nurture potential with well-designed market instruments.

are not the only options. There is a spectrum of possibilities, including regulated markets that balance laissez-faire efficiency with protective government standards and policies.[56] This last exhortation highlights an important lesson about nurturing indigenous technology. Although governments should try to avoid promoting winners among competing emergent technologies, governments should not necessarily view all technologies equally. A key tenet of developmental economics is that nations that seek to encourage indigenous development should prioritize industries and technologies that can leverage existing core competencies and link into existing economic activity.[57] Governments should not necessarily pick winners, but they should endeavor to support a pool of the most attractive contenders that possesses the greatest potential to succeed within their country's contextual environment.

NOTES

1. M. Friedman, *Capitalism and Freedom* (Chicago: University of Chicago Press, 2002), 196.

2. Robert H. Frank and Ben S. Bernanke, *Principles of Micro-Economics*, 3rd ed. (Boston: McGraw-Hill, 2007), 58–87; Marc Levinson, "Why Markets Matter," in *Guide to Financial Markets* (Princeton: Bloomberg Press, 2003), 1–13.

3. Charles Wheelan, *Naked Economics: Undressing the Dismal Science* (New York: W. W. Norton & Company, 2002), 28.

4. R. W. Bacon, "Privatization and Reform in the Global Electricity Supply Industry," *Annual Review of Energy and the Environment* 20 (1995): 119–143; J. Stern and S. Holder, "Regulatory Governance: Criteria for Assessing the Performance of Regulatory Systems—An Application to Infrastructure Industries in the Developing Countries of Asia," *Utilities Policy* 8 (1999): 33–50.

5. F. P. Sioshansi and W. Pfaffenberger, eds., *Electricity Market Reform: An International Perspective* (Amsterdam: Elsevier, 2006).

6. Bacon, "Privatization and Reform"; R. W. Bacon and J. Besant-Jones, "Global Electric Power Reform, Privatization, and Liberalization of the Electric Power Industry in Developing Countries," *Annual Review of Energy and the Environment* 26 (2001): 331–359; D. G. Victor and T. C. Heller, *The Political Economy of Power Sector Reform* (New York: Cambridge University Press, 2007).

7. Jean-Jacques Laffont and Mathieu Meleu, "Separation of Powers and Development," *Journal of Development Economics* 64 (2001): 129–145; J. Stern, "Electricity and Telecommunications Regulatory Institutions in Small and Developing Countries," *Utilities Policy* 9 (2000): 31–157.

8. J. Williams and R. Ghanadan, "Electricity Reform in Developing and Transition Countries: A Reappraisal," *Energy* 31 (2006): 815–844.

9. Greg Brannon, "Thoughts on Energy and the Environment" (June 2014), available at http://gregbrannon.com/energy-enviorment.

10. Hans Biebl, "Energy Subsidies, Market Distortion, and a Free Market Alternative," *University of Michigan Journal of Law Reform* 2 (December 2012): 43–44.

11. Douglas L. Norland and Kim Y. Ninassi, *Price It Right: Energy Pricing and Fundamental Tax Reform* (Washington, DC: Alliance to Save Energy, 1998).

12. Ulrich Beck, "From Industrial Society to the Risk Society: Questions of Survival, Social Structure and Ecological Enlightenment," *Theory, Culture, and Society* 9 (1992): 97–123.

13. Zachary A. Smith, *The Environmental Policy Paradox*, 5th ed. (Upper Saddle River, NJ: Prentice Hall, 2009).

14. R. J. Brulle and D. N. Pellow, "Environmental Justice: Human Health and Environmental Inequalities," *Annual Review of Public Health* 27 (2006): 103–124.

15. David W. Orr, *Earth in Mind: On Education, Environment, and the Human Prospect* (Washington, DC: Island Press, 1994), 172.

16. C. S. Holling and Gary K. Meffe, "Command and Control and the Pathology of Natural Resource Management," *Conservation Biology* 10, no. 2 (April 1996): 328–337.

17. Robert W. Hahn, "Economic Prescriptions for Environmental Problems: How the Patient Followed the Doctor's Orders," *Journal of Economic Perspectives* 3, no. 2 (1989): 94–114; Robert W. Hahn and Robert N. Stavins, "Economic Incentives for Environmental Protection: Integrating Theory and Practice," *American Economic Review* 82, no. 2 (1992): 464–468; T. H. Teitenberg, "Tradable Permit Approaches to Pollution Control: Faustian Bargain or Paradise Regained?" in *Property Rights, Economics, and the Environment*, ed. M. D. Kaplowitz (Stamford, CT: JAI Press, 2000), 175–201; A. Denny Ellerman, "A Note on Tradable Permits," *Environmental and Resource Economics* 31 (2005): 123–131.

18. R. J. Lazarus, "Super Wicked Problems and Climate Change: Restraining the Present to Leverage the Future," *Cornell Law Review* 94 (2009): 1153–1234.

19. Rita Tubb, "Millions Needed to Meet Long-Term Natural Gas Infrastructure Supply, Demands," *Pipeline and Gas Journal* 236, no. 4 (April 2009): 2–3.

20. Kevin Morrow, Donald Karner, and James Francfort, *Plug-in Hybrid Electric Vehicle Charging Infrastructure Review*, INL/EXT-08-15058 (Idaho Falls: Idaho National Laboratory, November 2008).

21. B. K. Sovacool and B. Brossmann, "Symbolic Convergence and the Hydrogen Economy," *Energy Policy* 38, no. 4 (April 2010): 1999–2012.

22. C. Ford Runge, "Environmental Protection from Farm to Market," in *Thinking Ecologically: The Next Generation of Environmental Policy*, ed. Marian R. Chertow and Daniel C. Esty (New Haven: Yale University Press, 2000), 200–215.

23. James T. B. Tripp and Daniel J. Dudek, "Institutional Guidelines for Designing Successful Transferable Rights Programs," *Yale Journal on Regulation* 6 (1989): 369–391.

24. Bonnie G. Colby, "Cap-and-Trade Policy Challenges: A Tale of Three Markets," *Land Economics* 76, no. 4 (2000): 638–658.

25. Vivien Foster and Robert W. Hahn, "Designing More Efficient Markets: Lessons from Los Angeles Smog Control," *Journal of Law and Economics* 38, no. 1 (April 1995): 19–48.

26. Richard J. McCann, "Environmental Commodities Markets: 'Messy' versus 'Ideal' Worlds," *Contemporary Economic Policy* 14, no. 3 (July 1996): 85–97.

27. Robert W. Hahn and Gordon L. Hester, "Marketable Permits: Lessons for Theory and Practice," *Ecology Law Quarterly* 16 (1989): 361–406.

28. Richard G. Newell, James N. Sanchirico, and Suzi Kerr, "Fishing Quota Markets," *Journal of Environmental Economics and Management* 49 (2005): 437–462.

29. Robert R. Nordhaus and Kyle W. Danish, "Assessing the Options for Designing a Mandatory U.S. Greenhouse Gas Reduction Program," *Boston College Environmental Affairs Law Review* 32 (2005): 97–168.

30. Jonathan Adler, "Free and Green: A New Approach to Environmental Protection," *Harvard Journal of Law and Public Policy* 24 (2001): 653–694.

31. Herman E. Daly, *Beyond Growth: The Economics of Sustainable Development* (Boston: Beacon Press, 1996).

32. Tom Tietenberg, "The Tradable-Permits Approach to Protecting the Commons: Lessons for Climate Change," *Oxford Review of Economic Policy* 19, no. 3 (2003): 400–419.

33. Maureen L. Cropper and Wallace E. Oates, "Environmental Economics: A Survey," *Journal of Economic Literature* 30, no. 2 (June 1992): 675–740.

34. Jan-Peter Voss, "Innovation Processes in Governance: The Development of 'Emissions Trading' as a New Policy Instrument," *Science and Public Policy* 34, no. 5 (2007): 329.

35. Jonathan Adler, "Free and Green: A New Approach to Environmental Protection," *Harvard Journal of Law and Public Policy* 24 (2001): 658.

36. Sam Napolitano et al., "The U.S. Acid Rain Program: Key Insights for the Design, Operation, and Assessment of a Cap-and-Trade Program," *Electricity Journal* 20, no. 7 (August/September 2007): 47–59; C. P. Carlson et al., "SO_2 Control by Electric Utilities: What Are the Gains from Trade?" *Journal of Political Economy* 108 (2000): 1292–1326.

37. Lauraine G. Chestnut and David M. Mills, "A Fresh Look at the Benefits and Costs of the US Acid Rain Program," *Journal of Environmental Management* 77 (2005): 252–266.

38. M. Mazzucato, "The Entrepreneurial State," *Soundings* 49, no. 49 (2011): 131–142.

39. E. F. Schumacher, *Small Is Beautiful: Economics As If People Mattered* (New York: Harper Collins, 2010 [1973]).

40. Joseph E. Stiglitz, "Information and the Change in the Paradigm of Economics," *American Economic Review* 92, no. 3 (June 2002): 460–501.

41. Agency for Natural Resources and Energy, *FY2007 Annual Energy Report* (Tokyo: Ministry of Economy, Trade, and Industry, 2008).

42. S. V. Valentine, "Japanese Wind Energy Development Policy: Grand Plan or Group Think?" *Energy Policy* 39 (2011): 6842–6854.

43. D. Thampapillai, *Environmental Economics: Concepts, Methods and Policies* (Melbourne, Australia: Oxford University Press, 2002).

44. Frank Ackerman and Lisa Heinzerling, *Priceless: On Knowing the Price of Everything and the Value of Nothing* (New York: New Press, 2005).

45. Veronica Franco, "Global Energy Industry to Grow 13.2% through 2014," *Market Wire*, February 28, 2011, available at www.marketwire.com/press-release/global-energy-industry-to-grow-132-through-2014-1402888.htm.

46. J. Hansen, "Global Warming Twenty Years Later: Tipping Points Near" (National Press Club, 2008), available at www.columbia.edu/~jeh1/2008/TwentyYearsLater_20080623.pdf.

47. Scott E. Atkinson and T. H. Teitenberg, "Market Failure in Incentive-Based Regulation: The Case of Emissions Trading," *Journal of Environmental Economics and Management* 21 (1991): 17–31.

48. Kenneth Rose, "Twelve Common Myths of Allowance Trading: Improving the Level of Discussion," *Electricity Journal* 8 (May 1995): 64–69.

49. Sarah K. Adair, David C. Hoppock, and Jonas J. Monast, "New Source Review and Coal Plant Efficiency Gains: How New and Forthcoming Air Regulations Affect Outcomes," *Energy Policy* 70 (2014): 183–192.

50. B. K. Sovacool, "The Policy Challenges of Tradable Credits: A Critical Review of Eight Markets," *Energy Policy* 39, no. 2 (February 2011): 575–585.

51. See John G. Clark, "Federal Management of Fuel Crisis between the World Wars," in *Energy and Transport: Historical Perspectives on Policy Issues*, ed. George H. Daniels and Mark H. Rose (London: Sage Publications, 1982), 135–147.

52. See John G. Clark, "The Energy Crisis of 1919–1924 and 1973–1975: A Comparative Analysis of Federal Energy Policies," *Energy Systems and Policy* 4, no. 4 (1980): 239–271.

53. Richard B. Mancke, "The Genesis of the U.S. Oil Crisis," in *The Energy Crisis and U.S. Foreign Policy*, ed. Joseph S. Szyliowicz and Bard E. O'Neil (New York: Praeger, 1975), 52–72.

54. Finland and the Netherlands have had a carbon tax since 1990, Sweden and Norway since 1991, Denmark since 1992, the United Kingdom since 1993, Costa Rica since 1997, Switzerland and Taiwan since 2008, India and Ireland since 2010, Australia and Japan since 2012 (though Australia's was repealed in 2014), France since 2013, and Chile since 2014. Moreover, as of late 2014, Canada, China, New Zealand, South Africa, and the United States all had proposals for a carbon tax pending debate in their legislatures.

55. B. K. Sovacool, "Energy Policy and Climate Change," in *The Handbook of Global Climate and Environment Policy*, ed. Robert Falkner (New York: John Wiley & Sons, 2013).

56. Smith, *Environmental Policy Paradox.*

57. D. Perkins, S. Radelet, and D. Lindauer, *Economics of Development*, 6th ed. (New York: W. W. Norton & Company, 2006).

II. ENERGY RESOURCES
AND TECHNOLOGY

Do Conventional Energy Resources Have a Meaningful "Peak"?

We can describe the debate over peak energy resources through an analogy, centering on a jar of jellybeans. One camp views the resource estimation process as a simple jellybean counting contest: someone fills a glass jar with jellybeans and asks passers-by to guess how many beans are in the jar; the closest guess typically receives some sort of prize. In estimating energy reserves, there is a belief that certain geological conditions are conducive to the presence of these resources. The challenge is to identify those likely locales, use past trends to estimate the likelihood of a discovery, and aggregate the data to establish a total number for resources available at that site. As in the jellybean counting contest, the estimate will rarely be exact, but a carefully commissioned analysis should yield a good approximation.

The opposing camp might agree that estimating conventional energy reserves is like a jellybean counting contest but would point out that the contest is complicated by a couple of facts: the size of the jar is not known (thus disputing the geological conditions conducive to resource formations), and the size of the jar is not fixed (because viable commercial access to energy resources depends on technological innovation and energy prices). Moreover, an unknown number of persons continue to eat the jellybeans out of the jar, so by the time an estimate is made, it's already inaccurate.

This difference of opinion might be an enjoyable diversion if ownership of a plush toy or some other trivial prize were at stake; however, this is far from the case. On the one hand, if we have passed the peak of known conventional energy resources and maintain our current global energy mix, humanity faces a near future of rapidly escalating energy prices. In this scenario, we will have endorsed a continued commitment to fossil fuel energy sources and exacerbated the risk of climate change for a pool of resources that we must transition out of before they are exhausted. Failure to make this transition will leave us fighting over the few jellybeans left at the bottom of the jar. On the other hand, if we are nowhere near reaching the peak of known conventional energy resources, renewable energy substitutes will have to become far more efficient if they are to compete commercially with these resources and if carbon capture and sequestration advocates are to find themselves with a viable seat at the energy policy table. Both camps pose some convincing arguments.

One Side: Resource Peaks Are a Serious Threat

The view that finite energy resources such as fossil fuels and uranium will soon be exhausted and that this poses a serious challenge to society rests on supply-side and demand-side assumptions deserving further examination.

No Finite Energy Resources Are Forever

The notion of a supply-side peak refers to a situation where the rate of extraction of a finite resource reaches a plateau, after which the rate of production enters a terminal, irreversible decline.[1] In 1956, the geophysicist Marion King Hubbert predicted that peak oil production in the United States would occur between 1965 and 1970. He probably would have toiled in relative obscurity save for one event: in 1970, oil reserves reached what appeared to be a national production peak. When the oil crises of the 1970s rolled in and American policymakers began to strategically assess oil reserves, Hubbert's work rose to prominence, forever linking the name Hubbert to the term "peak oil." Since then, proponents of production peaks have argued that global estimates of energy reserves and production speak for themselves, and they wield statistics that show the declining availability of high-quality oil, coal, natural gas, and uranium reserves. As physicist Kjell Aleklett put it, "Every oilfield reaches a point of maximum production, which advanced technologies can delay or extend, but not eliminate."[2]

This trend of production peaks, though contested, is perhaps best exemplified in the crude oil sector, as illustrated in figure 4.1. The United States recorded a peak in domestic oil production in 1970, after which production tailed off significantly,[3] until a turnaround with shale oil in recent years (panel A). Looking at peak production on a global scale introduces an added perspective: discovery peaks (panel B). Globally, oil production has yet to reach a confirmed peak. Part of the reason for this is that progressive oil field discoveries between 1950 and 1990 brought new capacity online, permitting oil production to expand to meet increases in demand. However, since the 1990s, new discoveries have declined precipitously, and some advocates contend that we have reached a global oil production peak. Finally, more than half of all currently produced oil comes from very small fields with production levels of 9,000 barrels of oil per day or less (panel C). The implication is that the largest reserves of oil have already been discovered and partially exhausted.

As a by-product of scarcity, oil is, as a whole, becoming more expensive to discover and produce. Many of the newest oilfields that entered production during the past decade were financially viable only at oil prices well above historical rates. To break even, the Kashagan oilfield in Kazakhstan's zone of

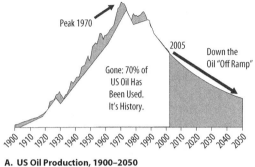

Peak 1970

2005

Down the
Oil "Off Ramp"

Gone: 70% of
US Oil Has
Been Used.
It's History.

1900 1910 1920 1930 1940 1950 1960 1970 1980 1990 2000 2010 2020 2030 2040 2050

A. US Oil Production, 1900–2050

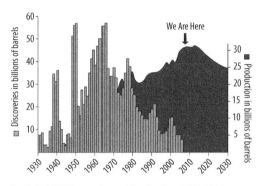

We Are Here

Discoveries in billions of barrels

Production in billions of barrels

1930 1940 1950 1960 1970 1980 1990 2000 2010 2020 2030

B. Global Oil Discoveries and Production, 1930–2030

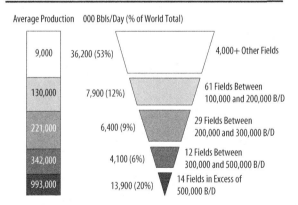

Average Production   000 Bbls/Day (% of World Total)

9,000	36,200 (53%)	4,000+ Other Fields
130,000	7,900 (12%)	61 Fields Between 100,000 and 200,000 B/D
221,000	6,400 (9%)	29 Fields Between 200,000 and 300,000 B/D
342,000	4,100 (6%)	12 Fields Between 300,000 and 500,000 B/D
993,000	13,900 (20%)	14 Fields in Excess of 500,000 B/D

C. The "Reverse Pyramid" of Current Oil Production

Figure 4.1. Trends in US and Global Oil Production. *Source:* Modified from Brian Hicks and Chris Nelder, *Profit from the Peak: The End of Oil and the Greatest Investment Event of the Century* (New York: John Wiley and Sons, 2008). *Note:* The "We Are Here" arrow points to 2008, the year of the study. 000 Bbls = thousand barrels; B/D = barrels per day.

the Caspian Sea, discovered in 2000, required a price of $121 per barrel. The Prirazlom oilfield located in the Pechora Sea, south of Novaya Zemlya, Russia, where drilling began in 2012 and 2013, required a breakeven price of $110 per barrel. The Al-Nasr field in Abu Dhabi, soliciting bids in 2014, required $109 per barrel. The Novoportovskoye, an oilfield in the Russian part of the Arctic Circle, required $108 per barrel. The Yurubcheno-Tokhomskoye field in east Siberia, where production began in 2009, required $102 per barrel.[4] As one commentator on oil markets exclaimed in 2014, "Not a single large project has come on stream at a break-even cost below $80 a barrel for almost three years."[5] The oil industry spent more than $10 trillion in exploration and production from 2000 to 2014 and saw its expenditures rise over this period by 300% per year, yet output is up only 14%.[6] In 2014, the biggest oil groups in Europe—BP, Shell, Total, Statoil, and Eni—spent $141 billion on operations but generated only $121 billion in cash flow, leading to a substantial deficit. As an analyst from Morgan Stanley concluded, "Oil development is so expensive that many projects do not make sense."[7]

As further evidence that global oil production has peaked, a study by an interdisciplinary team including the Sussex Energy Group, Imperial College London, and the Oil Depletion Analysis Center examined oilfield size, reserve growth and decline rates, and depletion rates for the entire industry. They concluded that, as a global average, "the reserve diminishment rate of post-peak fields is at least 6.5%/year and the corresponding decline rate of all currently producing fields is at least 4%/year. Both are on an upward trend as more giant fields enter decline, as production shifts towards smaller, younger and offshore fields and as changing production methods lead to more rapid post-peak decline. More than two thirds of current crude oil production capacity may need to be replaced by 2030, simply to keep production constant. At best, this is likely to prove extremely challenging."[8]

Numerous other studies suggest that oil peaks are imminent, if not already present. A study from the Oxford Institute for Energy Studies noted that, as of 2010, almost 30 major oil-producing nations, including Norway and the United Kingdom, registered declining domestic oil production (suggesting that they might be past their peaks) and that seven other countries, including China, India, and Russia, exhibited a plateau in production (suggesting that they have reached or are nearing their peaks).[9] As the author of that study noted, "Obviously, oil will decline and run out one day: just like gas and coal, it was formed in the distant geological past and its supply is finite. The issue is when."[10] Evidence in some countries and subregions indicates that they are decades past their peaks in oil production: Austria's occurred in 1955, Germany's in 1967, Canada's in 1974, Indonesia's in 1977, Alaska's in 1989, and Australia's in 2000.[11] A meta-study surveyed 29 separate, independent studies of peak oil globally

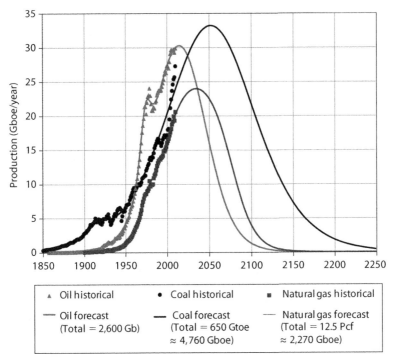

Figure 4.2. Projected Global Peaks in Production for Oil, Coal, and Natural Gas, 1850–2250. *Source:* Modified from G. Maggio and G. Cacciola, "When Will Oil, Natural Gas, and Coal Peak?" *Fuel* 98 (2012): 111–123. *Note:* Gb = gigabarrel; Gboe = gigabarrel of oil equivalent; Gtoe = gigaton of oil equivalent; Pcf = process classification framework.

and noted that the consensus among unbiased experts was that a worldwide production peak would occur somewhere between 2005 and 2035.[12]

Many who contend that we are on the leeward side of an oil peak acknowledge that this might make it possible to finance oil discovery efforts that were previously not commercially viable because, in the absence of a tail-off in demand, the cost of oil will continue to escalate. In this way, the appreciation of oil prices might be deferred through increased investment in discovery efforts and capacity expansion; in the long run, however, this strategy will only exacerbate the problem in much the same way that adding debt to a depreciating asset amplifies the financial impact of an ensuing bankruptcy.

Indeed, oil is not the only major energy resource that humanity is about to summit. A peer-reviewed assessment of the "most likely scenarios" estimated that global oil production would peak in 2015, natural gas production in 2035, and coal production in 2052—forming the bell-shaped curves in production illustrated in figure 4.2.[13]

Similar peaks have been confirmed by multiple independent analyses undertaken by some of the world's best geologists, economists, and energy analysts for oil, natural gas,[14] coal,[15] and even uranium.[16] In its 2014 "Statistical Review of World Energy," British Petroleum estimated reserve-to-production ratios for oil and natural gas and coal to sustain only 53.3, 55.1, and 113 years of continuous supply, respectively.[17] These estimates, of course, were based on current demand, which is not an accurate basis for such a calculation in a world of steadily increasing demand. Two of us (Sovacool and Valentine) investigated reserve-to-production ratios under various demand-side growth assumptions for each of these four conventional fuels, inclusive of shale gas, and estimated the ranges to be 40 to 43 years for oil, 37 to 60 years for natural gas, 61 to 137 years for coal, and 56 to 118 years for uranium.[18]

Fossil Fuels Face Social, Economic, and Environmental Limits, Too

As if the finite nature of conventional "supply-side" energy reserves were not enough to force a transition, there is ample evidence that the continued use of conventional energy fuels could also face growth impediments on the "demand side." One grave concern relates to the economics of conventional energy under conditions of scarcity; another is environmental restrictions on future fossil fuel use.

Proponents of this thesis argue that even if peak production estimates are off by two or three decades, humanity is about to experience an unavoidable period of energy price inflation. Those concerned with peak production believe that past the peak, only two developments can hold prices in check: an offsetting decline in demand or a renewed growth spurt in supply.

For example, the International Energy Agency (IEA) estimates that between 2011 and 2035, overall energy demand will increase by one-third under its New Policies Scenario. Demand is projected to increase for all forms of conventional energy: oil (+13%), coal (+17%), natural gas (+48%), and nuclear (+66%).[19] In short, the first possible development—declining demand—is not going to happen. Regarding the prospects of significantly ramping up production, the data presented earlier in the chapter and reinforced by the reverse pyramid (figure 4.1, panel C) suggest that enhancing oil production can come only at an increasingly high development cost. As Massachusetts Institute of Technology economist Morris Adelman said, "'Finite resources' is an empty slogan; only marginal cost matters."[20] When it comes to finding energy reserves in increasingly out-of-the-way places, marginal costs are bound to be higher. Overall, then, proponents suggest that the unavoidable era of high oil costs is going to catalyze an economic peak in conventional energy consumption. Evidence suggests that the heyday of conventional energy is behind us, whether we like it or not.

Many studies support such a contention. One research team predicted that the inflated prices for petroleum that are expected this century could almost bankrupt the iron, fertilizer, and air transport industries.[21] Citibank, a global financial firm, declared in 2013 that global oil demand was "approaching a tipping point" and that "the end is nigh" for growth, given the substitution trends of natural gas for oil coupled with improvements in the fuel economy of vehicles.[22] The net result of this substitution is reflected in figure 4.3. This projection (shown in panel A), we should note, assumes only vehicle efficiency gains and gas substitution. A sudden shift in major vehicle markets to electric vehicles, something explored in chapter 7, could expedite further declines. After all, to quote Sheikh Yamani, "The Stone Age didn't end because we ran out of stones." It ended because we developed better technologies.[23] Much oil could remain in the ground as countries transition to other fuels.

A separate thread of the "limits" argument centers on the social and environmental costs stemming from unabated climate change. This second major "demand-side" impediment relates to limits on environmental carrying capacity. Whether we choose to acknowledge this or not, humanity must undertake economic activities subject to a "carbon budget." At a certain level of greenhouse gas emissions, we cannot afford to utilize more fossil fuels, even if they were free.

One can make this claim based on either concentrations in the atmosphere or tons of carbon dioxide emitted. In terms of atmospheric concentration, beyond 350 parts per million (ppm), elevated radiative forcing causes atmospheric temperatures to rise, which heats our planet; if left unaddressed, this will lead to the melting of ice caps, sea level rise, and radically altered climate patterns. There is currently 401.3 ppm of CO_2 in the atmosphere, according to the Mauna Loa Observatory project.[24] At this level, the Intergovernmental Panel on Climate Change (IPCC) predicts a likely temperature rise of 1.5°C relative to the 1850–1900 period by 2100 under the high-policy-response scenario.[25]

In terms of emissions, one can think about a carbon budget of one trillion tons of carbon—that's the ultimate ceiling on what scientists say we can safely emit in total.[26] Right now humanity is emitting about 50 billion tons of CO_2 equivalent per year, or 14.1 billion tons of carbon (which is 3.67 times the weight of CO_2). Estimates suggest that human activities have already emitted a total of about 435 billion tons of carbon. That means we have, at most, 565 billion tons left that we can emit, or 40 years of emissions at today's rate.

Here's the rub: *five times* that amount of carbon—2,795 billion tons—is currently contained in all the proven coal, oil, and gas reserves of fossil fuel companies and countries. Essentially, this is the fossil fuel that many plan on burning, but it will completely blow our carbon budget.[27] As climate scientist

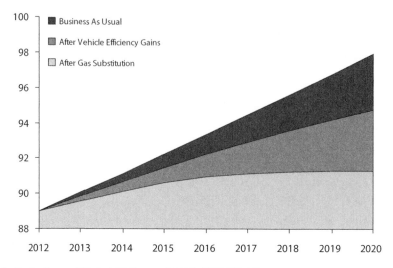

A. Projections of Global Oil Demand, 2012–2020 (Mb/d)

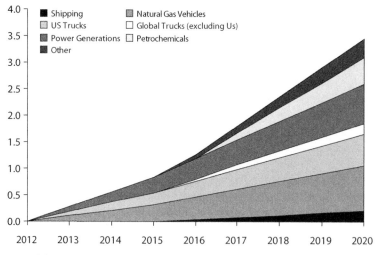

B. Potential Natural Gas Substitutions for Oil, 2012–2020 (Mb/d)

Figure 4.3. A Global Oil Peak on the Demand Side. *Source:* Citi Research, "Global Oil Demand Growth—The End Is Nigh" (March 26, 2013). *Note:* Mb/d = million barrels per day.

James Hansen and his colleagues note, "Burning all fossil fuels would produce a different, practically uninhabitable, planet."[28]

Thus, many barrels of oil, cubic meters of natural gas, and tons of coal will need to stay in the ground as "stranded assets." A study that examined the volumes of oil that "cannot be used" by 2035 due to carbon restraints projected that 500 to 600 billion barrels must be "unburnable" and 40% to 55% of new deep-water resources must not be developed.[29] Even if geological or economic peaks were avoidable, these folks argue, the climate change threat forces a retreat from fossil fuel consumption.[30] As one report concludes, "A precautionary approach means only 20% of total fossil fuel reserves can be burnt to 2050."[31] This translates into the need to "strand" a large amount of fossil fuel revenue. Under a global climate accord with an allowable 2°C atmospheric temperature rise, one investment analyst calculated that the fossil fuel industry would lose $28 trillion of gross revenues from 2015 to 2035 compared with business as usual, with the biggest loser being the oil industry—which would confront stranded assets of $19 trillion.[32]

Politics come into play, of course, under such circumstances. The risk of stranded assets in the fossil fuel industry is so serious that in late 2014, some banks were charged with investigating the risks of a "carbon bubble" in the economy, a situation where "if the world's governments meet their agreed target of limiting global warming to 2°C by cutting carbon emissions, then about two-thirds of proven coal, oil and gas reserves cannot be burned . . . With fossil fuel companies being among the largest in the world, sharp losses in their value could prompt a new economic crisis."[33] However, as we'll see, politicking is not the only weapon available to those who dismiss the peak energy resources argument.

The Other Side: Resource Peaks Are Irrelevant

The opposing thesis rests on three interconnected arguments: (1) we have been wrong about peaks before, (2) peaks can catalyze innovation and exploration, and (3) more efficient consumption can mitigate rates of depletion.

Predictions of Resource Scarcity Are Questionable

Claims of resource exhaustion and supply-side peaks do have a long history, and advocates of this thesis point out, correctly, that many projections of past peaks have been wrong—implying that current projections must be viewed with a degree of skepticism. Energy production has continued to surpass what was thought possible and achievable. We'll provide just a few examples. In 1874, the chief geologist of Pennsylvania, Peter Lesley, argued that if oil were used only to make kerosene for lighting, the state would run out of oil in a few years.[34] Davit T. Day, director of the Petroleum Division of the US Geological Survey

(USGS), predicted in 1908 that domestic oil reserves would be completely exhausted by 1927.[35] The US Department of Interior announced in 1939 that US oil supplies would run out by 1952. And the Department of State concluded in 1947 that no new oil reserves were left in the United States, yet 4.8 billion barrels of oil were discovered in 1948, the largest discovery in American history.[36]

One could argue that many of these examples are from times when the technical ability to estimate peaks was underdeveloped. However, there are more current examples as well. The IEA warned in 1979 that the world's 645 billion barrels of oil would be depleted by 1985, yet by 1990 only 320 billion barrels of that oil had been drawn down, and estimated global reserves had grown above 1 trillion barrels.[37]

Geologists and energy agencies have not been the only ones "off" in their projections of future shortages and crises. Authors of one study in 1972, at the peak of panic over the oil crisis, predicted that by 1976 the United States would find it necessary to permanently curtail heating and air conditioning in shopping centers, restaurants, public buildings, schools, homes, and hospitals to extend oil supplies. The authors further argued that by 1977, the government might have to take control of key industries to ration energy; by 1980, there could be widespread shortages of gasoline; and by 1985, a great depression on the scale of 1929 could ensue, followed quickly by world conflict over existing energy resources. "During the next two decades," the authors warned, "severe oil and gas shortages are inevitable."[38] A former director of the US Bureau of Mines also argued in 1972 that "we can anticipate that before the end of this century energy supplies will become so restricted as to halt economic developments around the world."[39] Given all of these erroneous projections, one could be forgiven for viewing current concerns about post-peak resources as offspring of a Chicken Little mindset.

One of the underlying explanations for erroneous historical predictions stems from the use of aggregate production rather than known reserves as an indicator of peaks. As Red Cavaney, president and chief executive officer of the American Petroleum Institute (an industry trade group), explains, "These forecasts were wrong because, nearly every year, we have found more oil than we have used, and oil reserves have continued to grow."[40]

In summary, then, production peaks are not accurate indicators of how much oil is left in reservoirs around the world. Using production peaks as a metric for how much oil exists in the world is like using monthly bank account withdrawals as a metric for the total amount of money deposited.

Peaks Can Drive Technological Innovation and Exploration

The inaccuracy of using production peaks as a metric for global fossil fuel reserves is further highlighted by understanding the role of innovation in extending fossil fuel supplies. The mere mention of scarcity in energy supplies is enough to inflate commodity prices, which drives fossil fuel firms to invest in initiatives to make better use of existing supplies and uncover new reserves. In other words, the threat of peaks, even when likely to be wrong, can accelerate advances in technology, which can dramatically increase the ability of energy companies to find and extract minerals and fuels.

In the oil and gas industries alone, several technical advances over the past few decades have enabled exploration in places that were previously unreachable. Wellbores used to be simple vertical holes drilled straight into the earth, but computer advancements and the development of sensors now allow companies to drill directional holes with great accuracy. Early wells could extend only 500 to 800 feet deep, but modern ones extend more than 8,000 feet into the earth. Three-dimensional seismic technology and enhanced computing power have enabled the industry to make more precise estimations of oil and gas deposits.[41]

Technological advances have also improved the efficacy of resource extraction. Oil that was once left in wells is now captured through advanced oil recovery techniques such as the injection of carbon dioxide. Coal-bed methane, which was previously left abandoned in mines, is now captured for use. In a number of refinement processes, the indiscriminate flaring of natural gas, previously thought to have little value, is now captured as an additional revenue stream.[42]

Historically, when oil prices spiked, oil companies were financially motivated to undertake exploration activities in more remote locations. For example, the oil crises of the 1970s justified building the $10 billion Trans-Alaskan Pipeline.[43] In modern times, the Canadian oil sands, once considered to be commercially unviable, are now being lauded by the Albertan government as a key contributor to provincial wealth and may justify building the Keystone XL pipeline.[44]

The USGS explains how, taken together, innovations have had a revolutionary effect on oil and gas production. It notes that:

- As drilling and production within discovered fields progressed, pools or reservoirs were found that were not previously known.
- Advances in exploration technology made it possible to identify new targets within existing fields.
- Advances in drilling technology made it possible to recover oil and gas not previously considered recoverable in the initial reserve estimates.
- Enhanced oil recovery techniques increased the recovery factor for oil and thereby increased the reserves within existing fields.

As a result of these advances, the USGS was prompted to increase its estimate of undiscovered, technically recoverable oil by 20% between 1990 and 2000.[45]

Table 4.1. Emergent and Potentially Disruptive Energy Technologies

Technology	Year technology becomes:		
	Scientifically viable	Mainstream	Financially viable
Fuel cells	2013	2015	2016
Lithium air batteries	2017	2018	2020
Hydrogen energy storage and transport	2019	2021	2022
Thermal storage	2022	2024	2027
First-generation smart grid	2014	2015	2016
Distributed generation	2017	2021	2022
Smart energy networks	2019	2020	2020
Tidal turbines	2015	2017	2017
Micro-Stirling engines	2020	2026	2027
Solar panel positioning robots	2014	2016	2017
Second-generation biofuels	2016	2017	2021
Photovoltaic transparent glass	2017	2020	2021
Third-generation biofuels	2022	2024	2025
Space-based solar power	2025	2027	2028+
Micro-nuclear reactors	2022	2023	2023
Inertial confinement fusion	2013	2021	2021
Thorium reactor	2025	2026	2027

Source: Michell Zappa, "17 Emerging Energy Technologies That Will Change the World," *Business Insider,* April 24, 2014.

Geographer Vaclav Smil remarks that judging the availability of oil and other energy resources by today's technology "may be akin to judging today's computer or aircraft performance by the standards of 1950."[46]

Although the notion that many of our key sources of energy may be extinguished within five decades leaves many with lower lips atremble, some are less fazed by this threat. The reason is that some advocates contend that current conventional energy sources are merely bridge fuels and that, if we bide our time, new, commercially viable technologies that can solve future energy needs will emerge. Table 4.1 lists 17 emerging energy technologies that energy researcher Michell Zappa, in a report for the Canadian government, identified as possible "game changers" that could revolutionize the way we conceive of and use energy.[47]

Advocates of this position do not necessarily dispute that there is a resource peak. Many in this camp accept the existence of finite energy resources but argue that fossil fuel scarcity, tightening emission standards, and geopolitical sensitivities will lead to inflated fossil fuel prices and will incentivize a transition—making many of these emerging technologies commercially viable in a very short period of time.[48] For more on the benefits and burdens of relying on one of these "bridge" fuels, shale gas, readers are invited to skip ahead to chapter 5.

Improved Consumption Efficiency Will Offset Depletion

The final related stream of logic underpinning the perspective that the peaking of resources is an irrelevant concern is that improvements in efficiency on the demand side can help buy humanity the time necessary to allow a less pernicious energy transition. If we install better home insulation or improve transportation systems to reduce the demand for oil, we can prolong the viability of our fossil fuel resources. If we consume one-third of the world's natural gas reserves but develop a natural gas combined-cycle turbine that produces one-third more energy per unit of fuel, we've similarly offset depletion.

To be sure, history is full of examples of humanity drastically improving the way energy is consumed through better end-use technology. A vivid illustration concerns lighting. Human beings have progressively transitioned from open fires to candles to incandescent lamps, compact fluorescent bulbs, and, now, LEDs. The economist William D. Nordhaus traced the pace of improvements in lighting efficiency from the oil lamps of the Paleolithic era up to today's modern technology and estimated that in terms of lumens delivered, the rate of progressive improvement increased from 0.0004% per year in ancient times to 0.04% per year in the nineteenth century and 3.6% per year from 1800 to 1992. Overall, Nordhaus calculated that the efficiency of lighting has increased by a factor of 1,200 from the times of Babylon until now.[49]

The historian Roger Fouquet makes an equally convincing point in his study of the history of heating, power, transport, and lighting in the United Kingdom from 1300 to 2000 (table 4.2)[50] Since 1750, due mostly to improvements in efficiency, the cost of lighting has decreased a thousandfold, transport fortyfold, and heating tenfold.

In sum, this line of thinking holds that the evolution of energy technology will help us to prolong the life of our finite fossil fuel reserves and present us with new opportunities in the future. As businessperson Duncan Clarke wrote, "What is known today is likely to be no more than a small fraction of what might be known tomorrow."[51] So why incur the prospects of economic

Table 4.2. Prices of End-Use Energy in Britain, 1300–2000

End use	Relative price (1900 = 100)								
	1300	1500	1700	1750	1800	1850	1900	1950	2000
Heat		225	275	300	140	110	100	80	28
Power	85	155	160	165	185	150	100	50	12
Transport	390	360	690	790	330	260	100	75	20
Lighting		950	1,115	1,170	570	300	100	6	1

Source: Modified from Roger Fouquet, *Heat, Power, and Light: Revolutions in Energy Services* (Cheltenham, UK: Edward Elgar, 2008).

disaster catalyzed by artificially inflated energy costs when technology and economic forces will eventually lead to a more natural transition?

Common Ground: Innovation Can Serve as a Failsafe

Though they may seem irreconcilable, these opposing theses can be synthesized. We can be certain of two things. First, the global demand for energy will not decrease anytime in the near future. Second, although fossil fuel energy reserves can never be known with absolute certainty, fossil fuel energy prices will certainly increase. If scarcity is not the main driver behind such increases, the ever-increasing cost of exploring increasingly remote places for fossil fuel resources will be the culprit.

Change will come. Inevitably, prices will rise to such a level that it is no longer economically prudent to commit a national energy strategy to a pool of technologies that are dependent on finite resources. To this energy market dynamic we must add the advancing perils of climate change. The costs associated with climate change invert the value of fossil fuels and turn these resources from things that ought to be explored, drilled, produced, and combusted to things best left where they are. We have entered an era where the "non-chrematistic value" (the value that cannot be measured materially) must be taken into consideration when cobbling together energy strategies.[52]

That said, we can never know precisely when supply or demand peaks will occur. Both are incredibly difficult to predict and perhaps can be determined precisely only in retrospect, after they have occurred. Many analysts have been wrong before. They have underestimated improvements in production or consumption efficiency or have simply miscalculated the timing of peaks entirely. However, the uncertainty of estimating peaks does not justify a strategy that treats peaks as nonexistent. When confronted by a tiger, one cannot vanquish it by closing one's eyes. Similarly, ignoring the inevitability of resource depletion does not make resource depletion more or less likely to occur.

Those who believe that the end of fossil fuels is still a comfortably distant concern might be right, for any one of the reasons enumerated in this chapter. However, those who stake such an ideological position are missing the point because they are busy counting jellybeans in the window of a shop that is on fire. The perils of climate change give rise to risks that are too grave to contemplate delaying a transition—a transition that will happen eventually anyway.

As such, perhaps the concept of a "peak" is less meaningful than some of the underlying factors identified in this chapter: issues of economic affordability and environmental desirability or rates of innovation and technical development. In other words, a simple metric or reserve-to-production ratio tells us little about fundamental market dynamics. As financial analyst James L. Smith writes, "In order to develop more effective forecasts of future resource

Resource Peaks and the Common Ground

Do conventional energy resources have a meaningful "peak"?

ONE SIDE: Resource peaks threaten global energy sustainability as oil and other fossil fuels are running out and social and environmental limits to their use become more apparent.

OTHER SIDE: Predictions of scarcity have been wrong before, and peaks can be offset by new systems and methods of energy supply and by changing patterns of consumer demand in a manner that will buy us time.

COMMON GROUND: We need to invest in efficiency and innovation, even if we cannot know the extent of peaks.

---

scarcity we need to look beyond the indicators to the production technologies, natural resource bases, and market structures that influence the indicators."[53] Put another way, although market fundamentals can determine the indicator of a peak, that indicator does not reveal the fundamentals about the market.

And so we come to the synthesis presented in box 4.1. We should continue to develop innovative energy systems (on both supply and demand sides) and work toward a transition out of carbon-intensive energy technologies, regardless of when peaks occur or we think they will occur. Constant innovation provides society with protection and surety in case resources are exhausted more quickly than expected. It also ensures that we keep energy resources available for future generations. As Cambridge engineering professor David J. C. MacKay argues, "Given that fossil fuels are a valuable resource, useful for manufacture of plastics and all sorts of other creative stuff, perhaps we should save them for better uses than simply setting fire to them."[54] However, if we're wrong—and if future breakthroughs or societal transitions lengthen the shelf life of existing energy resources—then our strategy of innovation has simply extended the value of those resources and given us more options.

Either way, civilization pursues some level of insurance against future, often unknowable, energy trends. Just as a participant in a jellybean counting contest would never assume that a covered jar of jellybeans is of infinite size, we need to recognize that our global coffers of fossil fuels are not limitless. A prudent strategy is to balance current economic costs and the prospects of ecological disaster. This is especially true if the shop that is holding the counting contest is aflame and the prizes we are competing for are about to be incinerated.

NOTES

1. R. G. Miller and S. R. Sorrell, "The Future of Oil Supply," *Philosophical Transactions of the Royal Society A: Mathematical, Physical and Engineering Sciences* 372, no. 2006 (December 2, 2013).

2. Kjell Aleklett, "Oil: A Bumpy Road Ahead," *World Watch Magazine*, January/February 2006, 10–12.

3. Brian Hicks and Chris Nelder, *Profit from the Peak: The End of Oil and the Greatest Investment Event of the Century* (New York: John Wiley & Sons, 2008).

4. Goldman Sachs, "330 Projects to Change the World" (April 6, 2011), available at www.energyventures.no/goldman-sachs-report-330-projects-to-change-the-world.

5. Ambrose Evans-Pritchard, "Oil and Gas Investment in the US Has Soared to $200b," *Daily Telegraph* (London), July 10, 2014.

6. Ibid.

7. Ibid.

8. Steve Sorrell et al., "Shaping the Global Oil Peak: A Review of the Evidence on Field Sizes, Reserve Growth, Decline Rates and Depletion Rates," *Energy* 37 (2012): 711.

9. David Buchan, *The Rough Guide to the Energy Crisis* (London: Rough Guides, 2010), 50.

10. Ibid., 51.

11. W. Zittel and J. Schindler, *Crude Oil. The Supply Outlook* (Ottobrunn, Germany: Energy Watch Group, 2007).

12. Ian Chapman, "The End of Peak Oil? Why This Topic Is Still Relevant Despite Recent Denials," *Energy Policy* 64 (2014): 93–101.

13. G. Maggio and G. Cacciola, "When Will Oil, Natural Gas, and Coal Peak?" *Fuel* 98 (2012): 111–123.

14. See Pedro de Almeida and Pedro D. Silva, "The Peak of Oil Production—Timings and Market Recognition," *Energy Policy* 37 (2009): 1267–1276; Ugo Bardi, "Peak Oil: The Four Stages of a New Idea," *Energy* 34 (2009): 323–326; Mikael Höök, Ugo Bardi, Lianyong Feng, and Xiongqi Pang, "Development of Oil Formation Theories and Their Importance for Peak Oil," *Marine and Petroleum Geology* 27 (2010): 1995–2004; Aviel Verbruggen and Mohamed Al Marchohi, "Views on Peak Oil and Its Relation to Climate Change Policy," *Energy Policy* 38 (2010): 5572–5581; R. A. Kerr, "Peak Oil Production May Already Be Here," *Science* 331 (2011): 1510–1511; James L. Smith, "On the Portents of Peak Oil (and Other Indicators of Resource Scarcity)," *Energy Policy* 44 (2012): 68–78; Sorrell et al., "Shaping the Global Oil Peak."

15. See L. Ruppert et al., "The US Geological Survey's National Coal Resource Assessment: The Results," *International Journal of Coal Geology* 50 (2002): 247–274; Zaipu Tao and Mingyu Li, "What Is the Limit of Chinese Coal Supplies—A STELLA Model of Hubbert Peak," *Energy Policy* 35 (2007): 3145–3154; R. A. Kerr, "How Much Coal Remains?" *Science* 323 (2009): 1420–1421; S. H. Mohr and G. M. Evans, "Forecasting Coal Production until 2100," *Fuel* 88 (2009): 2059–2067; Mikael Höök and Kjell Aleklett, "Historical Trends in American Coal Production and a Possible Future Outlook," *International Journal of Coal Geology* 78 (2009): 201–216; Bo-qiang Lin and Jiang-hua Liu, "Estimating Coal Production Peak and Trends of Coal Imports in China," *Energy Policy* 38 (2010): 512–519; Tadeusz W. Patzek and Gregory D. Croft, "A Global Coal Production Forecast with Multi-Hubbert Cycle Analysis," *Energy* 35 (2010): 3109–3122; Steve Mohr et al., "Pro-

jection of Long-Term Paths for Australian Coal Production—Comparisons of Four Models," *International Journal of Coal Geology* 86 (2011): 329–341; Robert C. Milici, Romeo M. Flores, and Gary D. Stricker, "Coal Resources, Reserves and Peak Coal Production in the United States," *International Journal of Coal Geology* 113 (2013): 109–115.

16. See Gavin M. Mudd and Mark Disendorf, "Sustainability of Uranium Mining and Milling: Toward Quantifying Resources and Eco-Efficiency," *Environmental Sciences and Technology* 42, no. 7 (2008): 2624–2630; Michael Dittmar, "The End of Cheap Uranium," *Science of the Total Environment* 461–462 (2013): 792–798.

17. BP, "Statistical Review of World Energy" (2014), available at www.bp.com/en /global/corporate/about-bp/energy-economics/statistical-review-of-world-energy.html.

18. B. K. Sovacool and S. V. Valentine, "Sounding the Alarm: Global Energy Security in the 21st Century," in *Energy Security*, ed. B. K. Sovacool (London: Sage Library of International Security, 2013), xxxv–lxxviii.

19. International Energy Agency, *World Energy Outlook 2013* (Paris: OECD, 2013).

20. M. A. Adelman, "Oil Resource Wealth of the Middle East," *Energy Studies Review* 4, no. 1 (1992): 7–8.

21. Christian Kerschner et al., "Economic Vulnerability to Peak Oil," *Global Environmental Change* 23, no. 6 (December 2013): 1424–1433.

22. Citi Research, "Global Oil Demand Growth—The End Is Nigh" (March 26, 2013), available at http://breakingenergy.com/tag/citi.

23. T. L. Friedman, *Hot, Flat, and Crowded* (New York: Macmillan, 2008), 241.

24. As updated regularly by CO_2Now.org, available at http://co2now.org.

25. Intergovernmental Panel on Climate Change (IPCC), *Climate Change 2013: The Physical Science Basis* (Geneva: IPCC, 2013).

26. Hal Harvey, Franklin M. Orr Jr., and Clara Vondrich, "A Trillion Tons," *Daedalus* 142, no. 1 (Winter 2013): 8–25.

27. Bill McKibben, "Global Warming's Terrifying New Math," *Rolling Stone*, July 19, 2012, 32–44.

28. James Hansen et al., *Climate Sensitivity, Sea Level, and Atmospheric CO_2* (New York: NASA Goddard Institute for Space Studies and Columbia University Earth Institute, 2013).

29. Christophe McGlade and Paul Ekins, "Un-burnable Oil: An Examination of Oil Resource Utilisation in a Decarbonised Energy System," *Energy Policy* 64 (2014): 102–112.

30. Aviel Verbruggen and Mohamed Al Marchohi, "Views on Peak Oil and Its Relation to Climate Change Policy," *Energy Policy* 38 (2010): 5572–5581.

31. James Leaton et al., *Unburnable Carbon 2013: Wasted Capital and Stranded Assets* (London: Carbon Tracker and the Grantham Research Institute, London School of Economics, 2013), 4.

32. Evans-Pritchard, "Oil and Gas Investment."

33. Damian Carrington, "Bank of England Investigating Risk of 'Carbon Bubble,'" *Guardian* (London), December 1, 2014.

34. Red Cavaney, "Global Oil Production About to Peak? A Recurring Myth," *World Watch Magazine*, January/February 2006, 13–15.

35. Patricia Nelson Limerick et al., *What Every Westerner Should Know about Energy* (Boulder, CO: Center of the American West, 2003), 6.

36. Jerry Taylor, *Energy Conservation and Efficiency: The Case against Coercion*, Policy Analysis No. 189 (Washington, DC: Cato Institute, March 9, 1993).

37. Ibid.

38. Lawrence Rocks and Richard Runyon, *The Energy Crisis: The Imminent Crisis of Our Oil, Gas, Coal, and Atomic Energy Resources and Solutions to Resolve It* (New York: Crown Publishers, 1972), 4.

39. John F. O'Leary, "Is the Energy Crisis Real?" *Chemical and Engineering News*, January 3, 1972, 4.

40. Cavaney, "Global Oil Production."

41. Ibid.

42. Robert L. Hirsch, Roger Bezdek, and Robert Wendling, *Peaking of World Oil Production: Impacts, Mitigation, and Risk Management* (Washington, DC: National Energy Technology Laboratory and US Department of Energy, February 2005).

43. D. Yergin, *The Prize: The Epic Quest for Oil, Money & Power* (New York: Free Press, 1993), 660.

44. Alberta Provincial Government, *Alberta's Oil Sands: Opportunity, Balance* (Edmonton, AB: Alberta Provincial Government, 2008), 20.

45. US Geological Survey (USGS), *2000 World Petroleum Assessment* (Washington, DC: USGS, 2000).

46. Vaclav Smil, "Peak Oil: A Catastrophist Cult and Complex Realities," *World Watch Magazine*, January/February 2006, 22–24.

47. For more on this topic see Michell Zappa, "17 Emerging Energy Technologies That Will Change the World," *Business Insider*, April 24, 2014, available at www.businessinsider.com/17-emerging-energy-technologies-2014-4.

48. This is summarized in "Yesterday's Fuel," *Economist*, August 3, 2013, available at www.economist.com/news/leaders/21582516-worlds-thirst-oil-could-be-nearing-peak-bad-news-producers-excellent.

49. William D. Nordhaus, "Do Real-Output and Real-Wage Measures Capture Reality?" in *The Economics of New Goods*, ed. Timothy F. Bresnahan and Robert J. Gordon (Chicago: University of Chicago Press, 1996), 27–70.

50. Roger Fouquet, *Heat, Power, and Light: Revolutions in Energy Services* (Cheltenham, UK: Edward Elgar, 2008).

51. Duncan Clarke, *The Battle for Barrels: Peak Oil Myths & World Oil Futures* (London: Profile Books, 2007).

52. L. Rival, "Ecuador's Yasuní-ITT Initiative: The Old and New Values of Petroleum," *Ecological Economics* 70 (2010): 358–365.

53. James L. Smith, "On the Portents of Peak Oil (and Other Indicators of Resource Scarcity)," *Energy Policy* 44 (2012): 68–78.

54. David J. C. MacKay, *Sustainable Energy—Without the Hot Air* (Cambridge: UIT Cambridge, 2008), 5.

Is Shale Gas a Bridge
to a Clean Energy Future?

In 2012, some communities in Pennsylvania were revitalized thanks to the rapid development of production of an unconventional energy resource known as shale gas. The state issued permits for 2,484 shale gas wells and completed more than 1,300. The total number of wells operating in the Pennsylvanian part of the Marcellus Shale formation yielded enough gas to meet the equivalent of Australia's national demand for that year. By one estimate, Pennsylvanians who allowed drilling on their land earned some $1.2 billion collectively in annual royalties.[1] The locals in Smithfield, Pennsylvania, were so enamored with this economic renaissance that they named a local delicacy the "frack burger" in honor of it.[2]

Such support contrasts with the less effusive perspective of one resident living 10 miles south of Smithfield. David Headley leads anyone who is interested to a shale gas wellhead, submerged in muddy rainwater, near his house. It continues to emit gases in a manner that Headley describes as a "witch's brew." He has published photographs of a stream beneath his house that became milky-white in 2012 after a drilling accident. The photographs include images of the well's collection tanks venting plumes of an unknown vapor into the air, dead fish bobbing on the surface of the stream, and a red rash on his son's leg that refused to go away. A spring with drinkable water that was once so clear you could see its bottom is now so full of methane that the water ignites when Headley applies a match to its surface.[3]

Despite the protestations of Headley and other Smithfield community members who are concerned about how this new energy resource is being exploited, we begin this chapter with a supportive view of shale gas. Advocates emphasize how the enormous stores of natural gas locked away in shale deposits across the globe provide much-needed bridges to a clean energy future. Thanks to breakthroughs in seismic imaging, horizontal drilling, and hydraulic fracturing, or "fracking," we are able to replace dirtier coal with cleaner gas. Moreover, this added bounty of natural gas raises the possibility of breaking the monopoly of petroleum in the transportation sector. It also enhances the potential role of natural gas as a strategic resource for supporting intermittent renewable energy sources such as wind and solar.

In contrast, critics argue that further exploitation of this new energy resource would only extend our addiction to fossil fuels by undermining investments in cleaner energy options and would give rise to massive ecological and
social problems. Foremost on the list of concerns is the knowledge that extracting, transporting, and burning natural gas all contribute significantly to
greenhouse gas (GHG) emissions and exacerbate global climate change. Those
harboring such views contend that "a golden age for gas is not necessarily a
golden age for the climate."[4]

One Side: Fracking Is Fantastic

To better comprehend the advantages and disadvantages of shale gas and to
understand what the fuss is all about, we first need a few definitions.[5] The term
shale gas refers to natural gas extracted from gas shales, porous rocks that hold
gas in pockets. Though the technology underpinning shale gas extraction
continues to evolve, two common steps are horizontal drilling and hydraulic
fracturing. In *horizontal drilling*, operators drill horizontally into the shale
formation for thousands of meters in different directions, using gas sensors to
ensure that the gas stays within the natural rock seam and does not seep out.
Hydraulic fracturing refers to a process that begins by perforating the concrete
casing of a horizontal pipe with small explosive charges. Water mixed with
sand, other propellants, and various chemicals is pumped through the holes at
up to 5,000 psi (pounds per square inch) to fracture the rock with hairline cracks
up to 1,000 feet from the pipe. The entire process can take 3 to 10 days.[6] One
group of experts proclaims that hydraulic fracturing is the most significant
energy innovation of this century.[7]

The thesis that shale gas is a boon to humankind is supported by three interconnected arguments: it is available and affordable, it is clean, and its production precipitates economic development. This thesis holds that natural gas
serves as a cost-effective bridge to a low-carbon future,[8] and that shale gas
can serve as a relatively clean and safe transitional fuel prior to an age of
renewable forms of energy.[9] According to Princeton physicist Freeman Dyson,
"Shale gas . . . makes an enormous difference to the human condition."[10] And
science writer Matt Ridley states that "a surge in gas production and use may prove
to be both the cheapest and most effective way to hasten the de-carbonization of
the world economy."[11] To gauge the verity of these claims, let's review the
premises of the main arguments in support of shale gas.

Shale Gas Is Abundantly Available and Eminently Affordable

One seemingly obvious advantage to shale gas is that there is a lot of it. The
US Energy Information Administration (EIA) assessed shale gas basins in 32
countries containing almost 70 shale formations and concluded that the

international resource is vast, with proven reserves at almost the same level as conventional natural gas.[12] The three largest energy consumers in the world—China, the European Union, and the United States—possess extensive shale gas deposits, fortuitously matching scale of supply with scale of demand. The EIA estimated that the 48 known shale gas basins (and their respective formations) hold about 6,609 trillion cubic feet of recoverable gas; these numbers are broken down in table 5.1.

Let's put some of these estimates in perspective. The Marcellus Shale formation, located in western New York State, Pennsylvania, and Ohio, is purported to possess a natural gas supply equivalent to 45 years of US national gas consumption; its reserves would be worth $500 billion at 2011 prices.[13] According to IHS, an energy-sector business information company, estimated recoverable shale gas nationwide could equal the total conventional gas discovered in the United States over the past 150 years, equivalent to about 65 times the current US annual consumption.[14]

Another advantage of shale gas, connected in part to its availability, is its affordability. Although the average cost of shale gas production varies from site to site, it has exhibited a historical trading range of $2 to $3 per thousand cubic feet of gas, about half to two-thirds the cost of production from conventional wells.[15] The United States is already producing so much shale gas that prices plummeted from $13 per million BTUs in the 1990s to $2 in 2012, rising only slightly to $4.50 in mid-2014.[16]

This cheap gas has translated into "cheap electricity," with North American factories paying half the going rate for electricity compared with Chile or Mexico. In 2012, New York State prices were the lowest they have been since the state established a wholesale market in 1999.[17] As figure 5.1 illustrates, the shale gas boom in the United States has depressed US natural gas prices significantly when compared with other major markets. However, as the figure further demonstrates, shale gas expansion in the United States has also served to slow the global escalation of natural gas prices. An interdisciplinary team at the Massachusetts Institute of Technology (MIT) calculated that without shale gas development, natural gas prices in the United States would be 2.5 times higher than they otherwise would be over the next decade.[18]

Global exploitation of shale gas reserves can suppress or even depress global natural gas prices and undermine the political clout of existing natural gas suppliers. Greater shale gas production could lessen European dependence on Russian gas, reducing Russia's ability to manipulate prices for political reasons.[19] Researchers at Rice University project that accelerated global shale gas production could depress prices in Asia as state-run gas monopolies lower their prices to match cheap American exports.[20] The implication of this is that cheap energy could catalyze a global economic renaissance.

Table 5.1. Projected Shale Gas Resources for Select Basins, 2011

	2009 natural gas market (trillion cubic feet, dry basis)			Proven natural gas reserves (trillion cubic feet)	Technically recoverable shale gas resources (trillion cubic feet)
	Production	Consumption	Imports (exports)		
Europe					
France	0.03	1.73	98%	0.2	180
Germany	0.51	3.27	84%	6.2	8
Netherlands	2.79	1.72	(62%)	49.0	17
Norway	3.65	0.16	(2,156%)	72.0	83
United Kingdom	2.09	3.11	33%	9.0	20
Denmark	0.30	0.16	(91%)	2.1	23
Sweden	—	0.04	100%		41
Poland	0.21	0.58	64%	5.8	187
Turkey	0.03	1.24	98%	0.2	15
Ukraine	0.72	1.56	54%	39.0	42
Lithuania	—	0.10	100%		4
Others	0.48	0.95	50%	2.71	19
North America					
United States	20.6	22.8	10%	272.5	862
Canada	5.63	3.01	(87%)	62.0	388
Mexico	1.77	2.15	18%	12.0	681
Asia/Australia					
China	2.93	3.08	5%	107.0	1,275
India	1.43	1.87	24%	37.9	63
Pakistan	1.36	1.36	—	29.7	51
Australia	1.67	1.09	(52%)	110.0	396
Africa					
South Africa	0.07	0.19	63%	—	485
Libya	0.56	0.21	(165%)	54.7	290
Tunisia	0.13	0.17	26%	2.3	18
Algeria	2.88	1.02	(183%)	159.0	231
Morocco	0.00	0.02	90%	0.1	11
Western Sahara	—	—		—	7
Mauritania	—			1.0	0
South America					
Venezuela	0.65	0.71	9%	178.9	11
Colombia	0.37	0.31	(21%)	4.0	19
Argentina	1,46	1.52	4%	13.4	774
Brazil	0.36	0.66	45%	12.9	226
Chile	0.05	0.10	52%	3.5	64
Uruguay	—	0.00	100%		21
Paraguay	—	—			62
Bolivia	0.45	0.10	(346%)	26.5	48
Total of above areas	53.1	55.0	(3%)	1,274	6,622
Total world	106.5	106.7	0%	6,609	

Source: US Energy Information Administration (EIA), *World Shale Gas Resources: An Initial Assessment of 14 Regions outside the United States* (Washington, DC: US Department of Energy, April 5, 2011).
   Note: Dashes indicate no data available, empty cells indicate a resource estimate of zero.

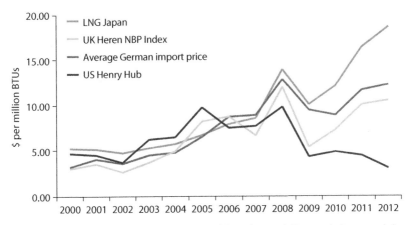

Figure 5.1. Natural Gas Prices in Europe (United Kingdom and Germany), Japan, and the United States, 2000–2012. *Source:* British Petroleum, "Statistical Review of Energy 2014." *Note:* LNG = liquefied natural gas.

Natural Gas Is the Cleanest Fossil Fuel

Advocates of shale gas argue that it possesses a cleaner environmental footprint than other fossil fuels. Compared with coal and oil, shale gas has lower emissions of sulfur oxides, nitrogen oxides, and mercury.[21] Already, the adoption of shale gas has lowered the overall emissions intensity of the coal-dependent US national grid and will continue to reduce emissions as coal- or oil-fired generation is displaced. Researchers at MIT compared future electricity scenarios for the United States with and without accelerated shale gas use and noted that intensified reliance on shale gas would reduce national emissions from the electricity sector by 17% compared with business as usual.[22] Many proponents have therefore argued that shale gas is the best way for the United States and other countries to meet their GHG reduction goals. William Press, a member of the US President's Council of Advisors on Science and Technology, writes that "America will only achieve the ambitious climate change goals outlined by President Barack Obama . . . by encouraging wide-scale fracking for natural gas over the next few years."[23]

Furthermore, the pro–shale gas argument runs, if natural gas can become a suitable substitute for gasoline and diesel in vehicles, it could break the transportation sector's dependence on oil. As Deutch speculates, if trends continue, "over time, natural gas will begin to replace oil, first in the power sector and then in the industrial and transportation sectors."[24] Globally, about 15 million natural gas–fueled vehicles already exist, supported by more than 20,000 refueling stations, including at least 2 million natural gas vehicles in (each of) Iran, Pakistan, and Argentina.[25] In the United States, one-fifth of buses and two out of every five garbage trucks run on natural gas.[26] Caterpillar and General Electric are in the process of building natural gas–powered trains, and TOTE, a marine shipping company, has begun to acquire container ships fueled by liquefied natural gas. General Electric argues that to the extent that sufficient infrastructure is built, such as large-scale pipelines and liquefied natural gas terminals, natural gas could replace diesel for rail and automobiles in North America.[27] As an added incentive for change, natural gas–powered vehicles produce much lower particulate matter emissions, about 60% fewer volatile organic compounds, 50% less nitrogen oxide, and 90% less carbon monoxide than petroleum-fueled vehicles—contributing to fewer deaths and hospital admissions due to smog, ozone, and haze.[28]

Finally, proponents emphasize that shale gas can have a symbiotic relationship with intermittent renewable sources of electricity such as wind and solar power, which need highly responsive peak-load generation support or backup generation. The low-capital, high-marginal-cost structure of natural gas combines well with the high-capital, low-marginal-cost structure of wind and solar. In the longer term, combined with other renewables, abundant natural gas can provide a path to a low-carbon, affordable energy system achievable with today's technologies, especially if configured as backup or coupled with compressed-air energy storage.[29] Most importantly, by reducing the cost of climate mitigation, natural gas could motivate more aggressive climate policies by world governments.

Thus, a Citigroup report concludes that rather than undermining renewables, shale gas could help them reach cost parity with conventional fuels by enabling effective peak-response and backup generation through flexible natural gas plants.[30] The Wuppertal Institute for Climate, Energy, and Environment in Germany contends that sufficient infrastructure already exists for natural gas to act as a complementary "bridge into a renewable age."[31]

Shale Gas Development Facilitates Prosperity

Some final benefits of shale gas stem from large-scale economic development of a new sector—the promise of jobs, infrastructure investment, profits, and tax revenues. Lower natural gas costs have begun to attract manufacturing back to the United States, offering hope of increased employment in the face of what so far has been largely a jobless economic recovery. As Andrew Liveris, chief executive officer at the Dow Chemical Company, noted, "The discovery of shale has really recreated the value proposition to build these facilities in what is the world's largest market."[32] Another commentator argues that "cheap gas costs—a third of EU prices and a quarter of Asian prices—have brought US industry back from near death."[33] The *Wall Street Journal* opined that the United States is on the cusp of a "real energy revolution" to become the world's "leading energy producer," a major coup for industry.[34] And Citi, a leading global bank, stated in one of its reports that the "potential re-industrialization of the US economy" made possible by shale gas is "both profound and timely."[35] The natural gas boom is a game changer, says economist Michael Porter. It is already generating a substantial part of US GDP, equivalent to a top-ten state like Ohio.[36]

There's evidence to support these giddy proclamations. Royal Dutch Shell committed to establishing a $2 billion chemical plant near Pittsburgh, Pennsylvania, in close proximity to the natural gas reserves of the Marcellus Shale.[37] BASF, the world's largest chemical company, is currently evaluating investments in a world-scale methane-to-propylene complex[38] and a world-class ammonia plant, both of which would be located on the US Gulf Coast to take advantage of plentiful, low-cost natural gas.[39] In 2008, Pennsylvania's shale gas boom created 29,000 new jobs, generated corporate revenues of $2.3 billion, and provided $238 million in tax revenue.[40] In 2009, gas production from the Marcellus Shale, which spans West Virginia and Pennsylvania, brought $4.8 billion in gross regional product, generated 57,000 new jobs, and created $1.7 billion in local, state, and federal tax collections.[41] Natural gas production at the Barnett Shale in Texas accounted for $11.1 billion in output in 2011, amounting to 8.1% of the entire region's economy and supporting 100,000 jobs— representing almost 10% of regional employment.[42] We should note that these numbers reflect only the direct economic benefits; indirect benefits arising from higher incomes and landowners' royalties would amplify the allure of shale gas.[43]

The Other Side: Fracking Is Folly

Before you rush out to purchase your own frack burger franchise, keep in mind that opponents of shale gas have an equally compelling case to put forward. They point out that extracting shale gas threatens water quality, pollutes the air, emits an excessive amount of GHGs, and will stymie efforts to invest in renewable energy. Moreover, when externalities are factored in, extraction is costly. In sum, opponents conclude that fracking is an unnecessary folly. As one former official at the Department of Conservation and Natural Resources in Pennsylvania put it, "We're burning the furniture to heat the house . . . in shifting away from coal and toward natural gas, we're trying for cleaner air, but we're producing massive amounts of toxic wastewater with salts and naturally occurring radioactive materials, and it's not clear we have a plan for properly handling this waste."[44]

Shale Gas Threatens Water Quality, Pollutes the Air, and Leads to Climate Change

Opponents argue that shale gas production is water-intensive and polluting. Most sites in the United States utilize between 2.7 and 3.9 million gallons of water per well—or 10 to 15 million liters of water. A drill site covering two hectares quickly turns into a "heavy industrial zone"—even if it's in suburban Fort Worth, Texas, with more than 100 large water tanker trucks coming and going each day to serve wells at a single site.[45] To catalyze the fracking process, drillers add as many as a dozen chemicals, such as biocides and benzene, constituting 2% or 3% of the total volume of fracking fluid, and the fluid invariably releases chemicals and salts from deep brines.[46] Roughly 200,000 liters of acids, biocides, scale inhibitors, friction reducers, and surfactants are pumped under high pressure into a typical well.[47] Moreover, shale gas production generates waste from drilling muds, flowback, and brines, all of which require proper treatment and disposal. The volume requiring treatment can be large, with 10% to 35% of initial chemical-water injections returning to the surface as flowback before production begins.[48]

Radiation from the release of naturally occurring radionuclides that surface during the production process poses another threat to water quality and availability. New York's Department of Environmental Conservation reported that 13 samples of wastewater from the Marcellus Shale gas reservoir contained levels of radium-226 as high as 267 times the safe disposal limit and thousands of times the safe consumption limit. The New York Department of Health also analyzed three Marcellus Shale production brine samples and found elevated gross alpha radiation, gross beta radiation, and radium-226 in the brine.[49] One study calculated that radon levels were 70 times higher than the average from the Marcellus gas field and determined that some shale gas deposits contain as much as 30 times the radiation in the normal background.[50] Other studies of Haynesville Shale (in Texas) and Marcellus Shale (in Pennsylvania) have similarly recorded high levels of radon, radium, and other radioactive substances.[51] Duke University researchers injected isotopic tracers into several shale gas basins across the United States—including the Utica, Marcellus, and Fayetteville fields—and confirmed "high levels" of salinity, toxic elements such as barium, and elevated levels of radioactivity.[52]

Such pollution is not confined to only the gas fields. In Pavilion, Wyoming, the US Environmental Protection Agency (EPA) documented that many drinking-water wells were contaminated by toxins often used in hydraulic fracturing fluids. The EPA cautioned that these toxins may be linked to abnormal rates of miscarriages, rare cancers, and central nervous system disorders, including seizures.[53] One medical journal concluded that pollution often extends into neighboring communities, noting that "the potential negative impact of natural gas well drilling on the environment and health of thousands of families and children . . . is disconcerting."[54]

Nor is such pollution limited to water. Most shale gas wells frequently employ diesel-powered pumps to inject and manage water, drills, compressors, and other machinery, leading to elevated levels of volatile hydrocarbons (such as benzene, toluene, and formaldehyde), ground-level ozone, and associated particulate pollution.[55] In Texas, benzene concentrations over the Barnett Shale area have exceeded acute toxicity standards to the point where they pose a risk of cancer due to chronic exposure.[56] Colorado has experienced increased ozone-forming pollutants along its Front Range, with measured ozone precursor emissions "twice the amount that government regulators . . . calculated should exist."[57] In Wyoming, air pollution has become a major challenge to shale gas production.[58] The Pennsylvania Department of Environmental Protection has recorded levels of nitrogen oxides and volatile organic compounds near shale gas–producing wells that are well above EPA standards.[59] Another study in Pennsylvania concluded that while the total emissions of a single well were less than those of a single coal-fired power plant, in areas where wells were concentrated, collective emissions were 20 to 40 times higher than permitted levels.[60]

Shale gas—this resource that proponents argue is an important piece in the climate change mitigation puzzle—is far from a "clean" energy source. Methane, the primary component of natural gas, is the second-most influential anthropogenic GHG, accounting for approximately one-third of global warming over the past century.[61] Methane emissions from natural gas production and distribution are one of the principal sources of anthropogenic emitted methane—and this was before all the excitement over shale gas. Cornell scientist Robert W. Howarth and his colleagues expect that fugitive methane emissions during natural gas production with fracking could be between 30% and 200% greater than those from conventional natural gas production.[62] If trends continue, shale gas may have a greater greenhouse effect than conventional gas and other fossil fuels used for both heating and electricity—conclusions consistent with figure 5.2.[63]

A few studies support Howarth's basic findings. Alvarez and colleagues reported that current leakage rates from natural gas production and delivery suggest that a switch from petroleum-fueled vehicles to natural gas–powered vehicles would have a net *negative* impact on climate change. Currently, at least 3.2% of methane is leaking from natural gas infrastructure, and Alvarez et al. warn that leakage rates for shale gas are probably worse.[64] Miller et al. studied methane emissions due to drilling, processing, and refining activities over the South Central United States (where oil and gas production are concentrated)

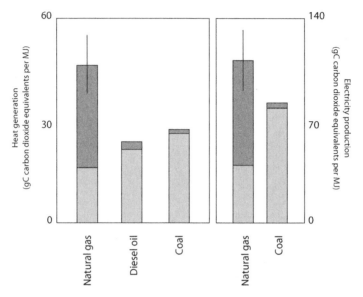

Figure 5.2. Greenhouse Gas Emissions from Natural (Shale) Gas, Diesel Oil, and Coal. *Source:* Robert W. Howarth, "A Bridge to Nowhere: Methane Emissions and the Greenhouse Gas Footprint of Natural Gas," *Energy Science and Engineering* (2014), 1–14, doi:10.1002 /ese3.35. *Note:* gC carbon dioxide = gram of carbon as CO_2; MJ = megajoule.

and concluded that emissions may be nearly five times larger than expected.[65] Brandt et al. analyzed more than 200 earlier studies of natural gas emissions spanning 20 years and confirmed that methane leakage rates are considerably higher than official estimates. They also cautioned that on a 20-year time-frame, methane is 87 times more potent than carbon.[66] Hultman et al. reached a more moderate, yet still troubling, conclusion. They found that for electricity generation, shale gas was 11% worse (from an emissions standpoint) than ordinary gas but not worse than coal.[67]

Shale Gas Production Is More Expensive Than It Appears

Many opponents of shale gas would argue that once the major problems associated with the extraction process—air pollution, water pollution, and contribution to climate warming—are economically quantified and incorporated into the cost of natural gas from this source, shale gas is no longer as economical as it currently appears to be. Also, in terms of investment attractiveness, fields in operation now represent low-hanging fruit that can never be picked again. Shale gas extraction is extremely site-specific. In Texas, for instance, wastewater can be managed by injecting it into deep wells that serve as natural

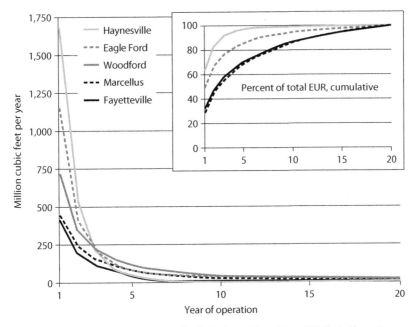

Figure 5.3. Average Production Profiles for Shale Gas Wells in Major US Shale Plays. *Source:* US Energy Information Administration, *World Shale Gas Resources: An Initial Assessment of 14 Regions outside the United States* (Washington, DC: US Department of Energy, April 5, 2011). *Note:* EUR = expected ultimate recovery.

depositories, but such formations do not exist in Pennsylvania or West Virginia, where wastewater must be pumped and treated externally.[68] In severe cases, shale gas development has contributed to increased seismicity and earthquakes.[69] Overall, one complicating factor is the lack of industrial knowledge and know-how. As one energy geologist explained, "At this stage of the game, we have very little experience with shale gas . . . Predictions of well performance over 15 to 20 years are based on 6 to 24 months of experience."[70]

There are indications that the metrics used for evaluating conventional natural gas investments are not valid for shale gas. While a conventional well can produce gas for upwards of 40 years, shale gas fractures peak within a matter of 30 to 40 months, as the production profiles in figure 5.3 indicate. In effect, this means that shale producers exchange a reduction in exploration risk for an increase in production risk.[71] In an analysis of 30 shale gas fields with 65,000 shale wells in the United States, geoscientist J. David Hughes noted that 5 fields produced 80% of the gas and that typical well output dropped 80% to 95% in its first three years.[72] This required producers to tap as many fields as possible to maintain stable output. As Hughes concluded, "Shale gas thus requires large amounts of capital from industry to maintain production. Governments and industry must recognize that shale gas and oil are not cheap or inexhaustible, [and that] 70% of US shale gas comes from fields that are either flat or in decline."[73]

An interdisciplinary team at MIT confirmed such dire findings. They estimated that most shale gas wells in the United States failed to make an expected 10% rate of return and that future costs of shale gas production could rise dramatically after drillers have depleted all of the "sweet spots" where geology has enabled easy extraction.[74]

Shale Gas Will Kill Investment in Cleaner Alternatives

Opponents further argue that shale gas can hurt cleaner alternatives such as renewable sources of electricity as much as it can help them. Economists at Resources for the Future have calculated that in conjunction with shale gas development, "some coal is displaced from the energy mix, but so are zero-emission nuclear energy and renewables. As a result, both energy consumption and carbon dioxide emissions increase slightly."[75] Similarly, complex long-term modeling published in *Nature* suggests that a shale gas boom would cut energy prices, depress incentives for adopting renewables, and, worse, increase overall carbon emissions by as much as 11% compared with business as usual.[76]

It gets worse, as there is evidence that shale gas operations are benefiting other fossil fuel technologies and pushing out cleaner alternatives such as renewable electricity. The lower cost of extracting gas through this method has convinced drillers in the United States to hunt for oil as well and to use fracking technology to liberate liquid fuel in addition to gaseous fuel. Indeed, the application of fracking to "tight oil" reserves in the Permian Basin and Eagle Ford Shale in Texas, the Bakken formation in North Dakota, and the Mississippian Lime between Oklahoma and Kansas enabled the United States to overtake Russia and Saudi Arabia to become the world's biggest producer of crude oil in 2013. The competitive attractiveness of shale gas has also driven coal producers into overseas markets where GHG emission standards are often more lax. There has been a net increase in the amount of coal that the United States exports to Asia and Europe, as coal companies shift their strategy to target international markets.[77] In Europe, specifically, a near-term swing back to coal rather than investment in new renewables is a perverse consequence of the abundance of cheap shale gas in the United States.[78]

Energy governance scholar Kirsten Westphal concludes that "unconventional fuels are no solution for global energy problems. At best they offer a viable bridge for conversion of the energy system, at worst they perpetuate existing use paths."[79] Shale gas might end up being a "bridge to nowhere," since "accommodating a large-scale shift toward natural gas assumes an investment of billions of dollars' worth of wells, pipelines, storage, and additional mid-stream infrastructure investments" which "will impede renewables unless steps are taken to prevent it."[80] As energy commentator John Farrell joked, "Natural gas isn't a bridge fuel, it's a gateway drug."[81]

Common Ground: Shale Gas Development Must Be Soundly Governed

Any attempt to synthesize these views must acknowledge that in addition to its discernible costs, shale gas production delivers tangible benefits when compared with coal, oil, and nuclear energy technologies. These benefits include a relative abundance of supply, prices that (in the near term) are highly competitive, less damage to the environment than coal and oil (under certain assumptions), and the potential for bolstering economic development in regions endowed with this resource. Nevertheless, the drawbacks associated with shale gas give cause for concern. These include the potential for costly leaks and accidents, damage to the environment and the atmosphere, geological instability, investment risk due to the uncertainty of future supply, and derailing of commitments to more sustainable forms of energy. As we described in the introduction to this chapter, shale gas production will lead to celebrations such as the naming of the "frack burger," but it will also give rise to the type of environmental tribulations experienced by David Headley.

Further obfuscating matters, these costs and benefits are not distributed evenly. Many of the benefits (such as improved regional economic development and fuel cost savings) are realized now, while many of the costs (such as contributing to climate change and the declining viability of drilling sites) will come to bear in the future. Thus, there is an irresistible short-term appeal of shale gas.

Costs and benefits also arise on different scales. The land, air, and human health impacts associated with fracking tend to be localized, whereas the climate change effects are globalized. Economic development benefits associated with shale gas have both local and global elements. Moreover, costs and benefits accrue to different actors: landowners and producers in shale gas regions benefit; companies promoting energy efficiency or renewable energy and people living adjacent to wells may suffer. Industries benefit from lower energy costs, but citizens will be adversely affected by amplified climate change patterns. This disaggregation of the winners and losers presents a collective action problem that inhibits consensus building.

To attenuate a contentious deadlock for parties on both sides of the debate, the end goal should be the same: ensure that exploitation of the resource occurs in a manner that minimizes environmental damage. For opponents of shale gas, the benefits of such a perspective allow a certain degree of concession in the face of an inevitable trend of further exploitation. For proponents of

BOX 5.1

Shale Gas and the Common Ground

Is shale gas a bridge to a clean energy future?

ONE SIDE: Shale gas is abundant, cheap, and clean, and its production facilitates economic prosperity and renewable energy.

OTHER SIDE: Shale gas production is complex, prone to accidents, bad for water quality, and dangerous to the climate, and trades off with low-carbon sources of energy.

COMMON GROUND: Shale gas production is beneficial only when properly governed by strong regulations and sound technologies and strategically planned as a bridge to cleaner energy.

shale gas, ensuring better environmental governance and closer ties to the renewable energy sector can position the shale gas industry as a meaningful part of a clean energy transition. This will reduce regulatory costs and create markets as nations strive to twin renewable energy systems and their associated stochastic energy flows with the peak-load benefits that natural gas has to offer. If the shale gas industry is not prepared to carry out further development with enhanced attention to mitigating environmental concerns, the sector might find itself in the unenviable position of a toxic competitor rather than a benign collaborator.

Given the powerful arguments on both sides of the shale gas debate, the synthesis for these two perspectives is straightforward: it has to be a bridge, and a short one at that. Like its cousins, oil and coal, natural gas of any sort is a finite resource. There might be abundant supplies of shale gas during these fledgling days of the new industry, but this will not be the case as demand accelerates and environmental regulations intensify to govern extraction and production. Our conclusion, illustrated in box 5.1, is that if you're going to develop and promote shale gas, its efficacy and utility will depend on sound governance principles and properly designed technology. That is, it can work with stringent safeguards in place and with the recognition that some fields just shouldn't be tapped because the environmental impacts would be too great. The publication of operational data on water use, the volumes and characteristics of wastewater, and methane and other air emissions, along with full mandatory disclosure of fracturing fluid additives and volumes, would help communities make more informed choices. This recommendation is endorsed by the Shale Gas Production Subcommittee of the Secretary of Energy Advisory Board in the United States[82] and by the International Energy Agency.[83] Governments—local and national—need to stipulate adequate regulations for the disposal of polluted water and for wastewater treatment. Companies and regulators should set restrictions on fracking in environmentally sensitive areas. They should select well sites that strategically reduce potential impacts on local communities, heritage, existing land use, individual livelihoods, and ecology. They should ensure the proper geological evaluation of gas fields to better assess the probability that deep faults or other geological features could generate earthquakes or permit fluids to pass between geological strata.[84]

In many ways, all of this is already happening. Best practices in technical design are emerging, including the use of multi-well drilling pads to minimize traffic and the need for the construction of roads; the elimination of flaring and venting and the isolation and sealing of leaks so that air pollutants and GHG releases are kept to a minimum; the reduction of air pollution associated with the extraction, production, and distribution processes, by substituting biofuels and electricity for diesel and deploying hyper-efficient pumps; and the requirement that wastewater be stored in tanks rather than open pits.[85] A full accounting of how the economic benefits and environmental costs of shale gas production are distributed needs to occur, modeled globally rather than only where shale gas is produced or utilized. When this is done, it will become apparent that, indeed, shale gas needs to be a bridge—perhaps a short and temporary one—and not a gateway drug.

NOTES

1. "Deep Sigh of Relief," *Economist*, March 16, 2013.

2. Peter Foster, "Now for the Downside of Fracking," *Daily Telegraph* (London), February 20, 2013.

3. Ibid.

4. Fatih Birol, " 'Golden Age of Gas' Threatens Renewable Energy, IEA Warns," *Guardian* (London), May 29, 2012.

5. Jan Bocora, "Global Prospects for the Development of Unconventional Gas," *Procedia: Social and Behavioral Sciences* 65 (2012): 436–442.

6. Matt Ridley, *The Shale Gas Shock* (London: Global Warming Policy Foundation, 2011).

7. Susan L. Sakmar, "The Global Shale Gas Initiative," *Houston Journal of International Law* 33 (2010–2011): 369–417.

8. MIT Energy Initiative (MITEI), "The Future of Natural Gas: An Interdisciplinary MIT Study" (Cambridge, MA, 2011), available at http://web.mit.edu/mitei/research/studies /natural-gas-2011.shtm.

9. Steffen Jenner and Alberto J. Lamadrid, "Shale Gas vs. Coal: Policy Implications from Environmental Impact Comparisons of Shale Gas, Conventional Gas, and Coal on Air, Water, and Land in the United States," *Energy Policy* 53 (2013): 442–453.

10. Dyson quoted in Ridley, *Shale Gas Shock*, 4.

11. Ridley, *Shale Gas Shock*, 7.

12. US Energy Information Administration (EIA), *World Shale Gas Resources: An Initial Assessment of 14 Regions outside the United States* (Washington, DC: US Department of Energy, April 5, 2011).

13. Madelon L. Finkel and Adam Law, "The Rush to Drill for Natural Gas: A Public Health Cautionary Tale," *American Journal of Public Health* 101, no. 5 (May 2011): 784–785.

14. Terry Engelder, "Should Fracking Stop? No" *Nature* 477 (September 15, 2011): 274–275.

15. John Deutch, "The Good News about Gas: The Natural Gas Revolution and Its Consequences," *Foreign Affairs* 90 (2011): 82–93.

16. EIA, "Natural Gas Weekly Update," available at www.eia.gov/naturalgas/weekly.

17. Economist, "Deep Sigh of Relief."

18. Henry D. Jacoby, Francis M. O'Sullivan, and Sergey Paltsev, "The Influence of Shale Gas on U.S. Energy and Environmental Policy" (MIT Joint Program on the Science and Policy of Global Change Report No. 207, November 2011), available at http://global change.mit.edu/research/publications/2229; H. D. Jacoby, F. M. O'Sullivan, and S. Paltsev, "The Influence of Shale Gas on U.S. Energy and Environmental Policy," *Economics of Energy and Environmental Policy* 1, no. 1 (2012): 1–22.

19. Deutch, "Good News about Gas."

20. Kenneth Barry Medlock, "Modeling the Implications of Expanded US Shale Gas Production," *Energy Strategy Reviews* 1 (2012): 33–41; Kenneth Barry Medlock et al., "Shale Gas and US National Security" (Rice University, Baker Institute, 2011), available at http://bakerinstitute.org/center-for-energy-studies/shale-gas-us-national-security.

21. A. Burnham et al., "Life-Cycle Greenhouse Gas Emissions of Shale Gas, Natural Gas, Coal, and Petroleum," *Environmental Sciences and Technology* 46, no. 2 (2012): 619–627.

22. Jacoby et al., "Influence of Shale Gas."

23. Press quoted in Robin McKie, "Fracking Is the Only Way to Achieve Obama Climate Change Goals, Says Senior Scientist," *Guardian* (London), February 16, 2013.

24. Deutch, "Good News about Gas."

25. See table 24.18 in Cutler J. Cleveland and Christopher G. Morris, *Handbook of Energy Volume I: Diagrams, Charts, and Tables* (London: Elsevier Science, 2013).

26. "Briefing: The Global Oil Industry," *Economist*, August 3, 2013, 20–22.

27. Peter C. Evans and Michael F. Farina, *The Age of Gas and the Power of Networks* (Schenectady, NY: General Electric Company, 2013), available at www.ge.com/sites/default /files/GE_Age_of_Gas_Whitepaper_20131014v2.pdf.

28. Ridley, *Shale Gas Shock*.

29. A. Gilbert and B. K. Sovacool, "Shale Gas: Better Modeling for the Energy Mix," *Nature* 515 (November 13, 2014): 198.

30. Citi Research, "Shale & Renewables: A Symbiotic Relationship" (2012), available at www.ourenergypolicy.org/wp-content/uploads/2013/04/citigroup-renewables-and-natgas -report.pdf.

31. Wuppertal Institute for Climate, Energy, and Environment, "Natural Gas: The Bridge into a Renewable Age" (January 2014), available at http://wupperinst.org/en /projects/details/wi/p/s/pd/307.

32. Liveris quoted in Daniel Gilbert, "Chemical Makers Ride Gas Boom," *Wall Street Journal*, April 18, 2012.

33. Ambrose Evans-Pritchard, "Oil and Gas Investment in the US Has Soared to $200b," *Daily Telegraph* (London), July 10, 2014.

34. "Saudi America: The U.S. Will Be the World's Leading Energy Producer, If We Allow It," *Wall Street Journal*, November 14, 2012.

35. Edward L Morse et al., *Energy 2020: North America, the New Middle East* (New York: Citi, March 2012), 1.

36. As reported by Chris Arnold, "America's Next Economic Boom Could Be Lying Underground" (National Public Radio, June 11, 2015), available at www.npr.org/2015/06/11/413395080/americas-next-economic-boom-could-be-lying-underground.

37. Gilbert, "Chemical Makers."

38. BASF, "Investments: Capital Expenditure Will Boost Future Organic Growth" (2014), available at www.factbook.basf.com/BASF-The-Chemical-Company/Strategy/Investments.

39. Ibid.

40. David M. Kargbo, Rong Wilhelm, and David J. Campbell, "Natural Gas Plays in the Marcellus Shale: Challenges and Potential Opportunities," *Environmental Science and Technology* 44 (2010): 5679–5684.

41. Stephanie Scott, "Who Shale Regulate the Fracking Industry?" *Villanova Environmental Law Journal* 24 (2013): 189–223.

42. Evan J. House, "Fractured Fairytales: The Failed Social License for Unconventional Oil and Gas Development," *Wyoming Law Review* 13 (2013): 6–67.

43. Thomas C. Kinnaman, "The Economic Impact of Shale Gas Extraction: A Review of Existing Studies," *Ecological Economics* 70 (2011): 1243–1249.

44. Quoted in Beren Argetsinger, "The Marcellus Shale: Bridge to a Clean Energy Future or Bridge to Nowhere? Environmental, Energy and Climate Policy Considerations for Shale Gas Development in New York State," *Pace Environmental Law Review* 29 (2012): 340.

45. Richard A. Kerr, "Natural Gas from Shale Bursts onto the Scene," *Science* 328 (June 25, 2010): 1624–1626.

46. Ibid.

47. R. W. Howarth, R. Santoro, and A. Ingraffea, "Methane and the Greenhouse-Gas Footprint of Natural Gas from Shale Formations," *Climatic Change* 106 (2011): 679–690.

48. House, "Fractured Fairytales."

49. Kargbo et al., "Natural Gas Plays."

50. Marvin Resnikoff, *Radon in Natural Gas from Marcellus Shale* (Brooklyn, NY: Radioactive Waste Management Associates, January 2012).

51. Y. K. Kharaka et al., "The Energy-Water Nexus: Potential Groundwater-Quality Degradation Associated with Production of Shale Gas," *Procedia: Earth and Planetary Science* 7 (2013): 417–422.

52. Avner Vengosh et al., "The Effects of Shale Gas Exploration and Hydraulic Fracturing on the Quality of Water Resources in the United States," *Procedia: Earth and Planetary Science* 7 (2013): 863–866.

53. Dianne Rahm, "Regulating Hydraulic Fracturing in Shale Gas Plays: The Case of Texas," *Energy Policy* 39 (2011): 2974–2981.

54. Lori S. Lauver, "Environmental Health Advocacy: An Overview of Natural Gas Drilling in Northeast Pennsylvania and Implications for Pediatric Nursing," *Journal of Pediatric Nursing* 27 (2012): 383–389.

55. Argetsinger, "Marcellus Shale."

56. Howarth et al., "Methane and Greenhouse-Gas Footprint."

57. House, "Fractured Fairytales."

58. Kargbo et al., "Natural Gas Plays."

59. Charles W. Schmidt, "Blind Rush? Shale Gas Boom Proceeds amid Human Health Questions," *Environmental Health Perspectives* 119, no. 8 (August 2011): 348–353.

60. Foster, "Now for the Downside."

61. J. Butler, "The NOAA Annual Greenhouse Gas Index (AGGI)" (2014), available at www.esrl.noaa.gov/gmd/aggi.

62. Howarth et al., "Methane and Greenhouse-Gas Footprint"; Robert W. Howarth et al., "Venting and Leaking of Methane from Shale Gas Development: Response to Cathles et al.," *Climatic Change*, January 10, 2012.

63. Robert W. Howarth and Anthony Ingraffea, "Should Fracking Stop? Yes," *Nature* 477 (September 15, 2011): 271–274.

64. R. A. Alvarez et al., "Greater Focus Needed on Methane Leakage from Natural Gas Infrastructure," *Proceedings of the National Academy of Sciences USA* 109 (2012): 6435–6440.

65. S. M. Miller et al., "Anthropogenic Emissions of Methane in the United States, *Proceedings of the National Academy of Sciences USA* 110, no. 5 (2013): 20,018–20,022, available at www.pnas.org/content/110/50/20018.

66. A. R. Brandt et al., "Methane Leaks from North American Natural Gas Systems," *Science* 343 (February 14, 2014): 733–734.

67. Nathan Hultman et al., "The Greenhouse Impact of Unconventional Gas for Electricity Generation," *Environmental Research Letters* 6 (2011): 044008. Hultman et al.'s study may be more fine-tuned than the studies cited above, given that it distinguishes between end uses such as heating and electricity, uses a greater variety of global warming potentials, and takes into account shale gas operators' learning from past failures and developing more efficient technology.

68. Laura C. Reeder, "Creating a Legal Framework for Regulation of Natural Gas Extraction from the Marcellus Shale Formation," *William and Mary Environmental Law and Policy Review* 34 (2010): 999–1026.

69. W. Ellsworth et al., "Are Seismicity Rate Changes in the Midcontinent Natural or Manmade?" *Seismological Research Letters* 83 (2012): 403; National Research Council, *Induced Seismicity Potential in Energy Technologies* (Washington, DC: National Academies Press, 2012); Richard A. Kerr, "Learning How to NOT Make Your Own Earthquakes," *Science* 225 (March 23, 2012): 1436–1437.

70. Kerr, "Learning How."

71. Jacoby et al., "Influence of Shale Gas."

72. J. David Highes, "A Reality Check on the Shale Revolution," *Nature* 494 (February 21, 2013): 308.

73. Ibid.

74. Jacoby et al., "Influence of Shale Gas."

75. Kerr, "Natural Gas from Shale."

76. Haewon McJeon et al., "Limited Impact on Decadal-Scale Climate Change from Increased Use of Natural Gas," *Nature* 514 (October 23, 2014): 482–485.

77. B. K. Sovacool, "Cornucopia or Curse? Reviewing the Costs and Benefits of Shale Gas Hydraulic Fracturing (Fracking)," *Renewable and Sustainable Energy Reviews* 37 (September 2014): 249–264.

78. Patrick Parenteau and Abigail Barnes, "A Bridge Too Far: Building Off-Ramps on the Shale Gas Superhighway," *Idaho Law Review* 49 (2013): 325–365.

79. Kirsten Westphal, "Unconventional Oil and Gas Global Consequences," *SWP Comments*, March 12, 2013, 1–8.

80. Ibid.

81. John Farrell, "Natural Gas Isn't a Bridge Fuel, It's a Gateway Drug," *Renewable Energy World*, February 3, 2014.

82. Brandon J. Murrill and Adam Vann, "Hydraulic Fracturing: Chemical Disclosure Requirements" (Congressional Research Service 7-5700, April 4, 2012).

83. International Energy Agency, *Golden Rules for a Golden Age of Gas* (Paris: OECD, November 2012).

84. Ibid.

85. B. K. Sovacool and V. Vivoda, "Enhancing the Energy Security and Governance of Shale Gas," *Oil Gas and Energy Law* 12, no. 3 (June 2014): 1–35.

Can Renewable Electricity Ever Be Mainstreamed?

The manner in which some energy systems are personalized and denigrated is astonishing. Fossil fuels are cursed for their multitude of negative externalities—air pollution, water pollution, climate change, unemployment, transfers of wealth to foreign nations, and more. The *Economist* calls fossil fuels "anti-heroes."[1] In a *Futurama* episode, they become greenhouse gas "villains" who cause global warming by capturing "Mr. Sunbeam" in the earth's atmosphere.[2] It is almost as if we're talking about not inanimate resources but living, breathing beings—malignant scoundrels terrorizing citizens. "So you want to run us out of town do ya?" asks Cole, leader of the fossil fuel villains. "Go ahead and try." "Yeah," cackles his oily sidekick, Slick. "Just try to make us leave!"

Fossil fuels, of course, are not malignant entities; they are resources that have been a boon to humankind. In truth, one would be hard-pressed to identify any other minerals that have provided greater utility. So, if fossil fuels were living entities, they wouldn't be astride motorcycles harassing families at rural fairs; they would be attending public galas, accepting lifetime achievement awards from grateful communities that wished them well in their retirement.

Retirement really is in the cards for fossil fuels, whether consumers wish to accept this inevitability or not. Even ignoring such ills as air pollution, water pollution, climate change, and terrorist links to oil, the era of fossil fuels' dominance of the energy supply may be drawing to a forced close. Renewable electricity—power supplied by wind turbines (onshore and offshore), solar energy (including solar photovoltaic panels and concentrated solar power), geothermal heat, biomass (including landfill gas, agricultural waste, trash, and energy crops), and falling and running water (from dams, pumped storage, and run-of-river turbines)—can generate energy with much less negative environmental and social impact. So in response to the question, "Can renewable electricity ever be mainstreamed?" the answer should be at least a partial yes, because humankind doesn't have many other viable options that don't involve our offspring having to manage massive amounts of hazardous nuclear waste. Nevertheless, before passing judgment on this issue, we need to examine the question from two competing perspectives.

One Side: Renewables Are Not Affordable on a Large Scale

Viewed from our lofty perch in 2016, it is understandable why many would view renewable electricity as merely a niche solution. There are valid grounds for trepidation about the potential of renewable energy technologies. One analyst joked that to force renewables to operate 24/7 "is like trying to make a pig fly: you won't succeed and you only make the pig unhappier."[3] Consider just three of the most prominent hurdles to the diffusion of renewable electricity: it requires backup, and not everyone has abundant resources; it is impeded by social and environmental obstacles; and it is expensive.

Renewable Energy Resources Are Intermittent and Inequitably Distributed

Many of the renewable technologies that possess the greatest utility-scale potential are plagued by energy generation profiles that, when plotted, resemble hand-sketched roller coasters, as figure 6.1 illustrates. Sometimes wind blows; at other times, it does not. Sometimes the sun shines; at other times, clouds and nightfall spoil the party. Wind power, solar power, some forms of hydropower, and even biomass power (due to seasonal changes) all suffer from the same curse: stochastic power profiles.[4] This makes these forms of energy a burden for utilities to manage. The task of providing power to energy users is hard enough, given variable demand patterns. When fluctuating supply patterns must be added to the planning challenge, the task of balancing electricity production with electricity demand can be expensive and can potentially undermine the quality of electricity services provided. For instance, in Germany in 2012, daily wind energy generation ranged from 0.007 to 0.53 TWh. The same holds true for renewable electricity supply in Australia, capacity factors in the United States, and the likely portfolio of a low-carbon German electricity sector in 2050 (panel C of figure 6.1). As an article in *Newsweek* stated, "Wind and solar are nice and clean—but the sun doesn't work 24/7 and the wind is fickle."[5]

As a result, advocates of this view argue that renewables need costly sources of backup or storage or pose a troublesome threat to grid reliability. The American Legislative Exchange Council, an industry group, proclaims that "wind is such an unreliable source of electricity that coal plants are required to operate around the clock as backup for wind power."[6] The Electric Power Research Institute, a think tank, notes that renewables work best only when "large-scale energy storage devices" can "accept electricity when it is produced by renewable resources and inject it into the grid during periods of peak demand."[7] Michael Webber of the University of Texas at Austin argues that if we aggressively promote solar panels, "it will actually mess up the power quality on the grid . . . you kind of put junk onto the line, so to speak, and that messes up grid reliability."[8]

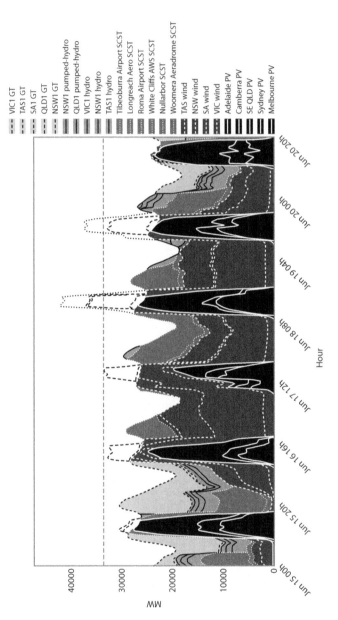

A. Variations in renewable electricity supply and demand for Australia, 2010

Legend:
VIC1 GT
TAS1 GT
SA1 GT
QLD1 GT
NSW1 GT
NSW1 pumped-hydro
QLD1 pumped-hydro
VIC1 hydro
NSW1 hydro
TAS1 hydro
Tibeoburra Airport SCST
Longreach Aero SCST
Roma Airport SCST
White Cliffs AWS SCST
Nullarbor SCST
Woomera Aeradrome SCST
TAS wind
NSW wind
SA wind
VIC wind
Adelaide PV
Camberra PV
SE QLD PV
Sydney PV
Melbourne PV

MW axis: 0, 10000, 20000, 30000, 40000

Hour axis: Jun 15 00h, Jun 15 20h, Jun 16 16h, Jun 17 12h, Jun 18 08h, Jun 19 04h, Jun 20 00h, Jun 20 20h

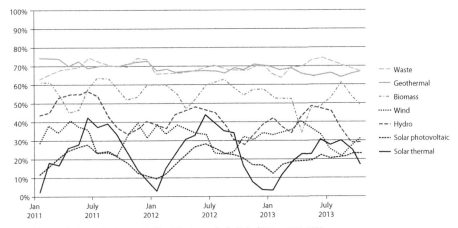

B. Monthly Capacity Factors for Renewable Electricity Sources in the United States, 2011–2013

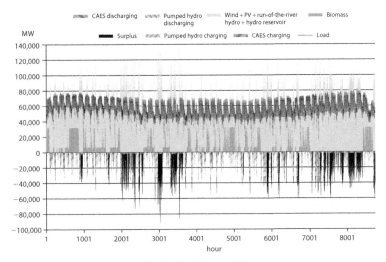

C. Projected Total Output from Renewable Electricity in Germany, 2050

Figure 6.1. Variations in Renewable Electricity Generation in Australia, the United States, and Germany. *Sources: A,* Mark Diesendorf, "Sustainable Energy Solutions for Climate Change, Air Pollution and Energy Security," Seminar to the School of Energy and Environment, City University of Hong Kong, 21 May 2014; *B,* US Energy Information Administration (EIA), *Electric Power Monthly* (Washington, DC: EIA, 2014); *C,* Olav H. Hohmeyer and Sönke Bohm, "Trends toward 100% Renewable Electricity Supply in Germany and Europe: A Paradigm Shift in Energy Policies," *WIREs: Energy and Environment* (online, June 24, 2014), doi:10.1002/wene.128. *Note:* CAES = compressed air energy storage; PV = photovoltaic. Given the rapidity of fluctuations in price and performance, such data continue to change rapidly.

The potential of many renewable technologies is also geographically constrained. Geothermal power is perhaps the most obvious example: many nations outside the Ring of Fire, a horseshoe-shaped area around the Pacific Ocean, do not have access to sufficient viable reserves of geothermal energy. Biofuel and biomass potentials are constrained by land availability and climate. Wind power is dependent on wind quality and wind speed, two factors that vary significantly depending on, among other things, latitudinal positioning, topography, climate, season, and atmospheric circulation patterns. Wave power and tidal power have significant potential that is largely dependent on coastal seabed topography. Solar power potential is undermined by latitudes, weather patterns, and land-use characteristics. Taken together, the argument runs, geographical constraints pose real and present hurdles in achieving high levels of diffusion of renewable technologies.

Social and Environmental Barriers Limit Diffusion

Renewable energy facilities typically require far more land and, as such, are sometimes perceived as more physically invasive and environmentally destructive than nuclear or fossil fuel power plants. One of the prime reasons for "not-in-my-backyard," or NIMBY, opposition to wind farms and utility-scale solar power installations is the invasive presence of these facilities.[9] Consider what your reaction would be to forty 3 MW wind turbines popping up on 100-meter-high towers near your community. According to the International Energy Agency (IEA), in 2011, global consumption of electricity amounted to 19,004 TWh.[10] To satisfy such demand only through wind power would necessitate more than 1.5 million 5 MW turbines (operating at 30% capacity). In any nation, nary a region would remain unaffected by the presence of their spinning blades.

A different but equally chaotic tale would unfold if solar photovoltaic (PV) systems were to provide all electricity at 2011 levels under current technological conditions. Assuming a 15% conversion efficiency and an average annual solar input of 0.1 kW/m^2, approximately 600,000 square kilometers—an area almost the size of France—would be required to provide enough space for solar panels. Conversely, a coal-fired or nuclear power plant is typically tucked away in a remote location, far from the prying eyes of the general public. Unless one lives in a city where the externalities from such conventional power plants are very severe (such as Chernobyl, Beijing, or Fukushima), the presence of these hidden threats to public health and the environment are seldom apparent.

Moreover, renewable sources of electricity, while clean, are not free from negative environmental or social impacts. Wind farms contribute to avian mortality, when birds collide with wind turbine blades.[11] Hydroelectric dams can destroy habitats, degrade water quality, and accelerate sedimentation and erosion patterns in rivers.[12] The use of energy crops and biomass waste streams to generate electricity can strip local ecosystems of needed nutrients and minerals and contribute to deforestation.[13] Geothermal plants require water during drilling and fracturing processes, and they release measurable amounts of hydrogen sulfide and carbon dioxide. They also produce toxic sludge containing sulfur, silica compounds, arsenic, and mercury.[14] Today's solar PV systems require silicon mining, which generates silica dust that causes lung diseases in miners, and the manufacturing of thin-film panels involves toxic materials such as cadmium and gallium that can pose environmental threats.[15]

Overall, opposition to renewable energy can come from disparate sources. Landowners may oppose a wind farm due to concerns that it will lower property values and increase electricity bills. Investors might be averse to investing in renewable energy because of concerns that project delays give rise to unacceptable risk levels. Politicians and regulators may be wary of renewable energy out of concern for job losses and public opposition to higher energy prices. Even environmentalists may reject certain forms of renewable energy—opposing wind power, for instance, because of the threat to bats—leading to what one scholar termed "green on green" conflict.[16] One researcher mused that modern resistance toward any energy project has become so strong that NIMBY has evolved into BANANA—build absolutely nothing anywhere near anything.[17]

Negative perceptions about renewable energy can significantly impair its realizable potential. To illustrate, Cambridge University energy scientist David MacKay broke down all energy requirements of the United Kingdom into kilowatt-hours per person per day—not only electricity, but also heating, transport, manufacturing, and even food and national defense—and assessed whether the nation could satisfy this demand exclusively through renewable electricity.[18] Promisingly, he found that the technical potential was close—demand was 195 kWh per person per day, and the technical potential of renewable energy was 180 kWh per person per day. However, as MacKay pointed out, technological incompatibilities (e.g., solar PV panels competing with solar thermal water systems on the same roof), costs, social constraints, and public opposition need to be incorporated into assessments to determine realizable potential. When this is done, realizable potential is reduced almost tenfold to 18.3 kWh.

Renewables Are Not Cost-Competitive

Perhaps the most prominent criticism of renewable energy diffusion focuses on cost. Most renewable energy technologies are seen as unable to compete commercially with coal-fired, gas-fired, or nuclear power, particularly when the externalities associated with conventional energy are not adequately incorporated into its cost. Due to adverse economics, even forward-thinking governments that ideologically support renewable energy eventually reach a stage, during efforts to expand renewable energy capacity, when economic concerns begin to stymie diffusion. Consider the current climate for renewable energy in Germany. The government is facing stiff criticism over the perceived economic inefficiencies associated with the nation's wind and solar power development program.[19]

There is also an enormous amount of capital and labor that must go into the design and implementation of a renewable energy–based system. With a colossal growth in energy consumption expected between 2008 and 2035, recent projections anticipate the need for trillions of dollars of investments in new energy infrastructure every year just to keep pace with growth.[20] However, if global society is to adopt technology to cut greenhouse gas emissions from the energy sector to 50% of 2005 levels by 2050—relying largely on renewable energy—the investment required becomes nearly unfathomable. The IEA, the energy planning arm of the Organization for Economic Cooperation and Development (OECD), estimates that a staggering $316 trillion of investment would be needed to achieve that goal: $270 trillion under "baseline" conditions and $46 trillion (17% more) for the BLUE emissions scenario.[21]

Advocates of fossil fuels argue that this is too high a price tag to expect global buy-in. It's an amount that is difficult to contemplate in a world trying to grow economies, provide jobs, and avoid recession. Canadian Prime Minister Stephen Harper argued that the costs of renewable energy would "destroy jobs and growth," and Australian Prime Minister Tony Abbott overturned renewable energy policies on the grounds that they are "job-killing."[22] As energy scholar David Victor summarizes, "The reality is that the policies you need to put into place and the technologies needed rapidly are completely beyond what any political system is going to do."[23] One economist estimated that it would be cheaper to colonize space and "terraform Mars" than to mitigate climate change on Earth with existing renewable energy technologies.[24] However, before you rush to dump your wind energy stocks and cancel your order for a solar panel kit, it might serve to explore the counter-position.

The Other Side: Renewables Are the Future

The disadvantages attributed to renewable energy represent only part of the story. To understand a fuller narrative, we need to examine the view that renewable energy can and probably will dominate the electricity sector in the not too distant future. In fact, in the long term, a transition to renewables will have to occur. As German parliamentarian Hermann Scheer put it, "Our dependence on fossil fuels amounts to global pyromania, and the only fire extinguisher we have at our disposal is renewable energy."[25] There are four key arguments that support this perspective: (1) the cost advantage of fossil fuel technologies is artificial; (2) there is sufficient renewable energy potential to meet demand;

(3) even without internalizing externalities, the cost profiles of renewable and conventional energy technologies are converging; and (4) renewable energy technologies have far greater potential for technological improvement.

The Economic Primacy of Conventional Energy Is Artificial

One argument in favor of renewable energy rests on the artificially inflated economic superiority of conventional energy. Fossil fuels and nuclear power have received levels of government support that might even make the occasional politician blush. Consider some statistics just for the United States. Between 1961 and 2008, nuclear power received $61 billion (in 2005 dollars) in research and development subsidies from the US Department of Energy.[26] If houses cost, on average, $100,000, this would be enough money to buy all the homes in the city of Boston. Over the same period, $26 billion was spent on fossil fuel research and development—a remarkable amount of money to throw at carbon-intensive energy technologies during a period of escalating climate change.

Thanks to such rich government support, the trajectory of technological development for fossil fuels and nuclear energy has been significantly altered for the better. Contrast this with what's happened for renewable energies. Between 1961 and 2008, all the renewable energy technologies in the United States plus all energy-efficiency research initiatives received a grand total of $26 billion—less than half of what the nuclear power sector received.[27] Biased support for conventional energy over renewable energy has falsely inflated the true technological potency of conventional energy systems. One can only speculate on what the economic fortunes of wind power technology might be if the same amount of government support channeled into nuclear power were funneled into wind power.

Globally, we see the same bias toward fossil fuels. The IEA estimated that fossil fuel consumption subsidies worldwide totaled $548 billion in 2013, with Iran, Saudi Arabia, India, Russia, and Venezuela topping the list.[28] The International Monetary Fund estimated that in 2013, energy subsidies across 176 countries amounted to $1.9 trillion. On a post-tax basis, this is equivalent to 8% of all global government revenue for that year.[29] A vast majority of these subsidies ($879 billion, or 46.3%) went to petroleum interests; $539 billion (28.4%) to coal interests; $299 billion (15.8%) to natural gas interests; and a meager $179 billion (9.4%) to electricity interests, including all types of renewable energy and other technologies such as storage, transmission, and distribution.

The other important characteristic of conventional energy technologies that fortifies their market advantage is that, generally, they have been base-load, centralized solutions to electricity generation. This has engendered the development of electricity grids that are, by and large, customized to fit the electricity generation profiles of these technologies. In other words, the stochastic profile of renewable energy flows is not necessarily an unavoidable technological problem. It is a technological problem only because of the nature of the systems currently in place for electricity load balancing.

This leads to a final factor that has bestowed false economic superiority on conventional energy: technological entrenchment. Consider the infrastructure that exists to deliver power from a conventional power plant to an end user, such as a residential electricity consumer. Land has to be procured and funds sunk into plant investment. Roads have to be built to allow fuel stock to be trucked into the plant or to rail links, and freight routes have to be developed to bring the fuel stock from the mine to the plant. Power lines and transformer stations have to be erected to transmit a stable flow of electricity from the plant to the grid. Grids have to be constructed, and load management systems have to be developed to deal with the electricity generation profiles of these technologies. At the residence where the electricity is being used, meters are installed and networks of wiring are put in place to accommodate a one-way flow of electricity.

Each of these entrenched features of the conventional electricity infrastructure is associated with vested financial interests that are committed to preserving profits and sunken investment. These are high-stakes economic activities, and threats to the economic well-being of this entrenched infrastructure will not be welcomed up to the veranda for a gin and tonic and a chat. Competitive threats will be met by aggressive market defense and, when necessary, by misinformation campaigns.[30]

Renewables Have Ample Technical Potential

Advocates of renewables contend that claims of insufficient renewable energy are, in the immortal words of Howard Cosell, poppycock. Consider, they say, what we know about the potential of some of the dominant renewables. A study by more than 100 researchers working with the Intergovernmental Panel on Climate Change (IPCC) contends that nearly 80% of the world's energy needs could be met by renewable energy by 2050, if the right public policies were put in place to support such an initiative.[31] According to a parallel study published in *Energy Strategy Reviews*, it would be feasible for 95% of the world's energy needs to be delivered by renewable energy under conditions of improved energy efficiency and electrification of vehicles.[32]

Even the United States has an enormous cache of renewable energy resources that it has only just begun to utilize. A comprehensive technical assessment undertaken by the US Department of Energy calculated that 93.2% of all domestically available energy within the country's borders was in the form of wind, geothermal, solar, and biomass.[33] The nonpartisan, independent National Research Council came to a similar conclusion regarding the importance of renewable energy: "Sufficient domestic renewable resources exist to allow renewable electricity to play a significant role in future electricity generation and thus help confront issues related to climate change, energy security, and the escalation of energy costs . . . Renewable energy is an attractive option because renewable resources available in the United States, taken collectively, can supply significantly greater amounts of electricity than the total current or projected domestic demand."[34] This support for renewable energy was replicated in a four-volume National Renewable Energy Laboratory report in 2012, which concluded that "renewable electricity generation from technologies that are commercially available today, in combination with a more flexible electric system, is more than adequate to supply 80% of total U.S. electricity generation in 2050 while meeting electricity demand on an hourly basis in every region of the country."[35]

In one of the most widely discussed studies, Mark Jacobson of Stanford University and Mark Delucchi of the University of California, Davis, contend that 100% of the world's energy for all purposes could be delivered by wind, solar, and water resources by 2030. They acknowledge that the plan calls for "millions of wind turbines, water machines and solar installations," but they also point out that "the scale is not an insurmountable hurdle; society has achieved massive transformations before. During World War II, the US retooled automobile factories to produce 300,000 aircraft and other countries produced 486,000 more."[36] In a follow-up study, the same authors conclude that "with sensible broad-based policies and social changes, it may be possible to convert 25% of the current energy system to WWS [wind, water, and sun] in 10–15 years, 85% in 20–30 years, and 100% by 2050."[37]

Renewables Exhibit Converging Cost Profiles

Advocates also take exception to the claim that renewable energy will never be as economically competitive as conventional energy. The International Renewable Energy Agency noted that in 2012, the cost profiles for onshore wind power, many biomass technologies, geothermal power, and all hydropower technologies fell within the cost range of fossil fuel power in OECD nations.[38] It also documented that all renewable energy technologies can provide commercially competitive decentralized power to remote locations ("island electricity prices"). Similarly, figure 6.2, which compiles data from a variety of sources to show the levelized cost of electricity, demonstrates how geothermal, biomass, wind, and even solar PV can all produce power more cheaply than conventional sources in certain configurations. In short, in response to the question of whether or not renewable energy technologies will be cost-competitive with conventional energy technologies in the future, the answer is that many already are.

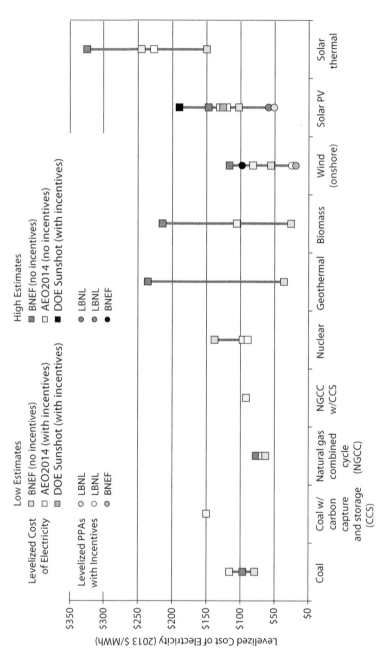

Figure 6.2. Levelized Cost of Electricity for New Power Plants, 2013. *Source:* World Resources Institute (WRI), *Seeing Is Believing* (Washington, DC: WRI, October 2014). *Note:* AEO = US Energy Information Administration's Annual Energy Outlook; BNEF = Bloomberg New Energy Finance; DOE = US Department of Energy; LBNL = Lawrence Berkeley National Laboratory; PPA = power purchase agreements; PV = photovoltaic.

The prospect for renewable energy technologies being able to compete with conventional energy technologies is increasingly optimistic. PV module prices dropped from more than $3 per watt in 2008 to less than $1 per watt in 2012. Admittedly, this was due in part to an oversupply that led to a significant reduction in prices for solar PV panels manufactured in China, which is now the dominant producer of solar PV equipment.[39] But prices are expected to fall even further as "soft costs" such as for permitting and labor for installation continue to decline.[40] The cost of wind power, which was estimated at approximately $50 to $60 per megawatt-hour in 2010, is projected to decline to $35 to $55 per megawatt-hour in 2030.[41] All the while, global conventional energy prices are expected to rise.[42] Figure 6.3 shows that unlike fossil fuels, renewable energy costs are set to decline in the future as technologies improve.

This promising future for renewable energy is supported by market data. From 2013 to 2014, for instance, renewable energy accounted for more than half (56%) of all new additions to global electricity capacity, and when hydro is included, renewables met about one-fifth of the world's total energy consumption.[43] According to a 2013 Bloomberg study, 70% of new electricity generation capacity installed between now and 2030 will be renewable.[44] The market is evolving so quickly that projections are outdated almost as soon as they are made. As testament to this, Bloomberg's projections are 35% higher than its same projections made a year earlier.

These cost numbers look at only part of the picture. Ironically, the greater use of renewable energy has been shown to depress fossil fuel prices, producing net social savings. In Germany, Italy, and the United Kingdom, accelerated deployment of solar PV significantly reduced peak electricity prices by more than 50% from 2007 to 2011.[45] In the United States, projections suggest that achieving a 20% national penetration rate for renewables would decrease consumer energy bills by an average of 1.5% per year and, overall, would save consumers a total of $49.1 billion on their electricity and natural gas bills by 2020.[46] Wind energy alone has been shown to have such immense benefits— by avoiding air pollution and improving public health—that spending $900 million on upgrading wind farms at Altamont Pass in California was estimated to avoid as much as $3.6 billion in human health–related and climate-related externalities.[47] In Europe, studies show that use of renewable energy decreases wholesale electricity prices and avoids the need for expensive energy imports, producing macroeconomic savings that extend well beyond their cost through a trend known as the "merit-order effect." In Spain, for instance, consumers pay total tariffs of about €1 billion for wind each year but see about €1.7 billion in avoided costs (for a net savings of more than €640 million). In Germany, customers spent €3.2 billion collectively on renewable energy in 2007 but saved €5 billion through the merit-order effect.[48] The implication seems to

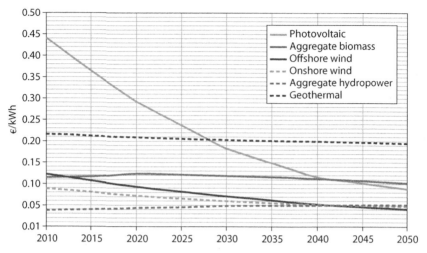

Figure 6.3. Expected Levelized Costs for Renewable Electricity, 2010–2050. *Source:* Olav H. Hohmeyer and Sönke Bohm, "Trends Toward 100% Renewable Electricity Supply in Germany and Europe: A Paradigm Shift in Energy Policies," *WIREs: Energy and Environment* (June 24, 2014), doi:10.1002/wene.128.

be that, yes, renewables may cost more in some situations, but the investment yields latent social, environmental, and economic benefits.

Performance and System Integration Are Improving

What about the other technological factors, outlined earlier, that are currently hampering the prospects of renewable energy diffusion: stochastic flows, geographical constraints, and NIMBY opposition? Proponents respond with several lines of evidence.

First, different renewable technologies exhibit differing ranges of intermittency or variability, as do traditional generators, but these can be managed. Coal plants suffer unexpected outages more commonly than one might think; nuclear plants need to shut down for more than a month when they refuel; natural gas plants often run in peaking mode, which means they operate at suboptimal levels. The US Department of Energy reports that the average capacity factor for all power plants across the nation—a reflection of both technical reliability and dispatching practice—is roughly 55%. That is, over a long period of time, an average power plant contributes to the electricity grid only 55% of its theoretical maximum output. As table 6.1 indicates, nuclear generators have boasted the highest capacity factors, occasionally exceeding 90%. Coal-fired plants rank near the middle, with a capacity factor of about 60% to 70%.

Natural gas combustion turbines have much lower capacity factors of between 4.1% and 6.0%, largely because they are used to meet demand during system peaks. Geothermal, biomass, and landfill gas systems already have higher capacity factors than the average coal plant and natural gas plant. However, no plants operate at full capacity 100% of the time, so accommodation always has to be made for times when energy generation technologies go offline, for whatever reason.

Moreover, over the past decade, concerns over how to manage the stochastic flows of renewable energy generation have weakened significantly due to improved energy portfolio management strategies, more effective plant-siting strategies, and technological advances within many of the renewable energy platforms. New approaches to load-balancing strategies are appearing in engineering journals,[49] and new strategies are being advanced to reduce stochastic variation through strategic siting of renewable energy plants and technical portfolio combinations such as twinning solar and wind energy.[50] Overall, improvements to renewable energy technologies and strategies for better employing storage technologies are beginning to make the stochastic flows of renewable energy generation far less of a concern.[51] As the supply of power becomes more variable, its demand can also become more flexible, meaning that building and industrial loads can be adapted into operating modes that provide a low-cost resiliency strategy for the grid that only enhances the potential of renewable capacity.[52] Given these developments, is it not possible, indeed probable, that advances will continue to mitigate the impact of intermittency?

Second, with regard to geographical constraints, it is undeniable that some nations are hard-pressed to find enough space to build sufficient capacity for renewable energy technologies that require significant swaths of land. Singapore, for example, is not going to adopt a 100% wind power diffusion policy anytime soon. However, research and evidence from practice indicate that most nations could achieve surprisingly high contributions of renewable energy to the electricity supply by focusing on technologies that best suit national circumstances and core competencies. For example, research in Australia demonstrates that it could accommodate an electricity grid powered by 100% renewable energy generation without a significant cost increase.[53] Other studies have noted the same for China,[54] Canada,[55] Denmark,[56] Germany,[57] Ireland,[58] Macedonia,[59] New Zealand,[60] and Portugal,[61] to name but a few. If researchers are already demonstrating the technical viability of electricity grids powered by 100% renewable energy, is it not possible, indeed probable, that technological advances will allow most nations to come close to fulfilling electricity demand through renewable energy in the near future?

Finally, it is true that NIMBY concerns have thwarted the realization of renewable energy projects in the past. The bad news is that there are many

Table 6.1. Annual Capacity Factors for Electricity Generators in the United States, 2008–2013 (%)

Period	Coal	Natural Gas				Petroleum		
		Natural gas–fired combined cycle	Natural gas–fired combustion turbine	Steam turbine	Internal combustion engine	Steam turbine	Petroleum liquids–fired combustion turbine	Internal combustion engine
2008	73.4	40.1	5.2	12.4	4.8	15.6	1.5	2.2
2009	65.1	39.8	4.5	11.2	4.8	14.5	1.6	2.3
2010	67.9	43.8	5.2	11.4	4.8	13.5	1.9	2.0
2011	63.7	43.6	5.1	12.4	7.3	12.0	1.2	2.2
2012	56.7	51.1	6.0	12.8	5.5	12.8	1.2	2.0
2013	59.7	46.5	4.1	10.7	21.5	11.7	0.9	6.7

Period	Nuclear	Hydro	Wind	Solar PV	Solar thermal	Landfill gas/MSW	Other biomass	Geothermal
2008	91.1	37.2	31.7	22.5	19.5	69.9	66.5	74.7
2009	90.3	39.6	28.1	20.6	23.6	70.2	62.1	73.3
2010	91.1	37.6	29.8	20.3	24.5	70.8	57.8	71.9
2011	89.1	45.9	32.1	19.1	23.9	70.0	56.3	71.8
2012	86.1	39.6	31.8	20.3	23.8	68.0	57.3	68.2
2013	90.1	38.1	32.3	19.4	17.8	69.6	50.8	66.0

Source: US Energy Information Administration (EIA), *Annual Capacity Factors for Electricity Generators in the United States, 2008–2013 (%): Electric Power Annual* (Washington, DC: EIA, 2014). *Note:* MSW = municipal solid waste; PV = photovoltaic.

facets to the NIMBY phenomenon—there is no single reason for people to be opposed to renewable energy projects. The good news is that much of the opposition can be reduced or eliminated altogether by better communication with those who are opposed. Often, opposition is based on misperceptions; NIMBY dissipates when these misperceptions are cleared up.[62] Transparency and communication can go a long way to minimizing the problem.[63]

One other aspect of NIMBY that does not receive enough attention is that opposition to new technology often dissipates as people become better acquainted with its associated benefits.[64] A vivid example of this is the transition from horse-drawn carriages to automobiles. In the early days of the automobile, this immature technology was the subject of many critical news reports highlighting the unreliability of automobiles, which often needed to be hauled back to the garage using the horses that they were intended to replace. Horse breeders, livery stable owners, and horse-drawn vehicle associations were all opposed to the new technology out of regard for vested interests, and this prompted political maneuvers that inflamed public sentiments about the dangers associated with the transition.[65] The point is that very few of us are now commuting to work on horses: opposition eventually subsides as the public realizes that the benefits exceed the costs.

BOX 6.1

Renewable Electricity and the Common Ground

Can renewable electricity ever be mainstreamed?

ONE SIDE: Renewable energy can be only a niche player. There is simply not enough technical potential, and the cost is prohibitively high.

OTHER SIDE: Many forms of renewable energy have sufficient technical potential to satisfy global electricity demand. Moreover, the allure of renewable energy will only increase as technology progresses.

COMMON GROUND: Electricity systems should be designed with a portfolio of energy technologies to maximize economic environmental, and security goals. Renewable technologies can be combined to create optimal portfolios.

Common Ground: Strategic Pieces Are Part of an Energy Mix Puzzle

In their current state of technological development, either wind power or solar PV could provide all of humanity's electricity needs, but this would come with costs certain to provoke dissent across financial board rooms and residential living rooms, at least in places where such technologies are not already cost competitive. Nevertheless, technically, either of these technologies could fully service the electricity needs of humankind. The implication of this observation is that renewable energy technologies can be far more than niche players in a nation's energy mix, but as for all other forms of energy, excessive reliance on any one technology comes at a cost. Therefore, renewable energy technologies should be seen as strategic pieces in an energy mix puzzle, a synthesis of perspectives illustrated in box 6.1.

Viewing renewable energy technologies as elements of a diverse energy mix portfolio offers numerous benefits. A diversified energy mix encourages technological competition among energy platforms, ensuring that progressive innovation takes place and costs are minimized.[66] A diverse portfolio of technologies also helps attenuate load imbalances caused by stochastic power flows and ensures that unexpected price increases associated with any given technology do not significantly affect the economics of the entire energy system.[67] Portfolio diversification allows nations to weather unexpected disruptions to energy supply, whether caused by economic shocks, natural disasters, terrorism, or geopolitical developments.[68] A diversity of technologies allows different regions of a nation to exploit local geographical or technical competencies by supporting renewable energy technologies that fit the community context.[69] Spreading out the supply of energy across a number of technological platforms minimizes the damage that can be caused by sole reliance on a single technology that, for whatever reason, suffers a competitive or technological setback.[70] Japan's nuclear power program comes to mind in this regard.

In conclusion, the notion of a world powered entirely by renewable energy is not an exercise in folly embraced by sects of environmental extremists; even using current technology, it is a viable possibility, one made even more attractive when an assortment of renewables is integrated into a diversified energy portfolio. Segue back to the fossil fuel protagonists that opened this chapter—Cole and Slick. They are retired now and seated side-by-side in rocking chairs on a veranda overflowing with bouquets of flowers adorned with messages of praise and gratitude from around the world. "Well, Slick, it was quite a run we had, don't ya think?" says Cole reflectively. "It sure was, buddy," rasps Slick, as he leans forward to adjust the speed setting on the solar-powered fan. "But ya know, those new whippersnappers in town—the renewables—they're gonna do just fine."

NOTES

1. "Special Report: Climate Change," *Economist*, September 7, 2006.

2. "Global Warming; Or: None Like it Hot!" in the *Futurama* episode "Crimes of the Hot," first broadcast November 10, 2002, available at www.youtube.com/watch?v=2ta ViFH_6_Y.

3. Quoted in B. K. Sovacool, "The Intermittency of Wind, Solar, and Renewable Electricity Generators: Technical Barrier or Rhetorical Excuse?" *Utilities Policy* 17, no. 3 (September 2009): 290.

4. G. Boyle, *Renewable Energy: A Power for a Sustainable Future*, 2nd ed. (Oxford: Oxford University Press, 2004).

5. Daniel Gross, "Solving 'Fission Impossible,'" *Newsweek*, October 29, 2007, E24.

6. American Legislative Exchange Council quoted in Union of Concerned Scientists, "Got Science?" (August 14, 2014).

7. Electric Power Research Institute quoted in Sovacool, "Intermittency of Wind, Solar, and Renewable."

8. Michael Webber, panelist, "American Perspectives on Energy Efficiency" (National Press Club, March 30, 2014), available at OurEnergyPolicy.org.

9. M. Wolsink, "Wind Power Implementation: The Nature of Public Attitudes—Equity and Fairness Instead of 'Backyard Motives,'" *Renewable and Sustainable Energy Reviews* 11, no. 6 (2007): 1188–1207.

10. International Energy Agency (IEA), *World Energy Outlook 2013* (Paris: OECD, 2013).

11. B. K. Sovacool, "Contextualizing Avian Mortality: A Preliminary Appraisal of Bird and Bat Fatalities from Wind, Fossil-Fuel, and Nuclear Electricity," *Energy Policy* 37 (2009): 2241–2248.

12. World Commission on Dams, *The Report of the World Commission on Dams* (London: Earthscan, 2001).

13. D. Pimentel et al., "Renewable Energy: Economic and Environmental Issues," *BioScience* 44 (1994): 42–48.

14. B. D. Green and R. G. Nix, *Geothermal—The Energy under Our Feet* (Golden, CO: National Renewable Energy Laboratory, 2006).

15. V. M. Fthenakis and E. Alsema, "Photovoltaics Energy Payback Times, Greenhouse Gas Emissions, and External Costs," *Progress in Photovoltaics: Research Applications* 14 (2006): 275–280; US Department of Energy, "SunShot Initiative" (October 2012), available at www1.eere.energy.gov/solar/sunshot/policy_reg_environment.html.

16. C. R. Warren et al., "Green on Green: Public Perceptions of Wind Power in Scotland and Ireland," *Journal of Environmental Planning and Management* 48 (2005): 853–875.

17. Arpad Horvath, "Construction Materials and the Environment," *Annual Review of Environment and Resources* 29 (2004): 182.

18. David J. C. MacKay, *Sustainable Energy—Without the Hot Air* (Cambridge: UIT Cambridge, 2008).

19. G. Wiesmann, "Germany Plans to Build Wind Power Grid," *Financial Times* (London), May 30, 2012.

20. IEA, *World Energy Outlook 2010* (Paris: OECD, 2010).

21. IEA, *Energy Technology Perspectives 2010: Scenarios and Strategies to 2050* (Paris: OECD, 2010), 4.

22. David Suzuki, "The Economics of Global Warming," *Nation of Change*, July 9, 2014, available at www.nationofchange.org/economics-global-warming-1404912339.

23. Bruce Lieberman, "David Victor: Views Examined on Climate Politics" (May 16, 2014), available at www.yaleclimateconnections.org/2014/05/david-victor-views-examined-on-climate-politics-communications.

24. Andrew Lilico, "We Can Terraform Mars for the Same Cost as Mitigating Climate Change: Which Would You Rather?" *Telegraph* (London), August 19, 2014.

25. Scheer quoted in Kate Connolly, "Endless Possibility," *Guardian* (London), April 16, 2008.

26. J. J. Dooley, *U.S. Federal Investments in Energy R&D: 1961–2008*, PNNL-17952 (Richland, WA: Pacific Northwest National Laboratory, 2008).

27. Ibid.

28. IEA, *World Energy Outlook 2014* (Paris: OECD, 2014), 320–321.

29. International Monetary Fund (IMF), *Energy Subsidy Reform: Lessons and Implications* (New York: IMF, January 2013).

30. N. Oreskes and E. M. Conway, *Merchants of Doubt* (London: Bloomsbury Press, 2010).

31. The press release on this study, "Potential of Renewable Energy Outlined in Report by the Intergovernmental Panel on Climate Change," can be accessed on the IPCC website

at http://srren.ipcc-wg3.de/press/content/potential-of-renewable-energy-outlined-report-by-the-intergovernmental-panel-on-climate-change.

32. Y. Y. Deng, K. Blok, and K. van der Leun, "Transition to a Fully Sustainable Global Energy System," *Energy Strategy Reviews* 1, no. 2 (2012): 109–121.

33. US Department of Energy (DOE), *Characterization of U.S. Energy Resources and Reserves*, DOE/CE-0279 (Washington, DC: DOE, 1989).

34. National Research Council (NRC), *Electricity from Renewable Resources: Status, Prospects, and Impediments* (Washington, DC: NRC, 2010), 4.

35. M. M. Hand et al., eds., *Renewable Electricity Futures Study*, NREL/TP-6A20-52409 (Golden, CO: National Renewable Energy Laboratory, 2012), 3, available at www.nrel.gov/analysis/re_futures.

36. M. Z. Jacobson and M. A. Delucchi, "A Plan to Power 100 Percent of the Planet with Renewables," *Scientific American*, November 2009, 58–65.

37. M. A. Delucchi and M. Z. Jacobson, "Providing All Global Energy with Wind, Water, and Solar Power, Part II: Reliability, System and Transmission Costs, and Policies," *Energy Policy* 39, no. 3 (2011): 1188.

38. International Renewable Energy Agency (IRENA), *30 Years of Policies for Wind Energy: Lessons from 12 Wind Energy Markets* (Abu Dhabi, UAE: IRENA, 2012).

39. DOE, *Revolution Now: The Future Arrives for Four Clean Energy Technologies* (Washington, DC: DOE, 2013).

40. K. Ardani and D. Seif, *Non-Hardware ("Soft") Cost-Reduction Roadmap for Residential and Small Commercial Solar Photovoltaics* (Golden, CO: National Renewable Energy Laboratory, 2013).

41. P. Hearps and D. McConnell, *Renewable Energy Technology Cost Review 2011* (Melbourne, Australia: Energy Institute, 2011).

42. IEA, *World Energy Outlook 2014.*

43. Renewable Energy Policy Network for the 21st Century (REN21), *Global Status Report 2014* (Paris: REN21 Secretariat, May 2014).

44. The report of this study is available at http://about.bnef.com/press-releases/strong-growth-for-renewables-expected-through-to-2030; it is expanded upon on the Think Progress website at http://thinkprogress.org/climate/2013/04/25/1916291/bloomberg-study-70-percent-of-new-global-power-capacity-through-2030-will-be-renewable.

45. Data based on figure 1 in Keith Barnham, Kaspar Knorr, and Massimo Mazzer, "Progress towards an All-Renewable Electricity Supply," *Nature Materials* 11 (November 2012): 908–909.

46. Alan Nogee, Jeff Deyette, and Steve Clemmer, "The Projected Impacts of a National Renewable Portfolio Standard," *Electricity Journal* 20, no. 4 (May 2007): 33–47.

47. D. McCubbin and B. K. Sovacool. "Quantifying the Health and Environmental Benefits of Wind Power to Natural Gas," *Energy Policy* 53 (February 2013): 429–441.

48. M. Mendonça, D. Jacobs, and B. K. Sovacool, *Powering the Green Economy: The Feed-in Tariff Handbook* (London: Earthscan, 2009).

49. D. S. Callaway, "Tapping the Energy Storage Potential in Electric Loads to Deliver Load Following and Regulation, with Application to Wind Energy," *Energy Conversion and Management* 50, no. 5 (2009): 1389–1400.

50. N. A. Ahmed, M. Miyatake, and A. K. Al-Othman, "Power Fluctuations Suppression of Stand-Alone Hybrid Generation Combining Solar Photovoltaic/Wind Turbine and Fuel Cell Systems," *Energy Conversion and Management* 49, no. 10 (2008): 2711–2719; Mark Diesendorf, *Sustainable Energy Solutions for Climate Change* (Sydney, Australia: UNSW Press; London: Routledge-Earthscan, 2014).

51. K. Hedegaard and P. Meibom, "Wind Power Impacts and Electricity Storage: A Time Scale Perspective," *Renewable Energy* 37, no. 1 (2012): 318–324; C. K. Ekman and S. H. Jensen, "Prospects for Large Scale Electricity Storage in Denmark," *Energy Conversion and Management* 51, no. 6 (2010): 1140–1147.

52. O. Ma et al., "Demand Response for Ancillary Services," *IEEE Transactions on Smart Grid* 4, no. 4 (December 2013): 1988–1995; J. Aghaei and M.-I. Alizadeh, "Demand Response in Smart Electricity Grids Equipped with Renewable Energy Sources: A Review," *Renewable and Sustainable Energy Reviews* 18 (2013): 64–72; S. Kiliccote et al., *Field Testing of Automated Demand Response for Integration of Renewable Resources in California's Ancillary Services Market for Regulation Products* (Berkeley, CA: Lawrence Berkeley National Laboratory, 2012).

53. B. Elliston, M. Diesendorf, and I. MacGill, "Simulations of Scenarios with 100% Renewable Electricity in the Australian National Electricity Market," *Energy Policy* 45 (2012): 606–613.

54. W. Liu et al., "Potential of Renewable Energy Systems in China," *Applied Energy* 88, no. 2 (2011): 518–525.

55. S. V. Valentine, "Canada's Constitutional Separation of (Wind) Power," *Energy Policy* 38, no. 4 (2010): 1918–1930.

56. H. Lund and B. V. Mathiesen, "Energy System Analysis of 100% Renewable Energy Systems—The Case of Denmark in Years 2030 and 2050," *Energy* 34, no. 5 (2009): 524–531.

57. Olav H. Hohmeyer and Sönke Bohm, "Trends toward 100% Renewable Electricity Supply in Germany and Europe: A Paradigm Shift in Energy Policies," *WIREs: Energy and Environment* (online, June 24, 2014), doi:10.1002/wene.128.

58. D. Connolly et al., "Modelling the Existing Irish Energy-System to Identify Future Energy Costs and the Maximum Wind Penetration Feasible," *Energy* 35, no. 5 (2010): 2164–2173.

59. B. Ćosić, G. Krajačić, and N. Duić, "A 100% Renewable Energy System in the Year 2050: The Case of Macedonia," *Energy* 48, no. 1 (2012): 80–87.

60. I. G. Mason, S. C. Page, and A. G. Williamson, "A 100% Renewable Electricity Generation System for New Zealand Utilising Hydro, Wind, Geothermal and Biomass Resources," *Energy Policy* 38, no. 8 (2010): 3973–3984.

61. G. Krajačić, N. Duić, and M. da Graça Carvalho, "How to Achieve a 100% RES Electricity Supply for Portugal?" *Applied Energy* 88, no. 2 (2011): 508–517.

62. S V. Valentine, "Sheltering Wind Power Projects from Tempestuous Community Concerns," *Energy for Sustainable Development* 15, no. 1 (2011): 109–114.

63. J. Firestone and W. Kempton, "Public Opinion about Large Offshore Wind Power: Underlying Factors," *Energy Policy* 35, no. 3 (2007): 1584–1598.

64. P. Devine-Wright, "Beyond NIMBYism: Towards an Integrated Framework for Understanding Public Perceptions of Wind Energy," *Wind Energy* 8, no. 2 (2005): 125–139.

65. H. Rao, "The Social Construction of Reputation: Certification Contests, Legitimation, and the Survival of Organizations in the American Automobile Industry: 1895–1912," *Strategic Management Journal* 15, suppl. 1 (1994): 29–44.

66. D. Helm, "Energy Policy: Security of Supply, Sustainability and Competition," *Energy Policy* 30, no. 3 (2002): 173–184.

67. S. V. Valentine, "Emerging Symbiosis: Renewable Energy and Energy Security," *Renewable and Sustainable Energy Reviews* 15, no. 9 (2011): 4572–4578.

68. B. K. Sovacool and I. Mukherjee, "Conceptualizing and Measuring Energy Security: A Synthesized Approach," *Energy* 36, no. 8 (2011): 5343–5355.

69. S. Awerbuch, "Portfolio-Based Electricity Generation Planning: Policy Implications for Renewables and Energy Security," *Mitigation and Adaptation Strategies for Global Change* 11, no. 3 (2006): 693–710.

70. X. Li, "Diversification and Localization of Energy Systems for Sustainable Development and Energy Security," *Energy Policy* 33, no. 17 (2005): 2237–2243.

Is the Car of the Future Electric?

Over the past century, people have become increasingly dependent on cars. The automobile is an icon of success. It is the second-biggest single expenditure made by families, after purchase of a home, and it connects people to work, schools, shopping, and recreation. The automobile is also a symbol of innovation and personal aspiration. As a 50-year-old *Book of Styling* published by Ford expressed it, "[We] dream of cars that will float and fly, or run on energy from a laser beam, or travel close to the ground without wheels."[1] Car races, car shows, car clubs, and car cruise-ins have become big business in the United States, Europe, and Japan, where people long to glide through town in a classic sedan or roar down a country road in a muscled-up sports coupe. NASCAR (National Association of Stock Car Auto Racing) events are broadcast to 150 countries, and it is the only sport in the United States with more fans than American football.[2]

People love cars. Expecting people to voluntarily give up their cars or their dreams of owning cars is a dubious proposition. Therefore, given the unlikelihood of private car ownership decreasing any time soon, many have concluded that we need to find a substitute for the internal combustion engine (ICE) vehicle, which is a prime source of air pollution (with its cardiovascular, pulmonary, and respiratory consequences) and greenhouse gas (GHG) emissions. We consider here the question of whether the electric vehicle (EV) will be the heir apparent—the dominant form of personal and public transportation in the future.

One Side: Electric Vehicles Represent the Platform of the Future

Let's begin with some statistics and definitions. The global fleet of EVs as of 2012 was estimated to exceed 150 million vehicles (including motorbikes). This may sound like a lot, but the world has more than 1 billion light-duty vehicles (LDVs, including passenger cars, sport utility vehicles, and small trucks).[3] EVs can be categorized into three types, as shown in table 7.1. Hybrid electric vehicles (HEVs) are passenger cars that get their propulsion energy from both liquid fuels and a battery. Plug-in hybrid electric vehicles (PHEVs) are capable

Table 7.1. Attributes of Electric Vehicles (EVs)

Type of vehicle	Type of propulsion energy and emissions	Technology readiness	Infrastructure requirements
Hybrid electric vehicle (HEV)	• Propulsion energy from both consumable fuels and battery • Battery recharged by regenerative braking and through internal combustion engine	• Numerous car models in the market today	• HEVs are compatible with existing refueling infrastructure.
Plug-in hybrid electric vehicle (PHEV)	• An HEV with a means of recharging its battery from an external power source • Pollution benefits depend on driving cycles and on energy resources used to generate electricity	• Several models on the market; recharging equipment needs to be cheaper	• PHEVs are compatible with existing grids in most industrialized countries.
Battery electric vehicle (BEV)	• Propulsion energy drawn entirely from battery • No tailpipe emissions, but total pollution benefits depend on energy resources used to generate electricity	• Battery improvements needed to extend driving ranges and shorten recharging times; battery costs must be cut.	• Substantial market penetration could exacerbate peak demand, requiring new power plant or solar PV investment, especially if smart grid technologies and pricing policies are not able to concentrate recharging into off-peak periods.

of recharging from an external power source, usually the grid. Battery electric vehicles (BEVs) draw their energy for propulsion strictly from a battery.

While most of the EVs currently in use are bicycles and scooters, predominantly found cruising streets in China, the electric car market is growing by leaps and bounds. More than 40,000 fully electric cars were sold in 2011 globally, dominated by the Nissan Leaf and Mitsubishi i-MiEV.[4] The market is burgeoning—about 460,000 HEVs, 87,000 PHEVs, and 4,000 full EVs were sold in the United States in 2012. This accounted for 3.85% of all new vehicle sales.[5] In total, 14.1 million EVs of some type were sold in the United States

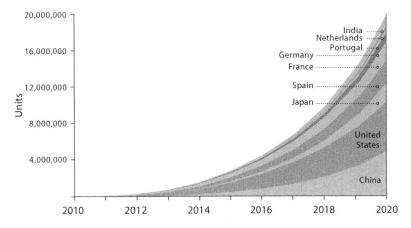

A. Global EV Stock Targets, 2010–2020

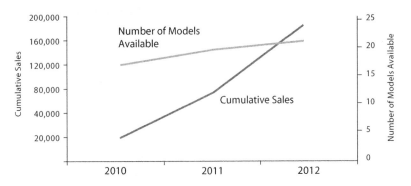

B. Models and Sales, 2010–2012

Figure 7.1. Worldwide Electric Vehicle Stock Turnover, Model Availability, and Sales. *Source:* International Energy Agency, *Global EV Outlook: Understanding the Electric Vehicle Landscape to 2020* (Paris: OECD, April 2013).

over the past 20 years, with EVs posting an average annual growth in sales of a healthy 26% between 1991 and 2010.[6] As figure 7.1 shows, 25 models were offered globally in 2012, and the International Energy Agency (IEA) projects that by 2020, more than 20 million BEVs will be on the road.[7]

Advocates of the electric car typically advance four arguments in favor of EVs: (1) lower pollution and GHG emissions, (2) enhancements to national energy security, (3) a symbiotic relationship with the electricity grid, and (4) the rapid pace of technical improvements.

EVs Offer Pollution and Carbon Benefits

Electric vehicles can reduce or eliminate tailpipe pollution and curtail GHG emissions, but the extent of the gains depends on the type of EV and the carbon intensity of the electricity used to recharge the batteries. PHEVs have greatly reduced tailpipe emissions, while BEVs emit no tailpipe pollution.[8] Grid-supported PHEVs and BEVs can drastically (and directly) mitigate GHG emissions and improve air quality in urban areas. A study at the Pacific Northwest National Laboratory (PNNL) estimated that for the United States as a whole, shifting roughly half the vehicles on the road in 2007 to EVs would reduce total GHG emissions by 27%.[9] Under such a scenario, PNNL further projected that pollution from total volatile organic compounds and carbon monoxide emissions would decrease by 93% and 98%, respectively, and total nitrogen oxides emissions would be reduced by 31%. In tempering their findings, the authors cautioned that total particulate matter emissions would increase 18% and sulfur oxides emissions would increase 125%, if the EVs were powered by electricity from coal-fired plants. However, the pollution would be transplanted from local urban areas to the more distant locations of power plants. The PNNL authors also pointed out that this disastrous scenario could be avoided and net gains made if electricity came from natural gas or renewable sources of energy.[10] Other studies, using a variety of "well-to-wheels" metrics, have reached similar conclusions, affirming the air pollution and carbon benefits of EVs compared with conventional LDVs.[11]

EVs could play a major role in supporting electricity industry reform. As technology evolves, PHEVs and BEVs will be able to provide storage support for intermittent renewable energy generators, allowing reduced CO_2 emissions and air pollution in the electricity sector. In a "vehicle-to-grid" configuration, the batteries in the vehicles could store surplus electricity produced by renewable energy technologies such as wind turbines and solar photovoltaics and then provide power back to the grid when needed. Integrated systems of this type can offset the need for additional generation capacity or storage, allowing intermittent renewable energy resources to integrate more effectively into the grid.[12] EVs could replace (or more likely supplement) large-scale pumped hydroelectric and compressed air energy storage systems in the United States, systems that have already proven effective for renewable energy technologies. Given that wind turbines produce most of their electricity at night, when PHEVs need to be recharged, integrated systems could greatly improve the efficacy of wind power installations and enhance the commercial appeal of a renewable energy technology that is already competitive with coal-fired power. One study estimated that a 50% penetration of PHEVs would provide sufficient ancillary storage to displace 2,800 MW of generation capacity, which in most nations is carbon-intensive.[13]

EV Penetration Can Enhance Energy Security

Advocates of EVs are quick to highlight the role these vehicles can play in reducing national dependence on oil—a dependence that has been acknowledged as detrimental to national security in many industrialized nations.[14] The numbers reinforce a compelling argument. A transition to EVs in the United States alone has the potential to displace 6.5 million barrels of oil equivalent per day, or more than 50% of the nation's entire oil imports over the past decade.[15] A large portion of that imported oil comes from countries that are either politically unstable or somewhat antagonistic toward liberal democratic states. This fact is highlighted by the observation that of the top 10 net oil-exporting countries, only 2 countries can be viewed as politically stable suppliers: Norway (ranked number 3 as a net oil exporter) and Mexico (ranked number 10). Exports from these two nations pale in significance compared with exports from nations such as Saudi Arabia, Russia, Iran, and the United Arab Emirates (which rank 1, 2, 4, and 5, respectively).

There are at least two obvious reasons that dependence on unstable nations for oil undermines national security. For starters, the supply of oil can be interrupted during times of political upheaval, and oil-dependent nations face the risk of oil price inflation due to events beyond their realm of control. As a result, oil-dependent nations are incentivized to meddle in the affairs of oil-rich nations in attempts to preserve the status quo. Indeed, in the 1970s, the Carter Doctrine held that the United States would adopt any means within its power to preserve its access to Middle Eastern oil. In recent years, an increasing number of high-profile critics, including former Federal Reserve chair Alan Greenspan, have embraced the obvious by acknowledging that access to foreign supplies of oil may have dictated the US government's decisions to wage war in the Middle East.[16] A second way in which oil dependence undermines national security relates to the link between terrorism and oil revenues. In 2007 alone, OPEC (Organization of Petroleum Exporting Countries) income amounted to $535 billion, funneled into the coffers of a small number of powerful oil sheiks and barons. Some of these individuals allegedly tap into these revenues for supporting terrorism. For example, the wealth of Osama Bin Laden's family came from government construction contracts that were financed by oil money.[17]

In the long term, reducing oil imports through greater EV penetration also promises a host of additional economic benefits. The most significant of these include a rise in domestic value-added economic activity due to the minimization of wealth transfers from oil consumers to producers and a reduced risk of economic disruption caused by wars, hurricanes, political unrest, or accidents that inflate the price of oil.[18] The National Defense Council Federation calculated that the direct economic costs of oil dependence in the United States alone included a loss of more than 820,000 jobs, $159.9 billion in gross national product, and $13.4 billion in federal and state revenues every year.[19] If reflected at the gasoline pump, these "hidden costs" would have increased the average price of a gallon of gasoline in the United States (in 2003, at the time of the study) from $2.12 to $5.28.[20]

EVs Can Save Money and Improve Electricity Reliability

Advocates claim that consumers will profit from the use of EVs because electricity is cheaper than gasoline for equivalent distances traveled. At 2014 petrol prices, for instance, the annual fuel cost of driving a conventional ICE car in the United Kingdom is about £1,440 ($2,390) for driving an average of 12,000 miles. The cost for running an EV for the same period of time over the same distances could be as low as £240 ($400).[21] Although infrastructure is not yet comprehensive enough to provide adequate charging services for people in all areas, most homes in industrialized countries already have the requirements to charge vehicles in at least one key locale: electrical outlets in the home or garage.

Even though current electric utility technologies lack the asynchronous capacities to support smart grids, there are still useful synergies to be harnessed. One synergy stems from the fact that many electricity generators and electricity grids operate in a suboptimal fashion because systems have a great deal of slack capacity to deal with demand peaks and troughs. In particular, in many countries, extreme overcapacity occurs in the late evenings and at night, when most people are sleeping. Fortuitously, recharging of EV batteries typically commences when commuters return home in the evening and can continue throughout the night. The result would be a system that could generate and deliver a substantial amount of energy to fuel the nation's vehicles at a very low cost. A study in 2007 suggested that 8% to 12% of peak demand occurs within just 80 to 100 hours during the year.[22] Because much of the generating capacity remains unused, about 84% of electrically powered cars, light trucks, and sport utility vehicles in the United States could be supported by the nation's existing electricity infrastructure, if they drew power from the grid at off-peak times.[23] In other words, if planned properly, vehicle recharging could take place today without any substantial expansion of electricity generation capacity. Moreover, utility companies would earn extra revenues from these sales.

Perhaps more exciting to EV advocates is the use of cars as suppliers of power to the grid, especially during times of peak demand, which offers a tantalizing opportunity to earn revenue and offset transportation costs. In a future supported by smart grid technology, EVs could be used as distributed storage devices, feeding electricity stored in EV batteries back into the grid (a process known as "vehicle-to-grid," or V2G). This could reduce electricity grid costs by providing affordable regulation of power flows, dampening peak power demands, and providing added reserve capacity without having to invest in power generation equipment.[24] The IEA estimates that the average storage capacity is 8 kWh for a PHEV and 30 kWh for a BEV. Thus, if 30,000 BEVs were available for V2G services 50% of the time, these batteries could replace a 100 MW peaking gas turbine unit.[25] In such a world, power companies could operate more like wholesalers—producing income from all existing equipment on the basis of economic attractiveness—and less like construction companies that constantly need to raise funds to build new generating resources.[26] These potential benefits for both consumers and electric utilities have already been widely studied and confirmed,[27] and as technology advances, more synergies will undoubtedly be identified.

EV Technology Is Advancing Rapidly
A key criticism of EVs has been that technological advances are required before they can become a practical option. Proponents of EVs argue that this is an increasingly unfounded criticism. In 2007, even the most advanced BEVs had limited charge ranges of about 100 to 160 kilometers, long recharge times of four or more hours, and high battery costs leading to comparatively high vehicle retail prices.[28] A mere seven years later, the 2014 Tesla Model S came equipped with a battery supercharger that can provide 170 miles (274 km) of range in as little as 30 minutes.[29] Most travel times and distances are relatively short, which suggests that for many commuters the current technology should suffice. Given that about 60% of vehicles travel less than 30 miles (48 km) per day, EV technology is not far from being able to effectively service the travel needs of most motorists.[30]

At 2014 gasoline prices, when compared with an ICE that gets 35 miles per gallon, a $21,000 investment in a Nissan Leaf will yield gas savings of almost $750 per year, based on 12,000 miles driven per annum.[31] Similarly, the Union of Concerned Scientists estimated that the lifetime charging costs for an EV in the United States amounted to a meager $5,200, compared with $9,800 in fuel costs for an HEV and a whopping $18,000 for a gas-powered passenger car—savings that more than offset the higher up-front cost of an EV.[32] All the while, research and development continues to improve EV performance and to make possible the integration of BEVs with other energy systems. As proponents of the EV would conclude, stay tuned for more exciting developments.

The Other Side: Electric Vehicles Are No Better
Than Conventional Cars

The case in favor of EVs is far from unchallenged. First of all, opponents argue that the promise of EVs has been touted by supporters for decades but has remained unrealized. More than a century ago, the *New York Times* declared that the electric car "has long been recognized as the ideal solution" because it "is cleaner and quieter" and "much more economical."[33] In the mid-1990s, General Motors EV-1 and three French BEV models from Citroen, Peugeot, and Renault were mass produced with great enthusiasm and high expectations but failed to realize commercial success.[34] In 2008, management scientists Struben and Sterman at the Massachusetts Institute of Technology argued that the reason for such slow diffusion is that the competitive playing field is far from level: "The current low functionality and high cost of alternatives, and low gasoline taxes are endogenous consequences of the dominance of the internal combustion engine and the petroleum industry, transport networks, settlement patterns, technologies, and institutions with which it has coevolved."[35] As a result, numerous attempts to promote EVs failed, as did initiatives to introduce compressed natural gas vehicles in Canada and New Zealand.[36]

Why would we expect a resurgence of EVs to take hold today? As economist Remy Prud'homme of the University of Paris concluded, "The idea that the electric car could be a general substitute to the fuel car is not acceptable. It can only, at best, be a niche market."[37] In addition to the sizable market barriers and the financial might enjoyed by the conventional automobile industry, four arguments challenge the promise of EVs. Criticisms entail (1) cost and range anxiety, (2) sizable environmental penalties, (3) the irreplaceability of liquid fuels, and (4) critiques of humanity's private vehicle paradigm.

Cost and Range Anxiety Will Constrain EV Growth

A key criticism of current EVs is that they are expensive and limited in range. Therefore, opponents argue, replacing ICE vehicles with EVs results in economic hardship and potential transportation inefficiencies.

There is a difference of opinion over which technology is more expensive. Conventional ICE vehicles sport cheaper sticker prices, and so the question of economic attractiveness comes down to competing assumptions about fuel and maintenance costs. A 2012 comparison between two five-door compact cars manufactured by Renault—the Clio diesel car and the Zoe electric vehicle—suggested that over a 15-year lifetime of operation, the BEV cars would cost consumers €4,000 to €5 000 more than their ICE equivalents.[38] Similarly, a 2014 Ford Focus EV purchased in California had a list price of $35,170. Its ICE counterpart, the 2014 Ford Focus four-cylinder 2.0 L model, had a list price of

$23,515.[39] With gasoline at $3.50 per gallon, it would take 12 years for the operational savings attributed to the Focus EV to offset the cheaper retail price of the ICE Focus.

Although EVs continue to close ground on their ICE counterparts and, depending on fuel cost assumptions, are now cheaper over an extended lifetime (12 years in the case of the 2014 Ford Focus), the disparity in sticker prices serves as a barrier to adoption. Initial outlay is weighted heavily in a purchase decision. Indeed, a survey among California households found that not one of them had estimated the net present value of fuel savings as part of a decision to purchase a new vehicle.[40] For those consumers that do consider fuel economy when purchasing a vehicle, the IEA found that buyers expected vehicle efficiency improvements to pay for themselves in the first three years or less.[41] Such high discounting of fuel savings may explain why hardly anyone purchased prototype EVs in the late 1980s and early 1990s, even though the vehicles demonstrated fuel economy as high as 71 miles (114 km) per gallon.[42]

A final, tangible, and serious obstacle is "range anxiety"—how far EVs can go before they run out of electricity.[43] To date, the cost and storage capacity of lithium-ion batteries have limited the range of travel per charge.[44] PHEVs have travel ranges as low as 10 to 40 miles, after which the vehicle shifts to gasoline operation. BEVs have a more limited travel range. The 2014 Nissan Leaf has a range of only 84 miles (135 km) before requiring a recharge, and this currently represents the state of the art. Anyone who has ever driven on a near-empty tank of gas will understand the nervousness associated with driving a car that is almost running on empty. Opponents of EVs argue that this state of anxiety is amplified by EVs because charging time is far more onerous than stopping at a local gas station.

EVs Engender Sizable Environmental Externalities

Opponents argue that EVs also give rise to a host of environmental ills. Most notable are GHG emissions from electricity use, toxic pollution from battery manufacturing and disposal, and elevated water consumption.

For EVs to deliver well-to-wheels CO_2 reductions, the carbon content of electric power must be low. Otherwise, EVs will simply shift the generation of air pollution away from urban areas and toward rural communities, which host the fossil fuel–fired power plants that recharge EV batteries. Beijing and other major Chinese cities exemplify this type of scenario. Many Chinese cities have adopted measures to encourage the use of electric motorcycles and, in return, have become cloaked in smoke from coal-fired power plants.[45]

Any overall GHG reduction benefit will depend on the carbon intensity of the electricity used for battery recharging. Even in nuclear power–dominated France, carbon intensity was found to range from 80 g/kWh for the average base-load generator to more than 600 g/kWh for peak generators fueled by oil and coal.[46] One expert called electrification of transportation "counterproductive" when used "in regions where electricity is primarily produced from lignite, coal or even heavy oil combustion."[47] Regrettably, as critics of EVs might point out, this is the case in a great many regions, given that in 2011, 41% of all electricity was generated by coal-fired power plants.[48]

Furthermore, the EV manufacturing process can be highly polluting. One study noted that electric car factories produced more hazardous waste than conventional car factories.[49] Anders Stromman of the Norwegian University of Science and Technology summarized many of the criticisms related to EVs in this regard: "Potential for effects related to acid rain, airborne particulate matter, smog, human toxicity, ecosystem toxicity and depletion of fossil fuel and mineral resources [means that] electric vehicles consistently perform worse or on par with modern internal combustion engine vehicles, despite virtually zero direct emissions during operation."[50] This statement implies that the heralded environmental benefits of an EV transition—fewer GHG emissions and improved air quality in urban environments—may come at the expense of greater environmental damage stemming from mining operations, increased pollution from factories making EV components, and toxic landfills and junkyards where obsolete models (and their batteries) end up.

The proliferation of EVs also raises a serious indirect environmental concern: the transition of LDVs from internal combustion engines to electric power is likely to increase the consumption of electricity and thus exacerbate water scarcity in some regions. This is because fossil fuel and nuclear power plants, which dominate the electricity generation sector, require large amounts of water for the production of steam and for cooling processes. One study projected that "in displacing gasoline miles with electric miles, approximately 3 times more water is consumed (0.32 versus 0.07–0.14 gallons/mile) and over 17 times more water is withdrawn (10.6 versus 0.6 gallons/mile), primarily due to increased water cooling of thermoelectric power plants to accommodate increased electricity generation."[51] The added water intensity associated with EVs makes it difficult to electrify transportation in regions where water is scarce—a prevalent condition in many large urban areas and arid regions across the globe.[52]

Liquid Fuels Lack Substitutes

A further argument against the universal proliferation of EVs is that electricity will not be able to displace petroleum or liquid fuels in many of the most needed applications. EVs are constrained by the reach of electricity networks. This might not be a problem in developed nations with comprehensive electricity grids, but it will pose a problem in regions such as Africa and Asia where 1.3 billion people reside in rural areas that are not serviced by electricity.[53] In these emerging economies, EVs remain virtually impossible to operate.

To compound the market diffusion challenge, the current state of EV technology might be able to support a degree of diffusion for LDVs, but the technology has not progressed sufficiently to facilitate a transition to electricity for heavy-duty vehicles such as buses, freight trucks, long-distance trains, and airplanes. These vehicles will most likely continue to operate on jet fuel, biofuel, diesel, and gasoline, at least over the next few decades. Even in Nordic parts of Europe committed to a low-carbon transportation future, EV penetration up to 2050 is expected to be limited to personal LDVs, some delivery trucks, and rail (figure 7.2, *A* and *B*).[54] This means that low-carbon transportation over the next few decades will increasingly rely on a diverse portfolio of fuels to meet energy demand, and electricity will be only one fuel source among many (figure 7.2, *C*). The role of EVs in this future system is not insubstantial, but it is far from universal.

The prospect of liquid fuel–powered vehicles outcompeting EVs for market dominance is further elevated when one considers current trends in the automobile industry. Although most major automobile manufacturers have released EVs, they also market an array of increasingly fuel-efficient models propelled by blended fuels or hybrid technology. Given the difficulty of transforming infrastructure systems toward the use of EVs, critics argue that similar CO_2 reductions can be gained by improving conventional ICE technology and shifting to flex fuels. As table 7.2 indicates, a number of emerging technologies are expected to significantly improve ICE vehicle efficiency over the next two decades.[55] In addition, reductions in vehicle size and weight and improved aerodynamics could bolster fuel efficiency. It has been estimated that a 20% reduction in weight for an average vehicle is possible over the next 25 years, producing a further 12% to 20% reduction in fuel consumption.[56] Proponents of conventional cars point out that all of these developments together could enable manufacturers of next-generation ICE vehicles to match the benefits that EVs offer.

Job concerns may also frustrate universal support for EVs. The cost of EV maintenance is typically minimal because the vehicles have fewer moving parts. Unlike ICE vehicles, EVs do not require oil changes, tune-ups, smog checks, or mandatory annual emission inspections. Given that automotive service technicians hold almost one million jobs in the United States and that most automobile dealers make their money not on new sales but on service, it is understandable that the industry may not want to support a transition away from conventional vehicles.[57]

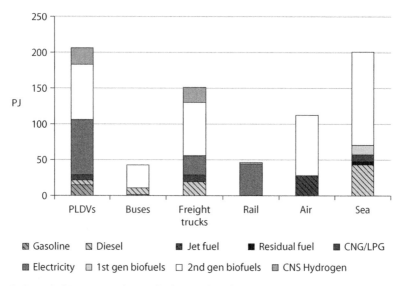

A. Expected Transportation Fuel Mixes and Modes in a Low-Carbon Nordic Region, 2050

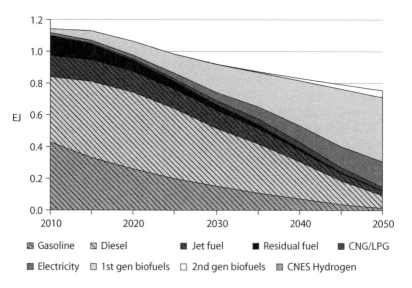

B. Expected Transportation Fuel Mix in a Low-Carbon Nordic Region, 2010–2050

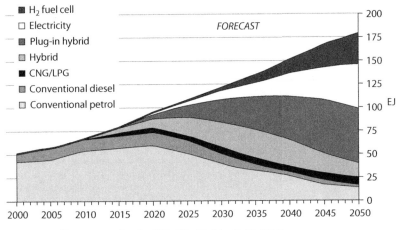

- ■ H₂ fuel cell
- □ Electricity
- ■ Plug-in hybrid
- ▨ Hybrid
- ■ CNG/LPG
- ▨ Conventional diesel
- □ Conventional petrol

FORECAST

C. Expected Transportation Fuel Mix Worldwide, 2000–2050

Figure 7.2. Transportation Modes and Fuel Mixes for Future Low-Carbon Societies. *Source:* International Energy Agency and Nordic Energy Research, *Nordic Energy Technology Perspectives: Pathways to a Carbon Neutral Energy Future* (Paris: OECD, 2013). *Note:* CNES = hydrogen; CNG = compressed natural gas; CNS = compressed hydrogen; EJ = exajoule; LNG = liquefied natural gas; PJ = petajoule; PLDV = personal light-duty vehicle.

EVs Still Promote Injury, Congestion, and Physical Inactivity

A final articulated downside to EVs stems from the macro-view that EVs are not much better than their conventional predecessors. There is a prevalent argument that the paradigm of private vehicle ownership, not the technology of private vehicles, needs to be changed. EVs depend on an assumption that transportation should be private rather than public, and motorized rather than human-powered; but private, motorized transport gives rise to a host of problems. It leads to increased congestion, a greater risk of accidents, and elevated demands on scarce natural resources.

For instance, private EVs can still play a leading role in traffic jams. According to the most recent "Urban Mobility Report," the congestion "invoice" for extra time and fuel in 2011 in 498 American urban areas showed a cost of $121 billion. Traffic congestion caused urban Americans to purchase an extra 2.9 billion gallons of fuel and to waste 5.5 billion extra hours (equivalent to the time US businesses and citizens spend each year on filing their taxes).[58] A shift to EVs would only prolong these problems. Potentially, EVs could also lead to more traffic accidents because their battery-powered engines make them stealthy. The World Health Organization (WHO) estimates that every year, 1.2 million people are killed and 50 million injured in traffic road crashes.[59] EVs will probably not help to solve this universal problem.

Table 7.2. Selected Technologies That Improve Conventional Motor Vehicle Efficiency

Technology	Average efficiency increase (%)	Description
Variable valve timing and lift	5	Improves engine efficiency by optimizing the flow of fuel and air into the engine
Cylinder deactivation	7.5	Saves fuel by deactivating cylinders when they are not used
Turbochargers and superchargers	7.5	Allows manufacturers to downsize engines without sacrificing performance or to increase performance without lowering fuel economy
Integrated starter/ generator (ISG) systems	8	Automatically turns the engine off when vehicle is idling
Direct fuel injection	12	Delivers higher performance with lower fuel consumption
Continuously variable transmissions	6	Employs an infinite number of "gears" to provide seamless acceleration and improved fuel economy
Automated manual transmissions	7	Combines the efficiency of manual transmissions with the convenience of automatics

Source: Modified from Cutler J. Cleveland and Christopher G. Morris, *Handbook of Energy, Volume I: Diagrams, Charts, and Tables* (London: Elsevier Science, 2013).

The failure to contribute to better societal health should also be considered as a knock against EVs. Private motorized vehicles of any sort contribute to less physical activity. Public transit is far superior in this regard, as it usually necessitates walking to the transit stop. The WHO projects that, annually, physical inactivity is responsible for 3.3% of worldwide deaths and 19 million disability adjusted life years. People who rely on private transportation have higher rates of diabetes, cardiovascular disease, and obesity than those who walk or take public transportation.[60] As the WHO's international team of health experts put it, "Increasing use of cars improves access for those individuals who are newly motorized but reduces access for others through danger and congestion."[61] In this context, private EVs are as bad as private conventional vehicles.

Advocates of a new transportation paradigm suggest that far better strategies exist—strategies that promote human fitness and better transportation efficiency and include a mix of walking, cycling, and public buses and trains. Figure 7.3 quantifies the enhanced efficiencies and reduced emissions associated with various modes of transportation, including walking, cycling, electronic bikes (ebikes), and light rail.

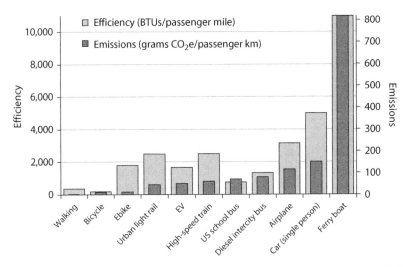

Figure 7.3. Energy and Carbon Intensity of Transportation Modes. *Source:* Modified from Cutler J. Cleveland and Christopher G, Morris, *Handbook of Energy, Volume I: Diagrams, Charts, and Tables* (London: Elsevier Science, 2013). *Note:* CO_2e = carbon dioxide equivalent.

Some nations are already making a noticeable shift toward this new transportation paradigm. Denmark is a case in point.[62] Danish cities feature "ample bike parking, full integration with public transport, comprehensive traffic education and training, and a wide range of promotional events intended to generate enthusiasm and wide public support for cycling."[63] The capital city, Copenhagen, received the European Green Capital Award in 2014 for planning to have 50% of commuters cycling to their workplaces in 2015 and to become carbon-neutral by 2025.[64] These ambitious goals are likely to be accomplished without EVs because the strategic focus is on encouraging nonmotorized and public transportation. In Delft, in the Netherlands, a "living street" lacks traffic lights, stop signs, lane dividers, or even sidewalks because motorized vehicles are prohibited or access is limited to public transit vehicles. Streets of this type encourage human interaction and exercise.[65]

Common Ground: Electric Vehicles Have Their Role But Can't Do It All

Both champions and critics of EVs have crafted persuasive arguments justifying their positions. In the end, both sides have merit and the two views can be reconciled, as shown in box 7.1. PHEVs deliver competitive benefits in the near

BOX 7.1

The Future of Electric Vehicles (EVs) and the Common Ground

Is the car of the future electric?

ONE SIDE: EVs are more carbon- and fuel-efficient, can displace costly petroleum dependence, save owners money, and improve the reliability of electricity infrastructure.

OTHER SIDE: EVs produce their own toxic pollution, continue to lock in private motorized modes of transport, are costly and limited in range, and cannot displace the use of liquid fuels in many applications.

COMMON GROUND: We should use plug-in hybrid electric vehicles in the near term and battery electric vehicles in the long term for passenger fleets, but leave other modes open to other fuel mixes.

term because they are fuel-flexible—running on electricity when it makes sense to do so, but reverting to petroleum when necessary. This not only gives the manufacturers of vehicles time to experiment with new technologies but also allows the sector to transition at a slower pace. This is of enormous benefit when one considers the scale and scope of the automotive industry and its support infrastructure. Over the longer term, as improvements in EV technology are made and travel ranges improve, BEVs might replace most ICE-propelled LDVs in nations where the electricity grid is comprehensive enough and is based on low-carbon generation sources. However, this will not happen everywhere and certainly not at once.

It is indeed counterproductive to encourage the proliferation of EVs in regions that rely on carbon-intensive electricity sources such as coal and petroleum, or in regions where there are no electricity grids, or for uses that cannot be supported by current technological levels (such as for airplanes, marine transport vessels, freight trucks, and trains—and even walking and cycling). Forced diffusion of this sort would simply push EVs into market segments where they replace more efficient and sustainable modes (such as mass transit or human power) or where the technology is too immature to succeed, leading to market opposition and elevating risks of accidents. In many regions (rural areas) and for many transportation needs (such as long-haul transportation and air travel), GHG reductions are likely to be more effectively achieved over the next few decades by improving ICE technologies, enhancing the adoption of flex-fuel systems, and employing alternative fuels such as cellulosic ethanol or natural gas.

In regions where adoption is appropriate (urban areas), the introduction of EVs propelled by low-carbon sources of electricity would significantly reduce emissions and financially facilitate a shift to smart grid technology. Certain transportation modes, most notably passenger travel in LDVs and by urban rail, are well suited to EVs. Conceivably, many households in developed nations could replace conventional cars with an EV with little to no disruption to quality of life. EVs could even be used in tandem with other "livable city" incentives to walk, cycle, and share public transit. As technology evolves, EVs could even enhance and augment standards of living by saving consumers money and by providing opportunities to offset transportation costs by selling battery storage capacity to utilities.

In the end, the pace and scale of EV diffusion depends on further technological breakthroughs in batteries, charging equipment, and support infrastructure and on consumer attitudes, societal conceptions of mobility, and policy support. EVs might represent the transportation platform of the future, but much development is still needed for them to become the dominant mode. This gives ICE stakeholders time to wage a competitive market defense that might confine EVs to specialized market niches for some time to come.

NOTES

1. Ford Motor Company, *The Ford Book of Styling* (Dearborn, MI: Ford, 1963), iv.

2. "All about NASCAR," *Shave Magazine*, March 15, 2011, available at www.shave magazine.com/cars/090601.

3. J. Sousonis, "World Vehicle Population Tops 1 Billion" (WardsAuto, August 15, 2011), available at http://wardsauto.com/ar/world_vehicle_population_110815.

4. Philippe Crist, "Electric Vehicles Revisited—Costs, Subsidies and Prospects" (International Transport Forum at the OECD, Discussion Paper 3, 2012), available at www.inter nationaltransportforum.org/jtrc/DiscussionPapers/DP201203.pdf.

5. Data for 2012 from Electric Drive Transportation Association, "Electric Drive Sales Dashboard" (updated monthly), available at www.electricdrive.org/index.php?ht=d/sp/i /20952/pid/20952.

6. Ibid.

7. International Energy Agency (IEA), *Global EV Outlook: Understanding the Electric Vehicle Landscape to 2020* (Paris: OECD, April 2013).

8. M. A. Kromer and J. B. Heywood, "Electric Powertrains: Opportunities and Challenges in the U.S. Light-Duty Vehicle Fleet" (Massachusetts Institute of Technology, 2007); Dong-Yeon Lee, Valerie M. Thomas, and Marilyn A. Brown, "Electric Urban Delivery Trucks: Energy Use, Greenhouse Gas Emissions, and Cost-Effectiveness," *Environmental Science and Technology* 47, no. 14 (2013): 8022–8030.

9. Michael Kintner-Meyer, Kevin Schneider, and Robert Pratt, "Impacts Assessment of Plug-in Hybrid Vehicles on Electric Utilities and Regional U.S. Power Grids, Part 1: Technical Analysis" (Pacific Northwest National Laboratory Report, 2007), available at www.pnl.gov/energy/eed/etd/pdfs/phev_feasibility_analysis_combined.pdf.

10. Ibid.

11. M. Duvall, "Comparing the Benefits and Impacts of Hybrid Electric Vehicle Options for Compact Sedan and Sport Utility Vehicles" (Electric Power Research Institute Final Report 1006892, July 2002), available at www.evworld.com/library/EPRI_sedan_options.pdf; William J. Mitchell, Christopher E. Borroni-Bird, and Lawrence D. Burns, *Reinventing the Automobile: Personal Urban Mobility for the 21st Century* (Cambridge, MA: MIT Press, 2010), 88; Paulina Jaramillo and Constantine Samaras, "Comparing Life Cycle GHG Emissions from Coal-to-Liquids and Plug-in Hybrids" (CEIC Working Paper 07-04, June 2007); Craig H. Stephan and John Sullivan, "Environmental and Energy Implications of Plug-in Hybrid Electric Vehicles," *Environmental Science and Technology* 42 (2008): 1185–1190; IEA, *Energy Technology Perspectives: Scenarios and Strategies to 2050* (Paris: OECD, 2010), available at www.iea.org/techno/etp/etp10/English.pdf.

12. Willett Kempton and Josna Tomic, "Vehicle-to-Grid Power Fundamentals: Calculating Capacity and Net Revenue," *Journal of Power Sources* 144 (2005): 268–279.

13. Willett Kempton and Amardeep Dhanju, "Electric Vehicles with V2G: Storage for Large-Scale Wind Power" (WindTech International, 2006), 3–7, available at www.udel.edu/V2G/docs/KemptonDhanju06-V2G-Wind.pdf.

14. K. M. Campbell and J. Price, eds., *The Global Politics of Energy* (Washington, DC: Aspen Institute, 2008).

15. Kintner-Meyer et al., "Impacts Assessment."

16. Bob Woodward, "Greenspan: Ouster of Hussein Crucial for Oil Security," *Washington Post*, September 17, 2007, A03.

17. T. L. Friedman, *Hot, Flat, and Crowded* (New York: Farrar, Strauss and Giroux, 2008).

18. David Greene and Sanjana Ahmad, "Costs of U.S. Oil Dependence: 2005 Update" (report to the US DOE, ORNL/TM-2005/45, 2005), available at http://cta.ornl.gov/cta/Publications/Reports/ORNL_TM2005_45.pdf.

19. National Defense Council Federation, *America's Achilles Heel: The Hidden Costs of Imported Oil* (Washington, DC: National Defense Council Federation, October 2003).

20. Ibid.

21. Tim Gibson, "Beat the Fuel Price Blues with an Electric Car," *Telegraph* (London), March 7, 2014, available at www.telegraph.co.uk/sponsored/motoring/ultra-low-emission-vehicles/10681607/lower-fuel-costs.html.

22. Amnad Faruqui et al., "The Power of 5 Percent," *Electricity Journal* 20 (October 2007): 68–77.

23. Kintner-Meyer et al., "Impacts Assessment."

24. T. Morgan, "Smart Grids and Electric Vehicles: Made for Each Other" (International Transport Forum at the OECD, Discussion Paper 2, 2012).

25. Ibid.

26. Christopher Cooper, James Rose, and Shaun Chapman, *Freeing the Grid: How Effective State Net Metering Laws Can Revolutionize U.S. Energy Policy* (New York: Network for New Energy Choices, November 2006).

27. Michael J. Scott et al., "Impacts Assessment of Plug-in Hybrid Vehicles on Electric Utilities and Regional U.S. Power Grids, Part 2: Economic Assessment" (Pacific Northwest National Laboratory Report, 2007), available at http://energytech.pnnl.gov/publications /pdf/PHEV_Economic_Analysis_Part2_Final.pdf; P. Denholm and W. Short, "An Evaluation of Utility System Impacts and Benefits of Optimally Dispatched Plug-in Hybrid Electric Vehicles" (National Renewable Energy Laboratory Technical Report NREL/TP -620-40293, October 2006), available at www.nrel.gov/docs/fy07osti/40293.pdf; Kempton and Tomic, "Vehicle-to-Grid Power Fundamentals."

28. B. K. Sovacool and R. F. Hirsh, "Beyond Batteries: An Examination of the Benefits and Barriers to Plug-in Hybrid Electric Vehicles (PHEVs) and a Vehicle-to-Grid (V2G) Transition," *Energy Policy* 37, no. 3 (March 2009): 1095–1103.

29. See the Tesla website at www.teslamotors.com/charging#/onthego.

30. US Department of Transportation, *NHTS 2001 Highlights Report*, BTS03-05 (Washington, DC: Bureau of Transportation Statistics, 2003).

31. The Nissan Leaf website provides specs on the cars and a tool for calculating gas savings at www.nissanusa.com/electric-cars/leaf/versions-specs/version.s.html.

32. Union of Concerned Scientists (UCS), *State of Charge* (Washington, DC: UCS, June 2012).

33. New York Times editorial quoted in Remy Prud'Homme, "Electric Vehicles: A Tentative Economic and Environmental Evaluation" (International Transport Forum at the OECD, Discussion Paper 22, 2010), 3.

34. Crist, "Electric Vehicles Revisited."

35. J. Struben and J. D. Sterman, "Transition Challenges for Alternative Fuel Vehicle and Transportation Systems," *Environment and Planning B: Planning and Design* 35, no. 6 (2008): 1070.

36. M. Hård and A. Knie, "The Cultural Dimension of Technology Management: Lessons from the History of the Automobile," *Technology Analysis and Strategic Management* 13, no. 1 (2001): 91–103.

37. Prud'Homme, "Electric Vehicles," 7.

38. Crist, "Electric Vehicles Revisited"

39. The data presented here were generated by the tool provided on the Department of Energy website at www.afdc.energy.gov/calc.

40. David L. Greene, John German, and Mark A. Delucchi, *Fuel Economy: The Case for Market Failure* (Knoxville, TN: ORNL/National Transportation Research Center, 2007).

41. Howard Geller and Sophie Attali, *The Experience with Energy Efficiency Policies and Programs in IEA Countries* (Paris: IEA, August 2005).

42. Richard B. Howarth and Alan H. Sanstad, "Discount Rates and Energy Efficiency," *Contemporary Economic Policy* 13, no. 3 (July 1995): 101–103.

43. A. Bandivadekar et al., *On the Road in 2035: Reducing Transportations Petroleum Consumption and GHG Emissions* (Cambridge, MA: MIT, 2008).

44. Ibid.

45. S. Ji et al., "Electric Vehicles in China: Emissions and Health Impacts," *Environmental Science and Technology* 46 (2012): 2018–2024.

46. Crist, "Electric Vehicles Revisited."

47. BBC World News, "Electric Cars 'Pose Environmental Threat'" (October 4, 2012).

48. IEA, "Key World Energy Statistics 2013," available at www.iea.org/publications /freepublications/publication/KeyWorld2013.pdf.

49. T. R. Hawkins et al., "Comparative Environmental Life Cycle Assessment of Conventional and Electric Vehicles," *Journal of Industrial Ecology* 17, no. 1 (2012): 53–64, doi:10.1111/j.1530-9290.2012.00532.x.

50. Stromman comment from BBC World News, "Electric Cars."

51. Carey W. King and Michael Webber, "The Water Intensity of the Plugged-in Automotive Economy," *Environmental Science and Technology* 42 (2008): 4306.

52. Tom Gleeson et al., "Water Balance of Global Aquifers Revealed by Groundwater Footprint," *Nature* 488 (August 9, 2012): 197–200; J. Alcamo and T. Henrichs, "Critical Regions: A Model-Based Estimation of World Water Resources Sensitive to Global Changes," *Aquatic Sciences* 64 (2002): 3523621.

53. IEA, "About Energy Poverty" (June 2014), available at www.iea.org/topics/energy poverty.

54. IEA and Nordic Energy Research, *Nordic Energy Technology Perspectives: Pathways to a Carbon Neutral Energy Future* (Paris: OECD, 2013).

55. Bandivadekar et al., *On the Road in 2035.*

56. Ibid.

57. Sovacool and Hirsh, "Beyond Batteries."

58. David Schrank, Bill Eisele, and Tim Lomax, "TTI's 2012 Urban Mobility Report (Texas A&M Transport Institute, 2012), available at http://d2dtl5nnlpfror.cloudfront.net /tti.tamu.edu/documents/mobility-report-2012.pdf.

59. James Woodcock et al., "Energy and Transport," *Lancet* 370 (2007): 1078–1088.

60. Ibid.

61. Ibid.

62. David Ogilvie et al., "Promoting Walking and Cycling as an Alternative to Using Cars: Systematic Review," *BMJ* 329 (October 2, 2004): 763.

63. John Pucher and Ralph Buehler, "Making Cycling Irresistible: Lessons from the Netherlands, Denmark and Germany," *Transport Reviews* 28, no. 4 (2008): 495.

64. "Environment: Copenhagen European Green Capital in 2014," *European Report* 3 (July 2012): 317594, available at http://ec.europa.eu/environment/europeangreencapital /winning-cities/2014-copenhagen.

65. R. Ewing and R. Cervero, "Travel and the Built Environment—A Meta-analysis, *Journal of the American Planning Association* 76 (2010): 265–294.

Can We Sustainably Feed
and Fuel the Planet?

If you drive northwest from Singapore into Malaysia, the first thing you see after the Tuas border checkpoint is not a shopping mall, a McDonald's restaurant, or a mosque but the start of some five million hectares of oil palm plantations.[1] Malaysia has been the top producer of palm oil for almost four decades. In 2013 it accounted for 39% of global production and 44% of global exports.[2] That same year, Malaysia produced some 19.2 million tons of palm oil worth $34 billion, responsible for 10% of national economic revenue. In total, 56% of all agricultural land, or 11.8% of the country's total land area, was dedicated to oil palm cultivation.[3] Oil palms are to Malaysia as sheep are to New Zealand—for every single person in Malaysia, the locals claim, there are almost two dozen oil palm trees.

Although it is a mainstay of the Malaysian economy, the palm oil industry is not without its foibles. It produces a considerable amount of waste. Every ton of oil obtained produces 6 tons of palm fronds, 5 tons of empty fruit bunches, and 100 tons of palm oil mill effluent, among other wastes.[4] Nationwide support for biofuel production has resulted in Malaysia becoming the second-fastest-growing emitter of greenhouse gases (GHGs) in the world (behind Vietnam), with an average annual growth rate of 7.9%. Emissions in a business-as-usual scenario were predicted to grow 74% from 2005 to 2020.[5] Moreover, Malaysia's impoverished people reside predominantly in rural areas where family farms have been displaced by conglomerate-owned oil palm plantations.

Has Malaysia made the right choice—becoming a major global player in the biofuel sector, while adding to the perils of the global climate? Can we truly facilitate a comprehensive replacement of petroleum by biofuel without exacerbating already alarming levels of poverty and environmental degradation?

The answer on one side of the debate is a resounding no. Proponents of this view argue that we cannot simultaneously feed the more than 7 billion people on this planet and power our transportation with biofuels such as ethanol and biodiesel. These fuels require orders of magnitude more water and land than does the production of petroleum fuels, and they compete directly with the growing of crops for food. Climate advocate Al Gore has confessed, "I shouldn't have supported corn-based ethanol."[6] As another commentator colorfully framed it, supporting global biofuel policy, given its implications for those in poverty, is akin to "burning the poor man's lunch."[7]

The other side suggests that governments worldwide are promoting the development of biofuel for sound reasons. These advocates argue that advances in biofuel development are progressing and, over time, will reduce dependency on oil imported from politically unstable regions of the world, spur agricultural development, and reduce the climate impacts of combusting fossil fuels. One article tells us that "biofuel could help poor nations modernize."[8] Scientists Lee Lynd and Jeremy Woods write that "bioenergy could help bring food security to the world's poorest continent."[9]

One Side: Biofuel Is a Bane

Before we can begin to consider the opposing perspectives on this important question, we must define some terms. By *biofuel*, we refer primarily to ethanol and biodiesel.[10] Ethanol is derived from fermenting grains, cereals, sugar crops, and other starches, predominantly corn and sugarcane. Crushing and soaking processes remove the sugar from these crops, and the sugar is fermented (to produce alcohol) using yeasts. Feedstocks for *biodiesel*, by contrast, tend to be oil-rich crops such as soybean, jatropha, oil palm, rapeseed (canola), and sunflower seeds. Nonagricultural sources of biofuel include animal fats and used cooking oil. Most biodiesel manufacturing plants employ transesterification, a process of mixing an oil feedstock with an alcohol through the action of a catalyst. Though definitions about "first" through "fourth" generations of biofuels persist,[11] table 8.1 summarizes a commonly accepted distinction between conventional and advanced biofuels.

So what could be wrong with something as simple as putting processed plant material into the gas pump? Plenty, say opponents of biofuel. They present a range of arguments against biofuel, but we focus here on the three that tend to dominate the discussion: (1) biofuels' comparatively high cost and poor energy payback, (2) land use and damage to the environment, and (3) adverse effects on food prices.

Biofuel Has High Costs and Poor Energy Payback Ratios

Using current technology, biofuel manufacturing is expensive and has low energy payback ratios. Ethanol, though higher in octane and easier on engines, possesses only 56% of the energy content of gasoline by volume, meaning that a driver operating a vehicle on ethanol will experience significantly reduced fuel economy.[12] Biofuel engines require more fuel per kilometer or mile traveled.

Low energy payback for most biofuel crops accounts in part for the high cost. Producing ethanol and biodiesel is an energy-intensive process. The payback depends on a number of variables, but in the United States, on average,

Table 8.1. Attributes of Conventional and Advanced Biofuels

Fuel classification	Stage	Fuel types/sources	Feedstocks/ processes
Conventional ("first" generation)	Commercially available or nearly available	*Ethanol:* Sugar- and starch-based plants or grains	*Ethanol:* Sugarcane, sugar beet, corn, wheat
		Biodiesel: Oil-based crops and straight vegetable oil	*Biodiesel:* Rapeseed (canola), soybean, oil palm, animal fats, used cooking oils
Advanced ("second," "third," or even "fourth" generation)	Basic and applied R&D or demonstration only	*Ethanol:* Cellulosic material, advanced or genetically modified energy crops	*Ethanol:* Switch-grass, *Miscanthus*, and other lignocellulosic biomass
		Biodiesel: Algal fuels (diesel from microalgae), biomass to liquid diesel, furanics, gasification	*Biodiesel:* Jatropha, hydrotreated vegetable oil, raceway ponds, photo-bioreactors

Sources: Modified from International Energy Agency, *Technology Roadmap: Biofuels for Transport* (Paris: OECD, 2011); US Agency for International Development (USAID), *Biofuels in Asia: An Analysis of Sustainability Options* (Washington, DC: USAID, March 2009); Royal Society, *Sustainable Biofuels: Prospects and Challenges* (London: Royal Society, 2008).

every unit of biofuel put into the production process barely yields more output,[13] and in certain cases the numbers can be negative—that is, it takes more energy to produce the fuel. This is especially true when low-energy feedstocks are used and/or distribution distances are long.[14]

Prospects for advanced-generation biofuel do not appear much rosier at the current time. The National Research Council cautioned in 2011 that advanced biofuel crops will not compete with petroleum "absent major technological innovation or policy changes" and calculated that oil prices would need to rise above $190 per barrel to make biofuels commercially viable.[15] Thus, as opponents will argue, in most places biofuel needs large subsidies to work—subsidies that can be better employed supporting more effective technologies.

Biofuel Induces Changes in Land Use and Environmental Damage

The cultivating and processing of biofuels require large amounts of land.[16] One report put it succinctly: the current business model involves taking a "large amount of biomass to get a small amount of energy."[17] The problem is that dry biomass has only one-third the energy density of gasoline (16 vs. 48 MJ/kg).[18] As one peer-reviewed assessment determined:

Producing a sufficient quantity of biomass to meet the biofuel volumes mandated in [the United States] will require millions of acres of land as well as extensive feedstock supply and logistic systems for biomass harvest, transportation, and processing. Recent studies have concluded that wide-scale harvesting of residues from existing cropland and timberland will alone be insufficient to meet biomass production targets. Other land would also need to be converted to the production of dedicated biofuel crops . . . Biomass production for bioenergy is therefore poised to become a major driver of land use change in the coming decades.[19]

Even then, this ramped-up use of biomass would offset only about one-third of current US petroleum consumption.[20]

The land-use constraints for biofuel crops globally are just as bleak. The International Energy Agency (IEA) reported that to achieve a 27% supply share of total transportation fuel in 2050, 65 EJ of biofuel feedstocks would need to be grown on 100 million hectares of land. The IEA concluded that "this poses a considerable challenge given competition for land and feedstocks from rapidly growing demand for food and fiber."[21] To reach slightly more than one-quarter of the world's demand for transportation fuel would require 16% of all arable land and 82% of all water that is currently drawn for human use.

The changes in land use required to cultivate biofuels give rise to serious concerns. Converting global rain forests, peat lands, savannas, and grasslands into biofuel plantations can incur "carbon debts" by replacing sinks (flora and soils that absorb CO_2) with sources of CO_2. The transition could result in releasing 17 to 420 times more CO_2 than the annual GHG reductions that these biofuel crops would offset by displacing fossil fuels. When ecologist Joseph Fargione and his colleagues assessed the production of biodiesel from palm oil and soybeans, ethanol from sugarcane and corn, and ethanol from prairie biomass in Brazil, Indonesia, Malaysia, and the United States, they discovered that in only one instance—prairie biomass ethanol made on marginal cropland—was no net carbon debt incurred.[22] Princeton's Timothy Searchinger and his colleagues projected that corn-based ethanol, instead of producing a 20% savings in emissions, would nearly double GHG emissions over 30 years and that biofuel from switchgrass, if grown on US corn lands, would increase emissions by 50%.[23] In severe situations, such as the direct deforestation of peat forests or tropical rain forests, it can take as long as 100 to 10,000 years to regenerate the flora to a state where the carbon debt can be paid back.[24] Similarly, Beer et al. estimated that compared with conventional diesel, biodiesel from palm oil increases GHG emissions 8 to 20 times, when carbon losses from rain and peat forests are considered.[25] Given this evidence, one can be excused for questioning why "biofuel" is even used in the same sentence as "clean energy."

The environmental arguments against biofuel extend beyond the themes of deforestation and land use. Many biofuel plantations depend on the profligate use of pesticides and fertilizers that contaminate habitats and sources of water. Biofuel operations can contribute to a degradation of water quality through the percolation of nitrogen and phosphorus from fertilizers into groundwater and other water sources.[26] Moreover, harvesting of biofuel crops tends to employ methods that have been known to erode soil at a faster pace than conventional farming.[27] Occupational diseases and health hazards are also associated with exposure to the pesticides and residues that accompany the harvesting and processing stages.[28] Paradoxically, small added amounts of air pollution from biofuel operations can have comparatively large health effects that exceed the impact of pollution from fossil fuels in certain respects.[29] This is because "the environmental health impacts of transportation depend in part on where and when emissions occur during fuel production and combustion."[30]

As table 8.2 suggests, with existing technology, conventional biofuel has the potential to do more harm than good; every biofuel option carries at least one major environmental disadvantage that places its efficacy into question. Most biofuel options are both water- and energy-intensive—not the type of attributes one would hope for in a water- and energy-constrained world.

Biofuel Production Adversely Affects Food Prices and Poverty

Opponents of biofuel argue that even if one could whisk away challenges related to cost, energy payback, land use, and the environment, the inexorable issue of food security remains. Economists C. Ford Runge and Benjamin Senauer illustrate the trade-off by pointing out that filling an automobile's 25-gallon tank with pure ethanol requires 450 pounds of corn, enough calories to feed a person for an entire year.[31] It is estimated that biofuel operations in the European Union now occupy enough land to feed 100 million people.[32]

There is also evidence that the displacement of food crops by biofuel crops is having an adverse effect on food prices. From 2004 to 2007, for instance, world maize, wheat, and vegetable oil prices climbed 86%, 110%, and 91%, respectively.[33] The World Bank concluded that increased use of feedstocks for biofuel explains 70% to 75% of these price hikes, with the remaining 25% to 30% caused by rising crude oil costs and depreciation of the US dollar.[34] Figure 8.1 suggests a loose correlation between increases in the production of biofuel and inflationary costs of corn (in the United States) and food baskets (across the globe).

Biofuel policy acknowledges the trade-off between biofuel and food. According to the Institute of Medicine, "It is exactly what is assumed in the major models used to predict the greenhouse gas effects of conventional biofuels."[35] Projections undertaken by the US Environmental Protection Agency, the

Table 8.2. Environmental Attributes of Biofuel Production in the United States

Crop	Fuel type produced	Greenhouse gas emissions	Use of resources during growing, harvesting, and refining				Cropland needed to meet half of US demand (% of existing cropland)
			Water	Fertilizer	Pesticide	Energy	
Corn	Ethanol	High	High	High	High	High	157–262
Sugarcane	Ethanol	Low	High	High	Medium	Medium	46–57
Switchgrass	Ethanol	Low	Medium to low	Low	Low	Low	60–108
Wood residue	Ethanol, biodiesel	Low	Medium	Low	Low	Low	150–250
Soybeans	Biodiesel	Medium	High	Low to medium	Medium	Medium to low	180–240
Canola (rapeseed)	Biodiesel	Medium	High	Medium	Medium	Medium to low	30
Algae	Biodiesel	Low	Medium	Low	Low	High	1–2

Source: Modified from Lisa Stiffler, "Bio-debatable: Food vs. Fuel," Seattle PI Reporter, May 2, 2008.

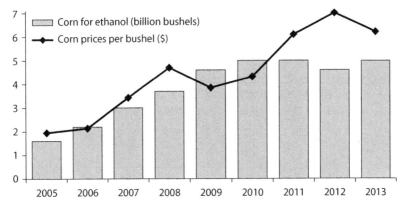

A. Corn Utilized for Ethanol in the United States and Corn Prices, 2005–2013

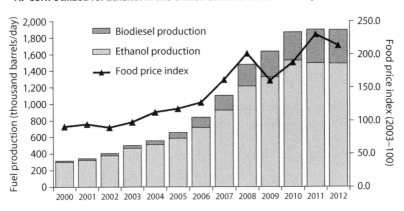

B. Weighted Food Price Index (including cereals, vegetable oils, meat, seafood, sugar, bananas, and oranges) and Global Biofuel Production, 2000–2012

Figure 8.1. Relationship between Biofuel Production and Higher Corn and Food Basket Prices. *Sources: A,* US data from US Department of Agriculture National Agricultural Statistics Service, "Prices Received: Corn by Year, U.S.," available at www.nass.usda.gov /Charts_and_Maps/Agricultural_Prices/pricecn.asp; and US Department of Energy, Alternative Fuels Data Center, available at www.afdc.energy.gov/data. *B,* global data from the FAO Food Price Index in 2014, in nominal dollars, and the US Energy Information Administration database of global biofuel volumes, available at www.eia.gov/cfapps /ipdbproject/IEDIndex3.cfm?tid=79&pid=79&aid=.

California Air Resources Board, and the European Union all assume that the calories from grains and corns diverted from human and animal consumption to biofuel are not replaced. Consequently, some of the GHG benefits presumed by these models depend on humans and animals eating less, expending less energy, producing less methane, and breathing out less carbon dioxide. As Searchinger and colleagues remark, "If you were to eliminate these savings, you would not have greenhouse gas savings according to all these models."[36]

Future conventional biofuel production—expected to expand greatly beyond today's levels—could force an even worse trade-off between biofuel and food security. The National Research Council projected in 2011 that producing 16 billion gallons of ethanol by 2022, as required by law in the United States, would require an additional 30 to 60 million acres of land and result in an increase of 20% to 40% in agricultural commodity prices. This would then cascade into a 1% to 2% increase in grocery prices.[37] This increase may not seem like much, but every percentage increase in the price of staple foods purportedly increases food insecurity for about 16 million people.[38]

Another study, which modeled the impact of ethanol expansion in the United States as a result of higher oil prices, concluded that "with increased ethanol expansion, the prices of both the agricultural feedstock commodities and their competing crops increase with implications for land allocations, food prices, and the environment."[39] Troublingly, the study also noted that price spikes would not be limited to grain: prices would also increase for chicken, other poultry, pork, and beef, and increases would spiral into other farm products such as butter, cheese, eggs, oats, and various oils. Many of these price increases would be greater than 10%. The study further noted that the impact of these escalations in price would be felt around the world, projecting that sub-Saharan Africa and Latin America would be particularly hard hit, with food basket costs rising 10% to 15% in Ethiopia, Guatemala, Mexico, Moldova, Mozambique, Rwanda, Tanzania, Uganda, and Zambia.[40]

A study undertaken by researchers at the Massachusetts Institute of Technology tackled the extreme case of estimating the amount of land needed for biofuel to replace petroleum fuel entirely by 2100, assuming that, in that year, a global biofuel industry would be producing 368 EJ of equivalent energy.[41] The study concluded, first, that the amount of land required was vast—between 1.4 and 1.77 billion hectares. By comparison, 1.6 billion hectares are currently used for *all* cropland worldwide. This bears repeating: to meet our transportation needs through biofuel, we would need more land than is currently dedicated to growing all of our food crops. It gets worse. The study pointed out that if the calculations were modified to take into account that not all of that land could be used simultaneously, 2.2 to 2.5 billion hectares would be needed; 2.5 billion hectares is "an amount greater than any other land cover category . . . including the area covered by natural forests."[42] Second, the study projected that tropical areas in Central and South America and Africa would become the biggest biomass suppliers, with grave implications for deforestation.[43] Third, worldwide prices for both agricultural and forestry products would rise 5% to 10% beyond business as usual. In short, satisfying world demand for petroleum through biofuel production would turn most of our usable land into biofuel plantations, completely wipe out biodiversity, and precipitate price increases for commodities grown on the few competing land tracts that would be left.

The Other Side: Biofuel Is a Boon

Advocates of biofuel have equally forceful reasons to support their view. The promise of biofuel is clearly revealed in market development statistics. Biofuel production has taken off globally, rising from a mere 18 billion liters per year in 2000 to 120 billion liters in 2012. The IEA expects the biofuel sector to attract $11 to $13 trillion worth of investment from 2010 to 2050.[44] More than 50 governments around the world have some type of goal or mandate for biofuel, and if these goals are met, biofuel will account for about 10% of the world's transportation fuels by 2020.[45] Table 8.3 profiles the largest producers of biofuel in 2012.

Again, there are many opinions on this side of the biofuel debate, and we focus on three critical arguments: (1) biofuel can satisfy the imperative of reducing dependence on oil, (2) the potential for commercializing advanced strains of biofuel can dramatically cut GHG emissions and/or rehabilitate degraded landscape, and (3) deteriorating food security has only a tenuous correlation with biofuel production, and it can be strategically avoided.

Table 8.3. Global Biofuel Production, 2012

Producer	Production (1,000 barrels/day)		
	Biodiesel	Fuel ethanol	Total biofuel
World	337.76	1,527.61	1,865.37
United States	22.40	867.44	889.85
Brazil	41.12	486.01	527.13
Germany	49.00	13.00	62.00
France	37.00	18.00	55.00
China	6.00	37.00	43.00
Argentina	36.00	2.10	38.10
Canada	2.40	24.00	26.40
Spain	16.00	8.00	24.00
Thailand	11.00	7.50	18.50

Biofuel Can Reduce National Oil Dependencies

Proponents of biofuel often argue vociferously for the benefits of energy self-sufficiency and independence, because biofuel crops can be grown in almost any country. As discussed in more detail in chapter 13, dependence on foreign sources of oil transfers trillions of dollars of wealth from importing to exporting countries. This is expected to intensify. One study estimated that between 2010 and 2030, many parts of the world will become increasingly dependent on imported petroleum. For example, India's import dependence will rise from 73% to 93%, and China's will rise from 50% to 75%.[46] "Why not meet demand for transportation fuel domestically?" the thinking goes. Doing so both minimizes dependence on the capricious behavior of supplier nations and keeps energy investments within national borders. Biofuel can even serve as a fungible commodity in the event of surplus.

Such benefits are already on display in nations that have embarked on large biofuel programs. An example is the Programa Nacional do Álcool, or Proálcool, in Brazil, started in 1975. While the net cost to the central government in terms of subsidies for Proálcool was about $11 billion from 1975 to 1985, the nation saved $55 billion in avoided oil imports. From 1975 to 2000, the entire ethanol program cost the government about $30 billion but saved about $750 billion by offsetting the need for oil imports. Two independent economists not affiliated with the Brazilian program estimated that the country's external debt was $100 billion less than it would have been in the absence of ethanol production.[47]

The Next Generation of Biofuels Will Have Enhanced Performance

Advocates emphasize that the next generation of advanced biofuels promises a greater array of benefits. For example, cellulosic ethanol has high levels of utilization efficiency as it can be made from compounds in green plants, including stalks and leaves, grasses, and even trees. Crops can be specifically engineered to maximize biofuel yield.[48] Engineered strains of algae and mold are technically capable of producing 10 to 30 times more oil per acre than crops currently in circulation. The science is already proven; it is just a matter of finding a commercially viable strategy for exploiting these advances.

In most cases, biofuel has GHG emission rates far lower than those of fossil fuels. Advocates argue that scenarios where a transition to biofuel exacerbates climate change are unusual and do not represent the norm. Figure 8.2 shows that GHG emissions for ethanol and biodiesel can be far lower than emissions from fossil fuels. The Organization for Economic Cooperation and Development (OECD) reviewed a large number of studies and determined that sugarcane ethanol, as it exists today, could reduce GHG emissions by at least 70% compared with gasoline. Even corn-based ethanol, a highly inefficient biofuel,

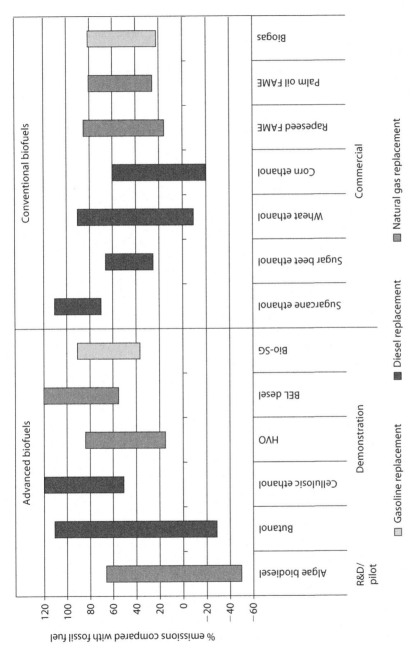

Figure 8.2. Greenhouse Gas Emissions of Biofuels Compared with Fossil Fuels. *Source:* International Energy Agency, *Technology Roadmap Biofuels for Transport* (Paris: OECD, 2011). *Note:* Bio-SG = biosynthetic gas; BtL = biomass to liquid; FAME = fatty acid methyl esters; HVO = hydrotreated vegetable oil.

could achieve 30% reductions.[49] As the IEA concludes, "Biofuels—liquid and gaseous fuels derived from organic matter—can play an important role in reducing CO_2 emissions in the transport sector, and [in] enhancing energy security."[50]

Biofuel's environmental benefits, its advocates elaborate, extend far beyond climate change mitigation. For example, ethanol contains oxygen, and when blended with gasoline it reduces the amount of carbon monoxide and soot that are created by burning fuel.[51] Ethanol biofuel is also low in sulfur dioxide, meaning that its combustion contributes far less to acid rain—and widespread biofuel use can therefore reduce ambient levels of hazardous air pollutants.[52] One report even suggested that biofuel "is an ideal source for highly polluted cities attempting to reduce toxic fumes."[53]

Biofuel crops can be used in tandem with agricultural crops to remediate soil. Compared with high-yield food crops that can leach nutrients from the soil, biofuel crops, if properly planned, can help stabilize soil quality, improve soil fertility, reduce erosion, and improve ecosystem health.[54] Switchgrass and sugarcane can grow without irrigation and can be harvested through a process similar to mowing the lawn. In some cases, switchgrass cultivation can create filter strips that trap and purify pollutants.[55] Algae, which require only carbon dioxide, water, and sunlight to grow, are tremendously efficient and have been called the fastest-growing species on the planet, growing 10 times faster than sugarcane. In laboratory experiments, certain strains of algae can yield oil at the rate of 2,000 gallons per acre per year, compared with only 18 to 635 gallons for corn, soybeans, and oil palm.[56] Algal fuels, produced in raceway ponds or photo-bioreactors, could also be grown on otherwise nonproductive or non-arable land, including deserts. These types of biofuel operations can recycle carbon dioxide and other waste streams and can thrive using saline water, brackish water, and wastewater.[57]

Advocates of biofuel argue that these benefits will multiply as technology improves and learning curves accelerate. Around the world, biofuel development is backed by massive research efforts, and these efforts are beginning to have an impact. In 2000, corn fields yielded less than 140 bushels per acre, but yield grew to 153 bushels per acre in 2007—a rate of increase of around 2% per year. In 2000, ethanol facilities averaged 0.37 liters of ethanol per kilogram of corn, but this jumped to 0.42 liters/kilogram in 2007.[58] Taken together, these advancements—improved yield per acre and improved fuel per kilogram of feedstock—have turned what was once a negative energy biofuel crop into a commercial contender. In Brazil, the average hectare of sugarcane produced less than 3,000 liters of ethanol in 1970, but this yield surpassed 6,500 liters/hectare in 2008 and is expected to reach almost 12,000 liters/hectare by 2018.[59]

These trends are likely to continue as more investments pour into the sector. DuPont is currently developing more than 130 seed hybrids. Monsanto is investing $1.5 million per day in research and development. Syngenta is channeling 10% of its sales back into research and is developing crop strains resistant to disease and drought.[60] These are all initiatives that are likely to lead to enhanced yields, which will be bolstered through advances in crop analytics, molecular breeding, plant protection and pest management, biotechnology, and harvesting.

Indeed, as figure 8.3 shows, current production techniques are far from optimal; adoption of industry "best practices" could result in outperforming "average" harvest and yield rates, in some cases by more than 100%. Moreover, recycling of residues and adoption of whole crop processing practices could see efficiency improvements jump by a factor of four (or more) for some feedstocks. As the US Agency for International Development adds, biofuel "cultivation can avoid significant environmental impacts through the adoption of agricultural best practices at every stage of production."[61]

This last point about the environmental impact of biofuel crops deserves a bit more attention. If well managed, biofuel can be produced "with much

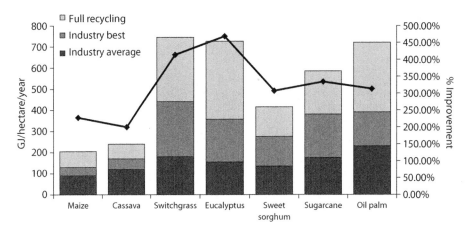

Figure 8.3. "Average" and "Best" Practices for Harvesting the Lignocellulosic Biomass of Selected Crops. *Sources:* Modified from Philips McDougall, "Resources for Biodiesel: Yield Per Hectare in Liters" (Bayer Crop Science Corporation, January 2008); European Commission and Food Agriculture and Natural Resources Policy Analysis Network, "Managing Arid and Semi-arid Ecosystems, Third Periodic Activity Report" (01.01.2009–31.12.2009, December 2009), Annex 4-3-3; and UK Department of Transport, "Renewable Transport Fuels Obligation (RTFO) Orders" (March 25, 2013). *Note:* GJ/hectare/year is displayed on the left axis with bars, % improvement on the right with lines.

lower life-cycle greenhouse-gas emissions than traditional fossil fuels and with little or no competition with food production."[62] How? Through strategic utilization of land and agricultural waste coupled with smart growing strategies. Employing perennial plants on degraded land or land that is not suitable for agricultural use minimizes competition (real or perceived) with food crops. Moreover, agricultural crop residues such as corn stover, rice husks, and straw could become substantial sources of biofuel as they are produced in abundance but rarely recycled. Similarly, residue from forestry operations can be utilized as biomass. Waste streams from industries rich in organic matter (including paper, cardboard, and yard wastes) can also be converted into liquid fuels. One study concluded that 500 million tons of such feedstocks were available in the United States alone, but this resource was barely utilized.[63] The final way in which biofuels can be cultivated in an environmentally benign manner is through mixed cropping systems, such as twinning of oil palm and jatropha crops. A similar strategy is to plant a fast-growing biofuel crop such as switchgrass between food crop seasons. If the right strains are used, the impact on soil health can be attenuated.

The Connection between Biofuel and Food Prices Is Tenuous

On the question of adverse effects on food prices, proponents of biofuel counter that it has had only a negligible effect. Only about 1% of the world's current supply of arable land was being used to manufacture biofuel in 2008 when the food security controversy reached its peak.[64] Robert Vos, director of Development Policy Analysis at the United Nations, attributed rising food prices to a multitude of issues, including:

- Underinvestment for many decades in the agricultural sectors of many countries
- Declining rates of productivity and yield growth rates of wheat, maize, and rice
- Accelerated urbanization and population growth causing mismatches between supply and demand
- Debilitating macroeconomic conditions, including higher interest rates and weakening of the dollar, which increased agricultural commodity prices
- Sharp increases in the cost of oil and fertilizer[65]

In short, Vos argued that the true culprits behind rising food prices had little to do with biofuel.

Further evidence indicates that the correlation between biofuel diffusion and a rise in food prices is not as strong as critics purport it to be. An independent study by the OECD suggested that the World Bank and the media "overestimated" the role of biofuels in food price spikes, and it concluded that bio-fuel expansion policies accounted for only 5% of the price increase in wheat, 7% of the overall increase in corn prices, and 19% of the overall increase in vegetable prices.[66] There is evidence that in certain areas, oilseed-based bio-diesel even suppressed livestock prices due to enhanced supply of cheap protein feed.[67] As has been widely reported, many of the causes of famine are political rather than technological. The world has had "enough" food, in terms of calories and nutrients, to feed everybody on the planet since at least the late 1970s, yet shortages persist and can occur despite good harvests and increasing prosperity.[68]

Common Ground: We Should Pursue Sustainable Biofuel Bridges and Niches

If biofuel were to substitute for fossil fuel for all uses, there would undoubtedly be cause for grave concern, and many of the criticisms leveled against biofuel would be valid. As chapter 7 indicates, however, electric vehicles should become a dominant form of transportation in the medium to long term, and one of the reasons for this is that biofuel, as the critics point out, cannot comprehensively substitute for gasoline without causing enormous environmental problems. Therefore, biofuel, in terms of private transportation, is really just a bridge. Its primary use in the future will be to substitute for petroleum in areas where electric vehicles are not well suited. This means that biofuel holds promise for shipping, long-haul railways, truck transportation, and aviation, and possibly for peak-load support for smaller, decentralized electricity systems. Under such a demand scenario, the two views presented in this chapter are easily recon-ciled because the role that biofuel is to play in the future will not be great enough to impinge on food production, provided that the transition is strategi-cally managed.

Much of the angst over biofuel has to do with the inefficiencies of first-generation crops, which were supported mainly for political reasons (US ethanol policy) or to generate jobs and economic growth (Indonesian and Malaysian palm oil policy). These crops were never intended to mitigate global CO_2 emis-sions or improve energy sustainability. Critiques of first-generation biofuels are legitimate. These fuels require intensive use of energy, water, and fertilizer, exhibit low or even negative payback ratios, and cause extensive environ-mental damage, especially when carbon sinks such as pristine rain forests and peat swamps are usurped to accommodate biofuel operations. However, these first-generation biofuels are akin to what the manual typewriter was to the computer: entry-level technological forays that do not denote the future of the sector.

As billions of dollars pour into second-, third-, and fourth-generation bio-fuel research, technologies can viably assume leading roles in certain niches—a

BOX 8.1
Biofuels, Food Security, and the Common Ground

Can we sustainably feed and fuel the planet?

ONE SIDE: Biofuels are costly, inefficient, environmentally destructive, and disruptive to global food security.

OTHER SIDE: Biofuels can strengthen national energy security, outperform fossil fuels and first-generation crops, and have a minimal impact on food prices.

COMMON GROUND: If produced and used smartly and efficiently, with improved feedstocks and sensitivity to sustainability concerns, biofuel offers a useful strategic bridge away from petroleum.

common ground depicted in box 8.1. Already, second-generation biofuel solutions such as cellulosic ethanol and algal fuel are exhibiting efficiencies that are orders of magnitude ahead of first-generation crops. The combination of strategic use of first-generation biofuels—by utilizing abandoned agricultural land, drawing on crop residues, recycling waste, and employing mixed cropping systems—and the promise of genetically engineered crops could allow these fuels to effectively support the uses described in this chapter for decades to come.

That said, the global biofuel industry is at a critical juncture. The industry can either position itself to sustainably occupy the competitive niches outlined here or try to ascend to the throne that petroleum currently occupies—in the process, solving one major problem with a host of others. Biofuel systems need to be smartly designed and managed to make the promise of sustainable development the norm rather than the exception.

NOTES

1. This introduction is adapted from B. K. Sovacool and I. M. Drupady, "Innovation in the Malaysian Waste-to-Energy Sector: Applications with Global Potential," *Electricity Journal* 24, no. 5 (June 2011): 29–41.

2. Malaysian Palm Oil Council, "Malaysian Palm Oil Industry" (February 23, 2013), available at www.mpoc.org.my/Malaysian_Palm_Oil_Industry.aspx.

3. See K. Y. Foo and B. H. Hameed, "Insight into the Applications of Palm Oil Mill Effluent: A Renewable Utilization of the Industrial Agricultural Waste," *Renewable and Sustainable Energy Reviews* 14 (2010): 1445–1452; Man Kee Lam and Keat Teong Lee, "Re-

newable and Sustainable Bioenergies Production from Palm Oil Mill Effluent (POME): Win-Win Strategies toward Better Environmental Protection," *Biotechnology Advances* 29 (2011): 124–141.

4. T. Y. Wu et al., "Palm Oil Mill Effluent (POME) Treatment and Bioresources Recovery Using Ultrafiltration Membrane: Effect of Pressure on Membrane Fouling," *Biochemical Engineering Journal* 35 (2007): 309–317.

5. Khazanah Nasional, *Opportunities and Risks Arising from Climate Change for Malaysia* (Kuala Lumpur: Khazanah Nasional, March 2010).

6. Emi Kolawole, "Al Gore: I Shouldn't Have Supported Corn-Based Ethanol" (Reuters, November 22, 2010).

7. "Why Burn the Poor Man's Lunch?" *Plague of Biofuels* (blog), April 12, 2008, available at http://nobiofuels.blogspot.com.

8. Natasha Gilbert, "Local Benefits: The Seeds of an Economy," *Nature* 474 (June 23, 2011): S18–19.

9. Lee R. Lynd and Jeremy Woods, "Perspective: A New Hope for Africa," *Nature* 474 (June 23, 2011): S20–21.

10. Britt Childs and Rob Bradley, *Plants at the Pump: Biofuel, Climate Change, and Sustainability* (Washington, DC: World Resources Institute, 2007).

11. V. Shunmugam, "Biofuels: Breaking the Myth of 'Indestructible Energy'?" *Margin: Journal of Applied Economic Research* 3 (2009): 173–189. See also F. X. Johnson and I. Virgin, "Future Trends in Biomass Resources for Food and Fuel," in *Food versus Fuel: An Informed Introduction to Biofuel*, ed. F. Rosillo-Calle and F. X. Johnson (London: Zed Books, 2010), 164–190.

12. E. S. Rubin and Cliff I. Davidson, *Introduction to Engineering and the Environment* (New York: McGraw Hill, 2001), 165.

13. Childs and Bradley, *Plants at the Pump*.

14. See Tad D. Patzek et al., "Ethanol from Corn: Clean Renewable Fuel for the Future, or Drain on Our Resources and Pockets?" *Environment, Development and Sustainability* 7 (2005): 319–336; R. Costanza et al., "Managing Our Environmental Portfolio," *Bioscience* 50 (2000): 149–155.

15. National Research Council (NRC), *Renewable Fuel Standard: Potential Economic and Environmental Effects of U.S. Biofuel Policy* (Washington, DC: National Academies Press, 2011), 44.

16. Mark A. Delucchi, "Impacts of Biofuels on Climate Change, Water Use, and Land Use," *Annals of the New York Academy of Sciences* 1195 (2010): 28–45.

17. Institute of Medicine, *The Nexus of Biofuels, Climate Change, and Human Health* (Washington, DC: National Academies Press, 2013), 12.

18. Rubin and Davidson. *Introduction to Engineering and Environment*, 165.

19. B. Keeler et al., "U.S. Federal Agency Models Offer Different Visions for Achieving Renewable Fuel Standard (RFS2) Biofuel Volumes," *Environmental Science and Technology* 47, no. 18 (2013): 10,095–10,101

20. V. M. Thomas et al., "Relation of Biofuel to Bioelectricity and Agriculture: Food Security, Fuel Security, and Reducing Greenhouse Emissions," *Chemical Engineering Research and Design* 87 (2009): 1140–1146.

21. International Energy Agency (IEA), *Technology Roadmap: Biofuels for Transport* (Paris: OECD, 2011).

22. Joseph Fargione et al., "Land Clearing and the Biofuel Carbon Debt," *Science* 319, no. 5867 (February 29, 2008): 1235–1238.

23. Timothy Searchinger et al., "Use of U.S. Croplands for Biofuels Increases Greenhouse Gases through Emissions from Land-Use Change," *Science* 319, no. 5867 (February 29, 2008): 1238–1240.

24. United States Agency for International Development (USAID), *Biofuels in Asia: An Analysis of Sustainability Options* (Washington, DC: USAID, March 2009).

25. T. Beer, T. Grant, and P. K. Campbell, "The Greenhouse and Air Quality Emissions of Biodiesel Blends in Australia" (Report No. KS54C/1/ F2.27, prepared for Caltex Pty Ltd., Department of the Environment and Water Resources, CSIRO, Collingwood, VIC, Australia, August 2007).

26. See Institute of Medicine, *Nexus of Biofuels.* See also K. Schilling et al., "Impact of Land Use and Land Cover Change on the Water Balance of a Large Agricultural Watershed: Historical Effects and Future Directions," *Water Resources Research* 44, no. 7 (2008), doi:10.1029/2007WR006644.

27. Food and Water Watch, Network for New Energy Choices, and Institute for Energy and the Environment at Vermont Law School, *The Rush to Ethanol: Not All Biofuels Are Created Equal* (Washington, DC: Food and Water Watch, 2007).

28. See K. Hagström et al., "Exposure to Wood Dust, Resin Acids, and Volatile Organic Compounds during Production of Wood Pellets," *Journal of Occupational Environmental Hygiene* 5, no. 5 (2008): 296–304; S. A. Sumner and P. M. Layde, "Expansion of Renewable Energy Industries and Implications for Occupational Health," *JAMA* 302, no. 7 (2009): 787–789.

29. NRC, *Renewable Fuel Standard.*

30. C. W. Tessum, J. D. Marshall, and J. D. Hill, "A Spatially and Temporally Explicit Life-Cycle Inventory of Air Pollutants from Gasoline and Ethanol in the United States," *Environmental Science and Technology* 46, no. 20 (2012): 11,409.

31. C. Ford Runge and Benjamin Senauer, "How Biofuels Could Starve the Poor," *Foreign Affairs* 86, no. 3 (May/June 2007): 41–54.

32. Bjørn Lomborg, "The Great Biofuels Scandal," *Telegraph* (London), December 16, 2013.

33. Rob Vos, "Green or Mean: Is Biofuel Production Undermining Food Security?" in *Climate Change and Sustainable Development: New Challenges for Poverty Reduction*, ed. M. A. Mohamed Salih (Cheltenham, UK: Edward Elgar, 2009), 233–250.

34. Donald Mitchell, *A Note on Rising Food Prices*, Policy Research Working Paper 4682 (Washington, DC: World Bank Development Prospects Group, July 2008).

35. Institute of Medicine, *Nexus of Biofuels*, 54.

36. Searchinger et al., "Use of U.S. Croplands for Biofuels," 1239.

37. NRC, *Renewable Fuel Standard.*

38. See Runge and Senauer, "How Biofuels Could Starve the Poor."

39. Amani Elobeid and Chad Hart, "Ethanol Expansion in the Food versus Fuel Debate: How Will Developing Countries Fare?" *Journal of Agricultural and Food Industrial Organization* 5, no. 6 (2007): 5.

40. Ibid., 6–7.

41. Angelo Gurgel, John M. Reilly, and Sergey Paltsev, "Potential Land Use Implications of a Global Biofuels Industry," *Journal of Agricultural and Food Industrial Organization* 5, no. 9 (2007): 1–34.

42. Ibid, 2.

43. Ibid., 3–4.

44. IEA, *Technology Roadmap.*

45. Institute of Medicine, *Nexus of Biofuels.*

46. M. A. Brown and B. K. Sovacool, *Climate Change and Global Energy Security: Technology and Policy Options* (Cambridge, MA: MIT Press, 2011), 44.

47. Childs and Bradley, *Plants at the Pump*, 26–27.

48. A. Farrell et al., "Ethanol Can Contribute to Energy and Environmental Goals," *Science* 311 (2006): 506.

49. Vos, "Green or Mean," 245.

50. IEA, *Technology Roadmap.*

51. Jose Goldemberg and Patricia Guardabassi, "Are Biofuels a Feasible Option?" *Energy Policy* 37 (2009): 10–14.

52. USAID, *Biofuels in Asia.*

53. Shunmugam, "Biofuels," 4.

54. T. See and F. M. Slater, "Ground Flora, Small Mammal and Bird Species Diversity in *Miscanthus* (*Miscanthus×giganteus*) and Reed Canary-Grass (*Phalaris arundinacea*) Fields," *Biomass and Bioenergy* 31 (2007): 20–29; T. Semere, and F. M. Slater, "Invertebrate Populations in *Miscanthus* (*Miscanthus×giganteus*) and Reed Canary-Grass (*Phalaris arundinacea*) Fields," *Biomass and Bioenergy* 31 (2007): 30–39; Union of Concerned Scientists (UCS), *How Biomass Energy Works* (Washington, DC: UCS, 2005); L. R. Lynd, "Overview and Evaluation of Fuel Ethanol from Cellulosic Biomass: Technology, Economics, the Environment, and Policy, *Annual Review of Energy and the Environment* 21 (1996): 405–465; J. Kort, M. Collins, and D. Ditsch, "A Review of Soil Erosion Potential Associated with Biomass Crops," *Biomass and Bioenergy* 14 (1998): 351–359.

55. M. A. Sanderson et al., "Nutrient Movement and Removal in a Switchgrass Biomass–Filter Strip System Treated with Dairy Manure," *Journal of Environmental Quality* 30 (2001): 210–216.

56. Gunter Pauli, *The Blue Economy* (Taos, NM: Paradigm Publications, 2010), 171–178.

57. See Morgan Bazilian et al., "The Energy-Water-Food Nexus through the Lens of Algal Systems," *Industrial Biotechnology* 9, no. 4 (August 2013): 158–162; NRC, *Sustainable Development of Algal Biofuels* (Washington, DC: Committee on the Sustainable Development of Algal Biofuels, 2012); Raphael Slade and Ausilio Bauen, "Micro-algae Cultivation for Biofuels: Cost, Energy Balance, Environmental Impacts and Future Prospects," *Biomass and Bioenergy* 53 (2013): 29–38.

58. A. E. Farrell et al., "Ethanol Can Contribute to Energy and Environmental Goals," *Science* 311 (January 27, 2006): 506–508.

59. Brown and Sovacool, *Climate Change*, 223.

60. Childs and Bradley, *Plants at the Pump.*

61. USAID, *Biofuels in Asia*, 13.

62. David Tilman et al., "Beneficial Biofuels: The Food, Energy, and Environment Trilemma," *Science* 325 (July 17, 2009): 270–271.

63. Oak Ridge National Laboratory and US Department of Energy, *Biomass as Feedstock for a Bioenergy and Bioproducts Industry: The Technical Feasibility of a Billion-Ton Annual Supply*, DOE/GO-102995-2135 (Washington, DC: US Department of Energy, 2005).

64. Shunmugam, "Biofuels."

65. Vos, "Green or Mean."

66. OECD, *Rising Food Prices: Causes and Consequences* (Paris: OECD, 2008).

67. Vos, "Green or Mean."

68. Mary Douglas, "The Food Problem," in *Food in the Social Order: Studies of Food and Festivities in Three American Communities*, ed. Mary Douglas (New York: Russell Sage Foundation, 1984), 1–39.

III. CLIMATE CHANGE

Is Mitigation or Adaptation the Best Way to Address Climate Change?

The biologist Garrett Hardin once used a lifeboat metaphor to characterize the consequences of unfettered population growth. He described a scenario in which a lifeboat is in the ocean with 50 people aboard and room for only 10 more. The lifeboat is surrounded by 100 swimmers. Hardin pondered the questions of when swimmers in distress should be taken aboard and how many should be offered salvation, given the capacity constraints of the lifeboat.[1] Climate refugees from the Maldives and Tuvalu represent a modern manifestation of the ethical conundrum put forth in Hardin's metaphor.[2] Should "safe" nations accept climate refugees and, if so, under what circumstances and with what limitations? Do the nations that have unwittingly been most responsible for the historical accumulation of anthropogenic greenhouse gas (GHG) emissions have a special responsibility to save these refugees? The problem, as explicated by Hardin, is that overloading the lifeboat will cause it to sink, precipitating disaster for all—wealthy and poor alike.

When applied to climate change, the lifeboat metaphor takes on a fresh nuance because the number of people at risk (the swimmers in the water) is not static—it increases as the impact of climate change intensifies. This gives rise to another intriguing question: given the limited time and resources available in the lifeboat that we call Earth, do we focus our attention on helping the swimmers who are currently in the water (i.e., adopt strategies to abate the impact of climate change) or on trying to limit the number of adrift swimmers who will need our help (i.e., adopt strategies to mitigate the progression of climate change)?

On the surface, mitigation might appear to be the logical choice. After all, if we fail to treat the cause of climate change, we will be faced with an increasing number of swimmers. On the other hand, mitigation requires collective action that, to date, has not materialized to any significant extent. As we noted in the introduction to this book, "adaptation" refers to the process of taking measures to reduce the adverse effects of climate change. It involves "managing the unavoidable" by minimizing the ineluctable harm associated with climate change. This, then, is the theme of this chapter: is mitigation or adaptation the best way of addressing climate change?

One Side: We Must Mitigate Now

Proponents of the need to prioritize mitigation draw on many lines of argument, but there is one common thread of logic: it is better to stop emissions than to deal with their likely effects. Proponents argue that mitigation lowers the overall cost of addressing climate change.[3] Critics highlighting the folly of adaptation draw parallels to insurance policies. Insurance policies are useful only to compensate for loss and, even then, only if the loss is within the financial limits of the policy.

What is a more effective strategy for a person who is a habitual smoker: buy life insurance or quit smoking? Clearly, for advocates of solving the root cause of a potential problem, mitigation is a better investment than compensating for disaster once the problem occurs. Advocates of mitigation hold that (1) mitigation is the best way to address climate change, (2) it offers benefits beyond stabilization of the atmosphere, (3) there is no guarantee that adaptation will work, and (4) adaptation can lead to its own set of pernicious drawbacks.

Mitigation Lessens the Future Impact of Climate Change

The argument in favor of prioritizing mitigation begins by noting that it is the only way to avert amplifying a problem that carries exponentially increasing costs as it progresses. Mitigation offsets adaptation costs because the more we mitigate, the less we will have to adjust to the effects of climate change down the road. As a US National Academy of Sciences report put it, mitigation is the best approach for "limiting the magnitude of climate change."[4] Proponents of mitigation would argue that when your house is on fire, the first action you should take is not to look for lumber but to put the fire out.

The primary climatic benefits of mitigation stem from the fact that immediate efforts can stop the buildup of GHGs in our atmosphere. As climate scientist James Hansen and colleagues argue, mitigation is humanity's best path for "preserving a planet similar to that on which civilization depended and to which life on Earth is adapted."[5] Immediate GHG emission reductions are needed to keep concentrations of these gases within a 500 to 550 ppm range. And as the Intergovernmental Panel on Climate Change (IPCC) implied in its 2014 global assessment, delaying mitigation invites disaster.[6]

A closely related theme in support of mitigation is that, because it directly curtails emissions, it fulfills the moral obligation of the polluting countries to rectify any damage done to the low-polluting nations. This is a matter of "corrective" justice. If a cluster of nations has engaged in wrongfully injuring another group, these nations should desist from their harmful actions and compensate the aggrieved nations for damages.[7] This ethical obligation extends both to rich countries that have exhausted the capacity of the atmospheric sink and to large developing economies that will become major future GHG emitters.[8] Some proponents of mitigation contend further that such inequalities must be "reversed" by imposing extra burdens on those entities responsible for inflicting and maintaining such inequalities.[9] This tenet is essentially the foundation upon which the Kyoto Protocol was based, wherein industrialized nations vowed to take the lead in mitigation efforts.

Lastly, advocates of mitigation point out that it's not just a matter of cost rationalization or of equity; it is also a matter of economic efficiency. Many things we are supposed to be doing to mitigate emissions are things we should be doing anyway because they can return new benefits to households and businesses. Figure 9.1 presents an array of items that can profitably displace emissions and generate revenues for investors and policymakers worldwide. Profitable initiatives include energy-efficiency investments in buildings and homes, better tillage and agricultural practices, industrial fuel switching, and transitions to various types of clean electricity generation, ranging from renewable sources to alternative sources such as landfill gas and small hydroelectric dams. Nicholas Paul Lutsey at the University of California, Davis, summarizes this approach: "There are many net-beneficial 'no regrets' climate change mitigation technologies—where the energy savings of the technologies outweigh the initial costs."[10] This could be why, in 2014, for the first time in four decades, CO_2 emissions from the world's energy industry stalled, even though economic expansion continued.[11]

Mitigation Offers Additional Benefits

Mitigation efforts can produce benefits beyond emission and cost savings. The imperative to mitigate can be premised as much on bolstering national competitive advantage or improving public health as on averting the perils of climate change.[12] As US President Barack Obama put it, "A transition to clean energy is good for business."[13] Similarly, the president of the World Bank Group, Jim Yong Kim, argues that there is no trade-off when it comes to transitioning to clean technology: "We believe it's possible to reduce emissions and deliver jobs and economic opportunity, while also cutting health care and energy costs."[14]

There are numerous real-world examples that support this contention. Investments in energy efficiency enhance industrial competitiveness and displace emissions at the same time.[15] Innovations that reduce the amount spent on energy allow capital to be invested more productively. Distributed generation, such as solar photovoltaic panels, landfill gas capture, micro-hydropower, and combined heat and power systems, provides more local jobs and reduces local pollution while also mitigating GHG emissions. Moreover, by offsetting the need for constructing large-scale electric transmission systems in capital-constrained developing countries, distributed generation offers a lower-cost, low-carbon infrastructure solution.[16]

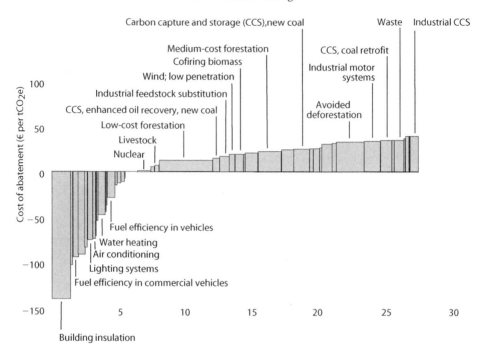

Figure 9.1. Cost Curve for Global Carbon Dioxide Abatement Options. *Source:* Modified from McKinsey & Company, *Global Greenhouse Gas Abatement Cost Curve*, Version 2.1 (Boston: McKinsey & Company, 2010), available at www.mckinsey.com/client_service /sustainability/latest_thinking/greenhouse_gas_abatement_cost_curves. *Note:* Items in the abatement curve below 0 have "negative cost"—they generate economic savings while displacing emissions. The horizontal side of each bar represents the amount of carbon saved. The idea is to start from the left of each curve, using the cheapest carbon-saving options first before moving up to the more expensive items. In theory, the savings from the negative-cost items can generate revenue to offset the positive-cost items. $GtCO_2e = $ giga-ton of carbon dioxide equivalent; $tCO_2e = $ ton of carbon dioxide equivalent. (To simplify the figure, not all entries have labels.)

Mitigation initiatives can also deliver dramatic improvements in health. The IPCC argues that the health co-benefits resulting from mitigation range from fewer deaths caused by heatwaves and forest fires to better food security and improved curtailment of disease epidemics.[17] These benefits can be roughly quantified at $40 to $198 of positive health value per metric ton of carbon dioxide mitigated by 2020.[18] A study from the International Institute for Applied Systems Analysis analyzed the co-benefits of pursuing a global mitigation strategy to keep temperature rise at 2°C by enacting policy initiatives in China, the European Union, India, and the United States.[19] The research team found that under a GHG emission mitigation strategy, expenditures on air pollution control would fall by €250 billion in 2050. The study also highlighted significant improvements in human health and average life expectancy as a result of these policy initiatives. In China alone, a mitigation strategy was projected to reduce concentrations of particulate matter by 50% and to improve average national life expectancy by 20 months. Decreases in ozone concentrations would also prevent nearly 20,000 premature deaths. The study concluded by identifying other indirect benefits arising from GHG emission reduction policies, such as reduced acidification of forests, improved water quality, and enhanced watershed health.[20]

There Is No Guarantee That Adaptation Will Work

If history is anything to go by, society will probably not make the investments in adaptation needed to truly bolster resilience. Many of the most vulnerable nations cannot finance sufficient adaptation, and the international community has shown little willingness to lend the needed support.

Adapting to climate change is like fishing for eels with greasy hands. Climate change is a dynamic event with consequences that scientists are still trying to grapple with. Some opponents argue that in an environment of such uncertainty, adaptation is of tenuous value. What is the point of building a 6-meter-high sea wall when storm surges turn out to be as high as 9 meters? Conversely, building a sea wall 9 meters high to confront storm surges of just 3 meters represents a waste of funds. For many adaptation initiatives, structural changes are required—changes that are often unalterable and irreversible and must be made under conditions of uncertainty.

The evidence suggests that adaptation is akin not only to a habitual smoker purchasing a life insurance policy, but to expecting a habitual smoker to purchase life insurance for a complete stranger. There is little incentive to follow through on such a commitment. During the Sixteenth Conference of the Parties in Cancun in 2010, industrialized countries pledged to mobilize $100 billion per year by 2020 to address the climate change adaptation needs of emerging economies.[21] A Green Climate Fund was created at this meeting to help poorer countries reduce emissions and prepare fortifications against climate change. However, the fund has remained little more than an organizing principle, and its fund-raising goals have fallen short.

The IPCC report *Climate Change 2014: Impacts, Adaptation, and Vulnerability* appealed to industrialized countries to commit the expenditure of $100 billion per year in assistance to developing countries for adaptation starting immediately, well before 2020.[22] However, the likelihood of this appeal

Table 9.1. Multilateral Adaptation Funds: Funds Disbursed to Date (millions of US dollars)

Fund	Goal	2007		2012		Mid-2014	
		Pledged	Disbursed	Pledged	Disbursed	Pledged	Disbursed
Strategic Priority for Adaptation (SPA)	Pilot projects that address local adaptation needs and generate global environmental benefits	50	15	—	50	—	50
Least Developed Countries Fund (LDCF)	Implementation of most-urgent adaptation projects in least-developed countries, based on National Adaptation Programs of Action (NAPAs)	163	12	537	334	880	726
Special Climate Change Fund (SCCF)	Activities aimed at adaptation and three other purposes: technology transfer, economic diversification, and support in key sectors	70	6	240	162	333	299
Adaptation Fund (AF)	Concrete adaptation projects in developing countries that are particularly vulnerable to the adverse effects of climate change	—	—	350	165	350	226
Total		283	32	1,128	711	1,563	1,301

Source: Modified from B. K. Sovacool, Energy and Ethics: Justice and the Global Energy Challenge (Basingstoke, UK: Palgrave Macmillan, 2013) and updated to 2014.

becoming reality is low, given the track record to date. As of 2014, wealthy nations have donated a paltry $10 billion per year to developing countries through the Green Climate Fund to help with adaptation, and they are backsliding even on those amounts.[23] As of mid-2014, developed countries had disbursed only $1.3 billion through four other major multilateral adaptation funds (see table 9.1). Another international adaptation program, pioneered by the Rockefeller Foundation and focusing on fostering resilient cities, was launched in 2013 to high expectations, but it has attracted only 32 participating entities to date.[24]

Adaptation Can Trade Off with Mitigation, Lead to Inequitable Outcomes, or Backfire

Adaptation, some critics say, is even worse than ineffective or underfunded mitigation efforts: it takes one's eye off the ball—diverting money and attention away from solving the problem. One reason that mitigation should not be underfunded is that a sizable portion of climate-related risk cannot be averted through adaptation measures alone. Adaptation, no matter how well designed, cannot be a substitute for actions to reduce GHG emissions in order to slow the rate of global warming. Although adaptation is an insurance policy against failed mitigation, adaptation measures also run the risk of displacing investments in mitigation initiatives that could make a difference. In a world with inviolable limits to the willingness of nations to commit funds to address climate change, adaptation and mitigation must compete for resources—producing, at times, a zero-sum game.[25]

Indeed, critics of adaptation are quick to point out that adaptation was initially considered an unthinkable strategy for addressing climate change. As climate scientists Susanne Moser and Max Boykoff explain, for many years, "some argued that it was immoral or un-strategic to discuss adaptation, when mitigation was the only way to avoid the terrible outcomes of climate change."[26] One magazine even joked that for these actors, mentioning adaptation was akin to "farting at the dinner table."[27] As testament to the lack of interest in adaptation, of the 728 pages of substantive text in IPCC reports in the 1990s, only 32 dealt with adaptation. There was an underlying fear that adaptation efforts "may weaken the social will to undertake greenhouse gas reduction and thus play into the hands of those that argue that any action is premature."[28] Today, critics of adaptation would tend to agree with this statement.

Some experts and climate advocates also oppose adaptation efforts because they can sometimes counteract mitigation efforts. A community adapting to warmer weather by using more refrigeration or air conditioning would use more energy and engender higher levels of GHG emissions.[29] Adaptation efforts to reduce disease vectors such as malaria that are influenced by changing rainfall patterns (e.g., wetlands conversion) might exacerbate GHG emissions and counteract associated mitigation efforts.[30] Reducing the spread of disease and providing potable water to drought-ridden communities (an adaptation initiative) can require more energy use for water treatment (counteracting mitigation).[31] The construction of desalination plants or inter-basin water transfers can help offset saltwater intrusion and shortages of potable water (an adaptation initiative), but only at the expense of higher energy consumption to operate the plants that pump, store, and transfer water (counteracting mitigation).[32] Relocating critical infrastructure away from coastal areas can enhance adaptation but would increase GHG emissions due to added construction demands, probable increases in urban sprawl, and associated emissions from transportation.[33] Farmers can adapt to climate change by increasing the use of nitrogen fertilizers to offset yield losses, but only at the expense of higher GHG emissions associated with fertilizer production, transportation, and use.[34] In many such instances, critics contend, pursuing adaptation initiatives can conflict with mitigation efforts, worsening climate change in the end. Literally and figuratively, adaptation efforts require energy.

Finally, there is evidence that even when adaptation measures are economically justified, such measures can create adverse social, political, and economic outcomes. A 2014 study that investigated eight adaptation projects in both developed and developing countries identified four features of such projects that can facilitate negative outcomes: enclosure, exclusion, encroachment, and entrenchment; these are summarized in table 9.2.[35] In enclosure, adaptation projects transfer public assets into private hands or expand the roles of private actors into the public sphere. In exclusion, adaptation projects limit people's access to resources or marginalize particular stakeholders in decision-making activities. In encroachment, adaptation projects intrude on biodiversity areas or contribute to environmental degradation and climate change. In entrenchment, adaptation projects disempower women and minorities or worsen wealth inequality within a community. In Mali, for instance, researchers discovered that some households were excluded from adaptation project discussions and interventions. In some instances, village leaders elected to implement particular measures—for example, flood defenses—to preserve their own cassava gardens.[36]

Some scholars have noted instances of maladaptation, where adaptation efforts backfire and lower overall resilience. Maladaptation can take on many forms. In the United Kingdom, planners established more stringent building codes to improve the energy efficiency of homes. At first blush, this was a positive intervention that simultaneously mitigated emissions and protected local inhabitants from extreme heat. Yet, when the United Kingdom experienced a period of heavy flooding and the houses were inundated by water, the cost of

Table 9.2. Examples of Adaptation Project "Losers" in Developed and Developing Countries

Process	Dimension	Description	Examples
Enclosure	Economic	Acquiring resources or authority: transferring public assets into private hands or expanding the role of private agents in the public sector	Wonthaggi Desalination Plant in Australia; disaster recovery in Honduras
Exclusion	Political	Marginalizing stakeholders: limiting access to decision-making processes and forums	Coastal protection in Norway; sea barriers in Alaska
Encroachment	Ecological	Damaging the environment: intruding on biodiversity-rich areas or other areas with predisposed land uses or interfering with ecosystem services	Marine Protected Areas in Tanzania; climate-proofing infrastructure in the Maldives
Entrenchment	Social	Worsening inequality: aggravating the disempowerment of women or minorities and/or worsening concentrations of wealth	Livelihood diversification in Burkina Faso; disaster relief in Kenya

repairing them exceeded the benefits of adaptation because of the higher expense of replacing the higher-quality insulation.[37] In Tanzania, adaptation efforts designed to lower the risk of drought ended up costing more than the damage they were trying to prevent.[38] In Australia, efforts to improve water availability around Melbourne relied on such capital-intensive, costly infrastructure that they worsened the economic vulnerability of the people that these measures were intended to benefit.[39] In Bangladesh, planners spent billions of dollars erecting a series of concrete dikes to protect coastal areas from storm surges and cyclones, but this infrastructure became a curse when floods overflowed the dikes, trapping water behind them and saturating agricultural fields with saline water. Bengali farmers had to break the concrete embankments to allow the water to escape.[40] What is tragic about these attempts is not only that they wasted money and resources but that, ultimately, doing nothing might have been better.

The Other Side: We Must Adapt Now

Proponents of adaptation point to negative experiences with climate change mitigation policies and conclude that adaptation is the best path forward for addressing climate change.[41] The crux of the argument is that although GHG emissions create a global problem, these emissions are driven by economic activities that occur at specific locales. Therefore, adaptation is an opportunity to match scalar costs with benefits.[42] Focusing on local adaptation can foster easier measurement, policy experimentation, democratic community engagement, and new platforms for sharing best practices among communities.[43] These efforts, in sum, can help enhance humanity's capacity to predict and effectively manage the expected impacts of climate change. To use an analogy, when confronted with a perpetual rainstorm, adaptation would be akin to building an ark, mitigation to using an umbrella.

Mitigation, enthusiasts for adaptation continue, interferes with climate resilience efforts. Kevin Watkins, director of the Overseas Development Institute in the United Kingdom, explained it this way: "There's a tension here between mitigation and adaptation. There are a large number of low income countries who will [shortly] be dealing with serious adaptation issues. There is a risk that mitigation could trump adaptation, and push out the world's poorest countries."[44]

Here we examine four streams of logic that support the contention that adaptation rather than mitigation should be prioritized: (1) without an international agreement there will be no global transition toward mitigation, whereas adaptation is more politically acceptable; (2) many "no-regrets" cost-positive adaptation options exist; (3) new technologies have the potential to significantly multiply the benefits of adaptation; and (4) adaptation can offer a backdoor for mitigating emissions while improving resilience.

Adaptation Is More Politically Feasible

Advocates of adaptation sometimes point out the futility of local, state, and national mitigation initiatives, given that the global community is unable to come to an effective agreement. The Kyoto Protocol was launched in 1997, to great fanfare, under the UN Framework Convention on Climate Change (UNFCCC). It set binding targets for 37 industrialized countries (and the European Community as a bloc) for reducing GHG emissions. These emissions commitments averaged 5% below 1990 levels over a five-year period from 2008 to 2012. A similar number of parties (including Australia, the 28 members of the European Union, Belarus, Iceland, Kazakhstan, Liechtenstein, Norway, Switzerland, and Ukraine) have agreed to binding targets for the second five-year commitment period, starting in 2013.

Data suggest, however, that claims about progress are tenuous. Many of the larger GHG emitters either have backtracked on climate change mitigation commitments or continue to sit on the sidelines. In December 2011, Canadian Prime Minister Stephen Harper informed the UNFCCC of his administration's intent to withdraw from the Kyoto Protocol, citing concerns that Canadian participation cost the nation $8.7 billion, while major emitters like China continued to operate without emissions targets.[45] Russia followed suit, despite being a beneficiary of first-round commitments as a result of lax emission reduction targets. This then led Japan, once a leader in mitigation efforts, to

declare that it would not sign on for further reduction targets, a position forti-
fied by the Fukushima disaster. These three nations accounted for 9.6% of all
GHG emissions in 2011. With the United States (14% of global GHG emissions in 2011) not ratifying the Kyoto Protocol and China (23.4% of global
GHG emissions in 2011) and India (5.4% of global GHG emissions in 2011) yet
to agree on targets, [46] nations responsible for 52.4% of global emissions in 2011
are not part of any worldwide agreement.[46] In short, the international community is regressing not progressing in terms of climate change negotiations.

Aside from insufficient national commitments, critics further argue that the
main tool driving GHG emission reductions—the establishment of carbon
markets—has exhibited an abysmal performance. Regional and state initiatives in the United States might look good on paper, but emissions still
grew 10.5% nationwide between 1990 and 2010, despite the existence of these
regional carbon markets and stagnant economic growth conditions that only
temporarily depressed emissions.[47] The European Union's Emission Trading
Scheme has all but collapsed, with credit prices plummeting from a high of
€30 per ton in 2006 to roughly €2 in 2012. Globally, there is strong evidence
that the Clean Development Mechanism (CDM) under the Kyoto Protocol
has been "gamed." Manufacturers of HFC-23, a by-product of refrigerants and
high-performance plastics, started producing excesses of the product just to
obtain CDM credits. Researchers at Stanford University calculated that payments to refrigerant manufacturers and carbon market investors for HFC-23
credits have exceeded $4.7 billion, whereas the costs of abating all HFC-23
emissions would have been about $100 million.[48] In short, there was 47 times
more money to be made in producing HFC-23 for CDM credits than in producing HFC-23 for its intended use.

In other instances, investments in mitigation projects have harmed communities. An independent assessment from the Center for People and Forests
warned that some governments have seized indigenous lands to capitalize on
forest carbon revenues and have coerced communities into accepting unfavor-
able contractual terms that sign away land-use rights or commit communities
to socially and environmentally invasive GHG mitigation projects.[49] The result has been diminished production of local food and deepening poverty. For
example, in the highlands of Ecuador, voluntary carbon-offset plantations participating under the REDD initiative proceeded without informing local communities and without impact assessments, delayed the delivery of promised
payments to villages, dismissed complaints and questions about company activities, and rejected appeals for an independent review of operations.[50] In
Uganda, two Norwegian companies purchased thousands of hectares of land
from the government with a 50-year lease to plant fast-growing eucalyptus and
pine trees.[51] The companies will receive millions of dollars of revenue from

selling credits but pay annual rents to the government of only $21,800. The companies have hired just a handful of local workers to tend the plantations (which are low-maintenance anyway) and evicted 8,000 people from 13 villages to make room for their projects.

The decades-long history of failure of negotiations on international climate change mitigation has led many to argue that actions on smaller scales may ultimately prove to be the faster path toward addressing climate change.[52] Mitigation is prone to a slew of collective action problems, essentially forcing politicians to either invest now for benefits in the distant future or sacrifice short-term national interests for long-term international gain (which politicians will rarely do).[53]

In contrast, advocates argue, adaptation efforts more effectively synthesize costs and benefits—attenuating the collective action problems. For the most part, adaptation projects are aimed either at avoiding near-term threats or at enhancing resilience. Since investments made today will provide some sort of payback in the near term, they possess political value. Moreover, costs incurred for adaptation initiatives largely benefit the citizens (voters) of the country that makes the investment.

There is ample evidence of progressive adaptation initiatives around the world as cities and municipalities continue to bolster adaptive resilience. Roughly 20% of cities and nations have developed or are in the process of implementing adaptation initiatives to address the effects of climate change. These include:

- Sea-level rise—restoring or planting mangroves in coastal embankments to reduce damage from storm surges (Bangladesh); coastal land acquisition programs (United States); building storm surge barriers (Netherlands); building higher bridges in anticipation of sea-level rise (Canada); and building seawalls to harden seacoast structures against sea-level rise (United Sates)
- Drought—expanding the use of traditional rainwater harvesting (Sudan); hydroponic gardening (Philippines); developing drought-resistant strains of wheat (Australia); and adjusting crop planting dates and crop variety (Mexico and Ethiopia)
- Global warming and extreme temperatures—opening designated cooling centers (Canada) and heatwave warning systems (United States)
- Permafrost melt—erecting barriers to protect against avalanches and debris flows stemming from permafrost thawing (Switzerland); constructing railway lines with insulation and cooling systems to minimize the amount of heat absorbed by permafrost (Tibet); and controlling glacial outburst floods (Bhutan and Nepal).

To these planners, adaptation is probably an easier investment to justify than incurring a domestic cost for international progress in climate change mitigation that may or may not be effective.

*Many "No-Regrets" Adaptation Pathways Are More
Cost-Effective Than Mitigation*

Adaptation, the thinking goes, is an essential, early investment to blunt the recurrence and severity of climatic events. Delay inflates the scale of the climate change problem, increasing the costs to future generations and the risk of climate-related damages.

Countries can significantly reduce the cost and impact of extreme weather caused by a changing climate. As table 9.3 indicates, adaptation measures can

Table 9.3. Types and Examples of Adaptation Pathways

Type of pathway	Explanation	Examples
Infrastructural	Assets, infrastructure, technologies, or "hardware" in place to ensure the delivery of services that could be disrupted by climate change (such as electricity or water)	Climate proofing of buildings; hardening of seawalls and other seacoast structures against sea-level rise; construction of reservoirs, irrigation systems, and wells to combat drought; maintenance of drainage systems; traditional rain and groundwater harvesting; water storage and demand management
Institutional	Response efficacy of an institution or set of institutions, usually government ministries or departments, in charge of planning and community and infrastructural assets	Improvement of environmental monitoring; early warning systems for floods and other natural disasters; climate prediction and weather forecasting; vulnerability mapping; published evacuation routes; public information on what to do during disasters
Community	Strategies to support the resilience and cohesion of communities and the livelihoods of the people that compose them	Development of agricultural systems and practices better suited to climate variability and change; land-use planning such as limiting development in low-lying coastal, wetland, and floodplain areas and other actions to limit precarious land uses; risk pooling at the regional or national level; use of social care networks to reach vulnerable groups; implementation of relocation systems for low-lying regions where storm surges may be severe
Ecosystem	Enhancing the ability of ecosystems, habitats, and species to survive or even thrive in the face of changing climate	Transplantation of species and establishing and maintaining gene banks to preserve biodiversity; mangrove conservation, restoration, and replanting to mitigate inundation related to rising sea levels

Source: Modified from B. K. Sovacool et al., "Expert Views of Climate Change Adaptation in Least Developed Asia," *Journal of Environmental Management* 97, no. 30 (April 2012): 78–88.

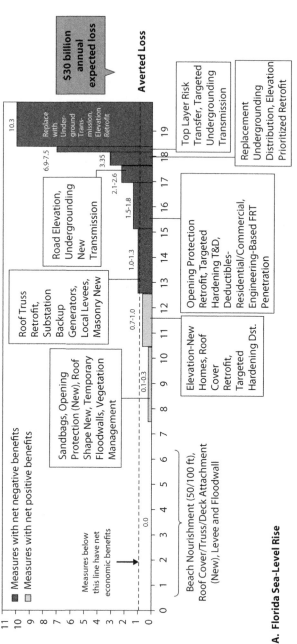

A. Florida Sea-Level Rise

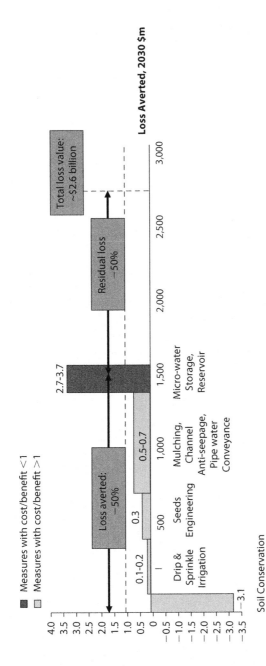

Soil Conservation

B. Drought in North and Northeast China

Figure 9.2. Adaptation Cost Curves for Florida and China. *Source:* Modified from Economics of Climate Adaptation Working Group, "Shaping Climate Resilient Development" (ClimateWorks Foundation, Global Environment Facility, European Commission, McKinsey & Company, Rockefeller Foundation, Standard Chartered Bank, and Swiss Re, 2009). *Note:* $m = million dollars; FRT = fault ride-through; T&D = transmission and distribution.

be carried out along infrastructural, institutional, community, and ecosystem pathways. Climate proofing of buildings and fortification of seawalls and other seacoast structures against sea-level rise are two critical infrastructural pathways. Institutional pathways tend to focus on improving the governance or management of climate-related risks, such as the creation of early warning systems for severe weather events. Community pathways focus on improving the resilience of communities and can include risk transfer or insurance measures.[54] Ecosystem pathways include initiatives for improving the resilience of ecosystems and employing environmental management strategies to foster adaptation, such as planting mangrove swamps to attenuate the damage caused by storm surges.

When put into action, many of these measures can save money, meaning that, economically, they are no-regret actions. A study looking at a mix of eight industrialized and developing countries found that adaptive measures could profitably offset 40% to 100% of the total expected loss from climate change.[55] Figure 9.2 shows how this methodology was used in evaluating possible measures for adapting to sea-level rise in Florida and drought in North and Northeast China. In the case of Florida, beach nourishment, roof covers and trusses, and levees and floodwall sandbags have cost-benefit ratios delivering the greatest climate protection and quickly generating returns on investment. In the case of drought in North and Northeast China, soil conservation, drip and sprinkle irrigation, seed engineering, and mulching are among the most cost-effective low-regret options, and they could cut the expected losses nearly in half. In a separate study, researchers in Vietnam calculated that investments in typhoon-resistant housing pay for themselves within a few years, with most investments seeing internal rates of return ranging from 11% to 21%, depending on the severity of climate effects.[56]

In aggregate, adaptation opportunities can be highly profitable and delay can be costly. The Asian Development Bank, for example, estimated that every $1 invested in adaptation in 2010 could yield as much as $40 in benefits by 2030.[57] Meanwhile, the Center for American Progress estimated that every year that the United States delays in cutting emissions will result in a $500 billion increase in requisite investment per year—a number that will balloon to $10 trillion by 2030— if the county does not change course.[58] By investing in adaptation, the cost of recovery from climate disasters can be at least partially constrained.

New Technologies Will Multiply Adaptation's Benefits

Another argument in favor of adaptation is that new technologies are emerging to make society even more resilient to climate change in a far more cost-effective manner. For instance, smart grid technologies allow deeper penetration and integration of distributed demand-side and supply-side energy resources.

Smart metering—technically termed "advanced metering infrastructure"—is one of the key technologies of the smart grid. In the United States, the penetration of installed smart meters increased fivefold in less than six years, reaching 46 million in 2013, from a base of only 7 million in 2007.[59] By 2015, installed smart meters are projected to surpass 65 million in the United States alone, which is approximately half of all US households. They are expected to be adopted by almost all households by the end of the decade. This new technological interface will lead to far more resilient electricity grids and minimize the probability of blackouts resulting from climate change–induced mega-storms.

New notions for planning technological development lend support to claims that technology for adaptation will continue to improve. Infrastructure ecology is an approach to planning urban infrastructure in a way that improves resilience to climate change, thereby reducing the cost of adaptation. By improving how flows of resources are utilized within complex urban systems and exploiting infrastructure synergies (e.g., with distributed energy–wastewater systems), it is possible to improve resiliency by reducing material and energy demands.[60] One application of this is being pilot-tested in the Dutch city of Rotterdam, where solar-powered floating pavilions were erected in 2008 as part of a climate resiliency project. The bubble-shaped domes are anchored off the city's waterfront and serve as a pilot for testing the viability of floating urban districts that will be able to rise with the changing sea levels.[61] This also comes with a social benefit, of course: if you don't like your neighbors, you simply weigh anchor and dock your house elsewhere.

Adaptation Can Enhance Mitigation

A final argument in support of adaptation measures, even those that come at a cost, is that many measures possess positive synergies with mitigation initiatives. Indeed, some forms of adaptation can act as mitigation measures, such as farming techniques that store moisture better (adaptation) while also reducing the need for fossil fuel–intensive fertilizers (mitigation).[62] Afforestation can create a carbon sink (mitigation) while protecting against flooding and providing income to communities (adaptation).[63] Restoring only 12% of the world's degraded forests could cut emissions from deforestation (mitigation) while feeding another 200 million people and raising farming incomes by $40 billion per year (adaptation).[64] More efficient space cooling and heating can reduce electricity consumption (mitigation) while making cooling more affordable for lower-income groups (adaptation). More efficient irrigation (adaptation) can lessen the electricity needs associated with water pumping (mitigation).[65] Restoring coastal wetlands can both increase carbon storage (mitigation) and create storm buffers against floods (adaptation).[66] Reducing the exploration and drilling of offshore oil can prevent GHG emissions from fossil fuel combustion (mitigation) while diminishing the risk of oil spills and consequent stress on ecosystems (adaptation).[67] The demand-side management programs discussed in chapter 2 can not only reduce energy use (mitigation) but also cut consumers' energy bills, translating into greater financial resilience to future shocks (adaptation).[68]

Indeed, one assessment of the Global Environment Facility's Least Developed Countries Fund—the largest international scheme to promote investment in adaptation in more than 60 of the world's poorest countries—found that the program mitigated emissions as an unintentional but positive by-product, for a cost of less than $2 per ton of carbon dioxide.[69] These initiatives displaced CO_2 2.35 to 500 times more cheaply than other mitigation options at that time.

Common Ground: We Should Optimize "No-Regrets" Adaptation with Mitigation

There are extreme differences of opinion on whether mitigation or adaptation initiatives should be prioritized to avert the costly consequences of climate change. Until recently, mitigation and adaptation were treated as distinct responses to climate change that, to some degree, required a financing trade-off.

We hold that mitigation and adaptation can be complements, not competitors. For these two strategies to be optimized, however, one must first acknowledge that adaptation and mitigation measures are not created equal. Some adaptation or mitigation measures come at high costs, while others offer no regrets from an economic perspective because they provide positive returns on investment. In reality, adaptation and mitigation can and must operate in tandem. Some nations are more at risk for sea-level rise and other effects of climate change. Many of these same nations are financially ill-equipped to adapt and require support from the international community. The need for support for adaptation cannot be ignored. However, mitigation is and always will be the key to averting future economic, social, and environmental catastrophe. Adaptation measures become costlier as the magnitude of climate change grows. Mitigation, if successful, lessens the financial burden of adaptation.

Alarmingly, the less we mitigate, the more the cost of adaptation grows over time. Therefore, premiums to protect systems against climate change and to expand resilience should be considered whenever infrastructure investments are being made and policies are being created. Not all adaptation measures will succeed, however. Success is undermined by the possibility of enclosure, entrapment, encroachment, and exclusion, so there are caveats to heed in many adaptation measures. Adaptation measures cannot be allowed to undermine mitigation efforts. Moreover, the threats associated with maladaptation are real and present with many projects.

BOX 9.1

Mitigation, Adaptation, and the Common Ground

Is mitigation or adaptation the best way to address climate change?

ONE SIDE: Mitigation is the most direct, cost-efficient way to fight climate change and catalyze technological innovation, whereas investments in adaptation can divert attention and resources.

OTHER SIDE: Mitigation cannot stop emissions already in the atmosphere; adaptation is a necessary insurance policy, made more affordable by early action and advanced technologies.

COMMON GROUND: Adaptation and mitigation should be co-optimized.

The common ground depicted in box 9.1 acknowledges that a balance is necessary, but the balance must slightly favor mitigation. Returning to the life-boat analogy, there is a moral imperative for all nations (both rich and poor) to ensure that the swimmers in the water (those most prone to climate change effects) are saved. But there is a greater imperative to ensure that conditions for those already in the water are not worsened and that no more swimmers are added to the ocean. We cannot afford abatement strategies if the damage that needs to be mitigated is as extensive as scientific projections suggest.

Yet, to a considerable degree, adaptation can attenuate many costs stemming from moderate degrees of climate change. Mitigation and adaptation should therefore be seen as partners in the response to climate change, not alternatives. Low-cost, no-regrets measures should be supported because these will result in net economic benefits. But designers of adaptation projects should ensure that costs and benefits are shared as equitably as possible.

Finally, to return to the smoking metaphor: We are like a growing society of habitual smokers. If we wish to improve our collective lot in life, the key isn't to purchase life insurance but to quit smoking—to mitigate our damaging behavior. But once we quit smoking, this will not guarantee us good health. We need to change our diet and lifestyle patterns—to abate the health threats that have accumulated due to a lifetime of smoking. Is abatement a worthy investment? Sure, but not if we keep smoking. A common ground ensures that we make cigarettes less harmful at the same time that we're trying to convince people to quit.

NOTES

1. G. Hardin, "Living on a Lifeboat," *Bioscience* 24 (1974): 561–568.

2. J. Cairns, "Life Boat Ethics Revisited: What Should Be the Reaction to the Maldives Crisis?" *Asian Journal of Experimental Biological Sciences* 24, no. 1 (2010): 17–20.

3. Curt Suplee et al., "What You Need to Know about Energy" (National Academy of Sciences, 2008), available at www.nap.edu/catalog.php?record_id=12204.

4. National Academy of Sciences, *Limiting the Magnitude of Future Climate Change* (Washington, DC: National Academies Press, 2011).

5. J. Hansen et al., "Target Atmospheric CO_2: Where Should Humanity Aim?" *Open Atmospheric Science Journal* 2 (2008): 217–231.

6. Intergovernmental Panel on Climate Change (IPCC), "Greenhouse Gas Emissions Accelerate Despite Reduction Efforts" (April 13, 2014), available at www.ipcc.ch/pdf/ar5/pr _wg3/20140413_pr_pc_wg3_en.pdf.

7. International Council on Human Rights Policy (ICHRP), *Climate Change and Human Rights: A Rough Guide* (Versoix, Switzerland: ICHRP, 2008).

8. Stephen M. Gardiner, "Ethics and Global Climate Change," in *Climate Ethics: Essential Readings*, ed. Stephen M. Gardiner et al. (Oxford: Oxford University Press, 2010), 3–35.

9. Simon Caney, "Cosmopolitan Justice, Responsibility, and Global Climate Change," *Leiden Journal of International Law* 18 (2005): 747–775. See also E. Neumayer, "In Defence of Historical Accountability for Greenhouse Gas Emissions," *Ecological Economics* 33 (2000): 185–192.

10. Nicholas Paul Lutsey, "Prioritizing Climate Change Mitigation Alternatives: Comparing Transportation Technologies to Options in Other Sectors" (doctoral dissertation, University of California, Davis, Spring 2008), 32.

11. John Schwartz, "Emissions by Makers of Energy Level Off," *New York Times*, March 14, 2015.

12. F. Reinhardt, "Bringing the Environment down to Earth," *Harvard Business Review*, July–August 1999, 149–157.

13. Renee Lewis, "Obama Pledges Action to Speed U.S. Transition to Clean Energy" (Associated Press, May 9, 2014).

14. Kim quoted in David Suzuki, "The Economics of Global Warming," *Nation of Change*, July 9, 2014, available at www.nationofchange.org/economics-global-warming -1404912339.

15. A. A. Niederberger et al., "Energy Efficiency in China: The Business Case for Mining an Untapped Resource," *Greener Management International*, 2005, 25–40. See also Marilyn A. Brown and Yu Wang, *Green Savings: How Policies and Markets Drive Energy Efficiency* (Santa Barbara, CA: Praeger Press, 2015).

16. B. K. Sovacool and S. V. Valentine, "Bending Bamboo: Restructuring Rural Electrification in Sarawak, Malaysia," *Energy for Sustainable Development* 15 (2011): 240–253.

17. K. R. Smith et al., "Human Health: Impacts, Adaptation, and Co-benefits," in *Climate Change 2014: Impacts, Adaptation, and Vulnerability. Part A: Global and Sectoral Aspects: Contribution of Working Group II to the Fifth Assessment Report of the Intergovernmental Panel on Climate Change* (Cambridge: Cambridge University Press, 2014), 709–754.

18. John M. Balbus et al., "A Wedge-Based Approach to Estimating Health Co-benefits of Climate Change Mitigation Activities in the United States," *Climatic Change* 127, no. 2 (November 2014): 199–210.

19. Peter Rafaj et al., "Co-benefits of Post-2012 Global Climate Mitigation Policies," *Mitigation and Adaptation Strategies for Global Change* 18, no. 6 (August 2013): 801–824.

20. Ibid.

21. Simon D. Donner, Milind Kandlikar, and Hisham Zerriffi, "Preparing to Manage Climate Change Financing," *Science* 334 (November 18, 2011): 908–909.

22. IPCC, *Climate Change 2014: Impacts, Adaptation, and Vulnerability*, Fifth Assessment Report of the Intergovernmental Panel on Climate Change (Cambridge: Cambridge University Press, 2014).

23. John Vidal, "Rich Countries 'Backsliding' on Climate Finance," *Guardian* (London), November 18, 2014.

24. 100 Resilient Cities, "About 100 Resilient Cities—Pioneered by the Rockefeller Foundation" (2015), available at www.100resilientcities.org/pages/about-us.

25. Richard S. J. Tol, "Adaptation and Mitigation: Trade-offs in Substance and Methods," *Environmental Science and Policy* 8 (2005): 572–578.

26. Susanne C. Moser and Maxwell T. Boykoff, *Successful Adaptation to Climate Change: Linking Science and Policy in a Rapidly Changing World* (London: Routledge, 2013), 132.

27. "Adapting to Climate Change," *Economist*, November 27, 2010, 80.

28. R. W Kates, "Cautionary Tales: Adaptation and the Global Poor," *Climatic Change* 45 (2000): 5.

29. R. J. T. Klein et al., "Inter-relationships between Adaptation and Mitigation," in *Climate Change 2007: Impacts, Adaptation, and Vulnerability: Contribution of Working Group II to the Fourth Assessment Report of the Intergovernmental Panel on Climate Change*, ed. M. L. Parry et al. (Cambridge: Cambridge University Press, 2007), 745–777.

30. B. K. Sovacool and M. A. Brown. "Scaling the Response to Climate Change," *Policy and Society* 27, no. 4 (March 2009): 317–328.

31. For more on these types of trade-offs, see T. J. Wilbanks et al., "Integrating Mitigation and Adaptation as Possible Responses to Global Climate Change," *Environment* 45, no. 5 (2003): 28–38; T. J. Wilbanks, "Issues in Developing a Capacity for Integrated Analysis of Mitigation and Adaptation," *Environmental Science and Policy* 8 (2005): 541–547; Richard J. T. Klein et al., "Integrating Mitigation and Adaptation into Climate and Development Policy: Three Research Questions," *Environmental Science and Policy* 8, no. 6 (December 2005): 579–588.

32. Susanne C. Moser, "Adaptation, Mitigation, and Their Disharmonious Discontents: An Essay," *Climatic Change* 111 (2012): 165–175.

33. Ibid.

34. Ibid.

35. B. K. Sovacool, B. O. Linnér, and M. E. Goodsite, "The Political Economy of Climate Adaptation," *Nature Climate Change* 5, no. 7 (July 2015): 616–618.

36. Sam Barrett, "Local Level Climate Justice? Adaptation Finance and Vulnerability Reduction," *Global Environmental Change* 23, no. 6 (December 2013): 1819–1829.

37. Moser, "Adaptation, Mitigation."

38. Kates, "Cautionary Tales."

39. J. Barnett and S. O'Neil, "Maladaptation," *Global Environmental Change* 20, no. 2 (2010): 211–213.

40. Don Belt, "Buoyant Bangladesh and the Coming Storm," *National Geographic* 219, no. 5 (May 2011): 58–83.

41. E. Ostrom, *Background Paper to the 2010 World Development Report: A Polycentric Approach for Coping with Climate Change* (Washington, DC: World Bank, 2009), available at http://elibrary.worldbank.org/doi/abs/10.1596/1813-9450-5095.

42. Matt Cox, Marilyn A. Brown, and Xiaojing Sun, "Energy Benchmarking of Commercial Buildings: A Low-Cost Pathway for Urban Sustainability," *Environmental Research Letters* 8, no. 3 (2013): 035018, available at http://iopscience.iop.org/1748-9326/8/3/035018.

43. Y. Mulugetta, T. Jackson, and D. van der Horst, "Carbon Reduction at Community Scale," *Energy Policy* 38, no. 12 (2010): 7541–7545.

44. Watkins quoted in Vidal, "Rich Countries 'Backsliding.'"

45. BBC News, "Canada to Withdraw from Kyoto Protocol" (December 13, 2011), available at www.bbc.com/news/world-us-canada-16151310.

46. These data are from the World Resource Institute's CAIT database, available at www.wri.org/resources/data-sets/cait-country-greenhouse-gas-emissions-data.

47. Data from figure ES1 in US Environmental Protection Agency, "Inventory of US Greenhouse Gas Emissions and Sinks" (April 15, 2012), available at www.epa.gov/climatechange/ghgemissions/usinventoryreport.html.

48. Michael Wara, "Is the Global Carbon Market Working?" *Nature* 445 (2007): 595–596; Michael W. Wara and David G. Victor, "A Realistic Policy on International Carbon Offsets" (Stanford University Program on Energy and Sustainable Development Working Paper 74, April 2008); Ben Pearson, "Market Failure: Why the Clean Development Mechanism Won't Promote Clean Development," *Journal of Cleaner Production* 15 (2007): 247–252.

49. Patrick Anderson, *Free, Prior, and Informed Consent: Principles and Approaches for Policy and Project Development* (Bangkok: RECOFTC and GIZ, February 2011).

50. Ibid.

51. Anne E. Prouty, "The Clean Development Mechanisms and Its Implications for Climate Justice," *Columbia Journal of Environmental Law* 34, no. 2 (2009): 513–540.

52. Ostrom, "Background Paper to 2010 World Development Report."

53. See David G. Victor, "Why the UN Can Never Stop Climate Change," *Guardian* (London), April 4, 2011; David G. Victor, *Global Warming Gridlock* (Cambridge: Cambridge University Press 2011).

54. Economics of Climate Adaptation Working Group, "Shaping Climate-Resilient Development" (ClimateWorks Foundation, Global Environment Facility, European Commission, McKinsey & Company, Rockefeller Foundation, Standard Chartered Bank, and Swiss Re, 2009), available at http://mckinseyonsociety.com/shaping-climate-resilient-development.

55. Ibid.

56. Phong Tran, Tuan Huu Tran, and Anh Tuan Tran, *Sheltering from a Gathering Storm: Typhoon Resilience in Vietnam* (Da Nang, Vietnam: Institute for Social and Environmental Transition–International, 2014).

57. Asian Development Bank (ADB), *The Economics of Climate Change in Southeast Asia: A Regional Review* (Manila: ADB, April 2009).

58. Jorge Madrid, Kate Gordon, and Tina Ramos, *America's Future under "Drill, Baby, Drill": Where We'll Be in 2030 If We Stay on Our Current Oil-Dependent Path* (Washington, DC: Center for American Progress, May 2012).

59. Institute for Electric Efficiency, "Utility Scale Smart Meter Deployments" (August 2013), available at www.edisonfoundation.net/IEE.

60. M. Xu et al., "Gigaton Problems Need Gigaton Solutions," *Environmental Science and Technology* 44, no. 11 (2010): 4037–4041.

61. A. Lisa, "Rotterdam's Solar-Powered Floating Pavilion Is an Experimental Climate-Proof Development," *Inhabit*, January 6, 2013, available at http://inhabitat.com/rotterdams -floating-pavilion-is-an-experimental-climate-proof-development.

62. Economist, "Adapting to Climate Change," 79–82.

63. Klein et al., "Inter-relationships between Adaptation and Mitigation."

64. *New Climate Economy, Better Growth, Better Climate* (New York, 2014), available at http://newclimateeconomy.report/wp-content/uploads/2014/08/NCE_GlobalReport.pdf.

65. Sovacool and Brown. "Scaling the Response."

66. Moser, "Adaptation, Mitigation."

67. Ibid.

68. Ibid.

69. B. K. Sovacool, *Energy and Ethics: Justice and the Global Energy Challenge* (Basingstoke, UK: Palgrave Macmillan, 2013), 181–182.

Should Geoengineering Be Outlawed?

On September 16, 1991, a small group of scientists sealed themselves inside Biosphere II, a glittering 3.2-acre (1.3-hectare) glass and metal dome in Oracle, Arizona, built by some of the world's finest engineers. Two years later, when this radical attempt to replicate the earth's ecosystems through technology ended, the engineered environment was dying and the inhabitants of this designed world were mired in interpersonal conflict. Biosphere II had failed to generate breathable air, drinkable water, and adequate food for just eight people.[1]

This experience was possibly prophetic. Our best available technology, at least in the early 1990s, could not simulate the ecosystem services that human beings need to survive. Failure of the Biosphere II experiment suggests that we do not understand our natural systems sufficiently well to replicate them. Moreover, even the cost of a failed effort is expensive: the eight researchers inside Biosphere needed $200 million worth of high-tech equipment to produce the ensuing chaos. Applied to the context of the climate crisis, the prediction is dire: we may not be able to design our way out of the problem.

A number of scientists would disagree with such a pessimistic outlook. Many would argue that technical solutions to climate change are fairly straightforward. We do not need to replicate the earth's biosphere—as was the case in the experiment—we merely need to refurbish one facet of it. Therefore, the challenge is more akin to remediating a polluted river, something that we have successfully done in many parts of the world. The logic behind "climate engineering" or "geoengineering," the term we use in this chapter, is that applying new technology represents the least disruptive and most industrially attractive approach to addressing climate change. As political scientist David G. Victor explains, geoengineering is "planetary scale, active interventions in the climate system to offset the build-up of greenhouse gases."[2] Pursuing such a path, the thinking goes, will allow us to enjoy the fruits of continued reliance on fossil fuel energy without disrupting our lifestyles. Some of the strategies being put forth would not be out of place in an episode of *Star Trek*.

Opponents of this view consider the prospect of geoengineering ludicrous. As environmental sociologist Thomas A. Heberlein wryly observes, "Solutions to global warming are not hard. All we have to do is shoot 800,000 Frisbeesized ceramic disks into space from 20 electromagnetic guns every 5 minutes for 10 years to create a giant sunshade. Or we could have 1,500 ships spray sea water to create white clouds to reflect the sun. Or we could pump tons of sulfur dioxide into the upper atmosphere to imitate the cooling effect of volcanic eruptions. Don't laugh. These geo-engineering projects were actual proposals— some by Nobel Laureates, no less."[3]

Critics argue that geoengineering is akin to the Biosphere II debacle: prohibitively expensive and destined to fail. Such fears led to the 193 signatories to the 2010 UN Convention on Biodiversity agreeing to a de facto moratorium on geoengineering until more about its risks can be known.[4] Is effectively outlawing geoengineering really justified?

One Side: We Can Geoengineer Our Way out of Trouble

The advocates of geoengineering embrace the idea that solving environmental problems is possible by employing technology to modify aspects of our biosphere. They argue that the challenge of climate change lies not in abating greenhouse (GHG) emissions but rather in mitigating atmospheric warming. The geoengineering route would have us build a new climate or atmosphere that safely reflects sunlight or stores carbon in oceans and other places. Some proponents of these strategies contend that special interests will continue to thwart mitigation and abatement efforts and that the best way to avert global catastrophe is to accept these realities and design strategies that will guarantee success. As a 2013 article put it, "It's not the sexiest sounding topic, but a small group of scientists say [geoengineering] just might be able to save the world."[5]

We examine here the rationale put forth in support of geoengineering. The main premises are: (1) it's too late to fully mitigate emissions, (2) technologies can serve as ecosystem substitutes, (3) geoengineering can safely counter the effects of climate change, and (4) geoengineering can be cheaper and more feasible than both mitigation and adaptation.

It's Too Late to Mitigate or Adapt; We Need to Try Something Else

Proponents of the geoengineering option point to historical trends that suggest we may be past the point of no return already. Carbon dioxide emissions from fossil fuels and industrial processes were barely discernible in 1750, but they have grown since the industrial revolution, reaching more than 30 billion tons/ year in 2010.[6]

It appears that we are unable to stop this trend. To avoid the most serious effects of climate change, many scientists believe, a 70% reduction of GHG emissions from 2000 levels is required by 2100.[7] Yet, despite the development over the past several decades of policies to mitigate GHG emissions, the rate of emissions is accelerating—it grew 2.2% per year between 2000 and 2010,

Table 10.1. Increasing Evidence of Anthropogenic Influence on Global Warming

1990	"The unequivocal detection of the enhanced greenhouse effect from observations is *not likely* for a decade or more."
1995	"The '*balance of evidence*' suggests a discernable human influence on global climate."
2001	"There is *new and stronger evidence* that most of the warming observed over the last 50 years is attributable to human activities."
2007	"Most of the observed increase in globally averaged temperature since the mid-20th century is *very likely* due to the observed increase in anthropogenic greenhouse gas concentrations."
2013	"It is *extremely likely* that more than half of the observed increase in global average surface temperature from 1951 to 2010 was caused by the anthropogenic increase in greenhouse gas concentrations and other anthropogenic forcings together."

Sources: Quoted from Intergovernmental Panel on Climate Change reports, various years, available at www.ipcc.ch. Emphasis added.

compared with 1.3% per year between 1970 and 2000.[8] So despite widespread alarm over the perils attributed to climate change, over the past decade, things got worse. As the Intergovernmental Panel on Climate Change (IPCC) remarked, "Emissions grew more quickly between 2000 and 2010 than in each of the three previous decades."[9] Making matters crueler, GHG emissions are projected to increase between 990 billion and 6.1 trillion tons of CO_2 equivalent from 2012 to 2100.[10] It is clear that human activity set this runaway train in motion. As table 10.1 indicates, the five IPCC climate science reports since 1990 have expressed an increasing confidence that human activity is culpable for global warming.

Scientists contend that even if we stopped emissions today (that is, if we achieved 100% mitigation), climate change would still continue because GHG molecules do not simply disappear once they are in the atmosphere. Once emitted, a ton of CO_2 takes a very long time to process through the atmosphere—according to the latest estimates, one-fourth of all fossil fuel–derived CO_2 emissions will remain in the atmosphere for several centuries, and complete removal could take as long as 30,000 to 35,000 years.[11] Put another way, the climate system is like a bathtub with a very large faucet and a very small drain.[12] Some emissions already in the atmosphere could endure longer than Stonehenge, Mozart's compositions, and perhaps even high-level nuclear waste.[13] Given the knowledge that emission trends are still increasing and existing concentrations will continue to warm the atmosphere for decades, no matter what humanity does now, mitigation, the rationale goes, is an exercise in futility. Money can be better spent on engineering our way out of a wickedly warmer world.

What's worse, advocates of geoengineering continue, is that making investments in mitigation or adaptation is no guarantee that societies will be resilient enough to withstand climate change. The Asian Development Bank warns that because of unique geographical characteristics, countries such as Indonesia, the Philippines, Thailand, and Vietnam are expected to lose 6.7% of combined gross domestic product (GDP) by 2100 ($86 billion) if climate trends continue as usual. This is more than twice the estimated rate for global GDP loss.[14] The study noted that no amount of adaptation could eliminate some of the risks that these countries face, such as flooding and rising sea levels. Another study, which used satellite imagery and historical records to create detailed projections of tropical cyclones, droughts, floods, landslides, and sea-level rise, concluded that all of the Philippines, the Mekong River Delta, all of Cambodia, North and Eastern Laos, and the islands of Sumatra and Java in Indonesia will be "especially vulnerable" to climate change, even with international investments in adaptive capacity.[15] Alarmingly, yet another study identified 10 nations with the highest number of people living in low coastal areas: China, India, Bangladesh, Vietnam, Indonesia, Japan, Egypt, United States, Thailand, and the Philippines.[16] With the exception of the United States and Japan, it is doubtful that the nations on this list possess the financial capacity to effectively deal with seawater inundation of major urban areas, let alone protect rural areas that are home to the most destitute population clusters.

The plights of two countries in particular—the Maldives and Bangladesh—illustrate the improbability that adaptation or mitigation measures can eliminate the risks of climate change. The Republic of Maldives could lose 80% of its land due to rising sea levels and has already started purchasing land in Sri Lanka for its "climate refugees."[17] Recent evidence that the West Antarctic ice shelf is melting faster than expected has added a new, dark twist to this story and is likely to result in an elevated flow of climate refugees.[18] As one official in the Maldives put it, "No amount of adaptation, no level of improved adaptive capacity, could save [us]."[19]

Similarly, if sea levels rise as predicted under these scenarios, practically no amount of adaptation or investment in resilience can save Bangladesh from social, environmental, and economic catastrophe. As a Bangladeshi government official told one of the authors (Sovacool), "The challenge Bangladesh now faces is to cope with changes in climate already happening every year. We are strengthening coastal embankments, yes, but the intensity of erosion and frequency of storms are also increasing and I feel like we are often in a race against time where time is running out. We have developed saline-tolerant rice varieties but the concentration of salinity is going up. We can't keep on producing crops when land is flooded and water salty; it's practically not possible at the moment. Adaptation has its limits."[20] If the situation worsens, or if adaptation investments are not able to keep pace with vulnerabilities and risks, Bangladesh may have to switch to retreat strategies such as forcibly relocating communities to higher ground. In an already densely populated country, this type of adaptation response may be a recipe for disaster.

These aren't the only two countries at extreme risk. Under the most alarming projections, the Arctic Ocean could be ice-free by as early as 2037,[21] and if the Greenland Ice Sheet melts, sea levels could rise a whopping 6 meters—enough to inundate almost all low-lying island states, as well as coastal areas from San Francisco and New York to Amsterdam and Tokyo. Climate scientist James Hansen and his colleagues see little hope for adaptation or mitigation if fossil fuel energy continues to dominate, warning that "burning all fossil fuels would produce a different, practically uninhabitable, planet."[22] Kevin Anderson, past director of the Tyndall Centre for Climate Change, worries that "a 4 degrees C [temperature rise] future is incompatible with an organized global community, is likely to be beyond 'adaptation,' is devastating to the majority of ecosystems, and has a high probability of not being stable."[23] As another climate scientist cautioned, if global temperatures rise by more than 4°C, "the limits for human adaptation are likely to be exceeded in many parts of the world, while the limits for adaptation for natural systems would largely be exceeded throughout the world."[24]

Despite the direness of these projections, the international community continues to plod along toward loose consensus on lax targets. With no expeditious solution in site, arguments that it is too late to adequately mitigate are gaining traction and attention.

Technologies Can Substitute for Ecosystem Services

Before you rush out to buy snorkels and fins in anticipation of sea-level rise, be aware that geoengineering sponsors point to a solution: advanced technology. The biosphere provides each of us with a plethora of goods and services, though we don't directly pay for them—including, for instance, protection from harmful cosmic influences, the recharging of aquifers, and breathable air—services that enable humanity and other species to flourish.[25] In an attempt to roughly quantify environmental endowments in economic terms, ecological economist Robert Costanza and his colleagues infamously set out to estimate the economic value of 17 ecosystem services across 16 biomes in the 1990s. They concluded that, at the time of their study, the 17 services they evaluated were worth some $33 trillion per year, almost twice as much as annual gross world product at that time.[26] A 2014 update to their analysis estimated that total global ecosystem services ranged between $125 and $145 trillion per year, meaning that they contribute "more than twice as much to human well-being as [does] global GDP."[27] In short, their findings—which, by the way, also ignored many environmental endowments—suggest that the "environment" is worth far more to humanity than our economy is.

Optimistic engineers and technologists, however, have reframed this argument to note that if one breaks down ecosystem services into critical provisioning services (food, timber, fiber, fuel and energy, freshwater), regulating services (air and water quality, climate, erosion, diseases, pests, natural hazards), cultural services (fulfilling spiritual, religious, and aesthetic needs), and supporting services (soil formation, photosynthesis, nutrient cycling), then each can be replaced or augmented by technology. Table 10.2 illustrates how pervasive technology can be in providing substitutes for the services provided by Mother Nature.[28]

Geoengineering Can Safely Address the Effects of Climate Change

It's not too much of a jump from the technocratic ideology put forth above to arrive at the contention that humans can design technologies to "solve" our climate change woes. We can take deliberate actions to manipulate the earth's environment and thereby offset the adverse consequences of climate change.[29]

Relevant geoengineering strategies include interventions to remove CO_2 from the earth's atmosphere ("carbon dioxide removal," or CDR) and actions to increase the earth's reflection of sunlight ("solar radiation management").[30] CDR strategies focus on manipulating natural processes to reconfigure how much CO_2 winds up in the atmosphere.[31] Solar radiation management focuses on how much solar energy reaches the planet's surface; it involves strategies that manipulate the radiation budget of the earth to ameliorate the main effects of GHGs. Put another way, as figure 10.1 illustrates, CDR deals with managing short-wave solar radiation, and solar radiation management deals with managing long-wave radiation.

Carbon dioxide removal options, sometimes referred to as "carbon cycle engineering" or "post-emission carbon management,"[32] incorporate strategies to reduce atmospheric CO_2 concentrations by manipulating biological processes. Two prominent strategic themes focus on increasing absorption rates in bodies of water and on land. Examples of specific initiatives include the industrial extraction and geological sequestration of CO_2 from ambient air, technologies to accelerate chemical weathering by carbonate or silicate rocks on land, and spreading of crushed silicate rocks over vast areas of land surface to enhance terrestrial absorption. Reforestation and afforestation are also CDR strategies; however, such initiatives are considered to be far slower than many of the other hi-tech solutions being put forth. In general, expedient solutions come at far greater risks because many of the biological interventions give rise to other environmental threats. For example, manipulating oceanic absorption rates will influence acidity levels, which, if altered over a short period of time,

Table 10.2. Ecosystem Services and Their Technological Substitutes and Augmentations

Category	Service	Product	Substitutes or augmentations
Provisioning services	Food	Crops	High-yield agriculture; precision agriculture; genetically modified (GM) crops
		Livestock	Cloning; breeding; artificial insemination; GM animals; fortified feeds; high-lysine feed
		Capture fisheries	Aquaculture; fish hatcheries; GM fish; crop-based feeds
		Wild foods	Agriculture
	Timber	Wood	High-yield tree crops; GM trees; aluminum, steel, and plastics
	Natural fiber	Cotton, silk, jute, flax, coir, hemp	Synthetic fibers; plastics
		Furs, skins	Synthetic fibers
	Fuel	Energy services	Fossil fuels; photovoltaics; higher-efficiency wind and solar; geothermal; nuclear; high-yield biofuel crops; cellulosic ethanol
	Transportation and work	Mobility	Bicycles; mechanized transport (i.e., trucks and cars); airplanes; tractors
	Genetic resources	Information	Polymerase chain reaction; gene banks; zoos and botanical gardens
	Biochemicals, medicines, pharmaceuticals	Medical care	Synthetic drugs and pharmaceuticals; GM strains; biofactories
	Fresh water	Water	Water purification and treatment; recycling and reuse technologies; desalination; water pricing and marketing; property rights for water

Regulating and supporting services	Air-quality regulation	Pollution control and regulation	Scrubbers, fabric filters, and electrostatic precipitators for traditional air pollutants; emissions trading
	Climate regulation at local, regional, and global levels	Temperature and weather control stability	Carbon sequestration on land, oceans, geological formations; conservation tillage; geoengineering; modification of land cover and albedo
	Water regulation	Water	Water purification and treatment; recycling and reuse technologies; desalination; water pricing and marketing; property rights for water
	Erosion regulation	Soil	No- or low-till agriculture; hydroponic cultivation; cover crops
	Water purification and waste treatment	Water	Chlorination; wastewater treatment; filtration; reduction in oxygen demand
	Disease regulation	Health care	Chlorination; drugs and pharmaceuticals; insecticides
	Pest regulation	Productivity	Insecticides; integrated pest management; GM crops
	Pollination	Fecundity	Managed pollination by nonnative/cultured pollinators (e.g., European honeybee in the USA); hand/mechanical pollination; electrostatic enhancement
	Natural hazard regulation	Resilience and disaster management	Artificial or restored wetlands and mangroves; dams, sea walls, levees, and dykes; concrete and steel houses
Cultural services	Spiritual/religious values	Entertainment and fulfilment	Photographs; movies; high-definition and holographic television; virtual reality
	Aesthetic values	Recreation and ecotourism	Constructed or augmented landscapes or ecosystems; artificial reefs; zoos and arboretums; photographs, movies, and videos; virtual tourism

Source: Modified from Indur M. Goklany, "Technological Substitution and Augmentation of Ecosystem Services," in *The Princeton Guide to Ecology,* ed. Simon A. Levin et al. (Princeton: Princeton University Press, 2009).

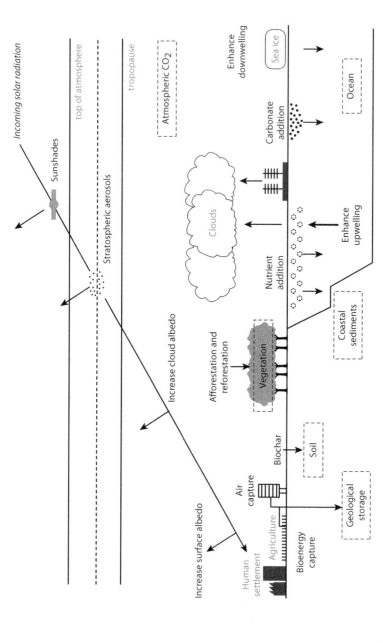

Figure 10.1. An Array of Geoengineering Options. *Source:* M. A. Brown and B. K. Sovacool, *Climate Change and Global Energy Security: Technology and Policy Options* (Cambridge, MA: MIT Press, 2011). *Note: Upward arrowhead* indicates short-wave radiation; *downward arrowhead*, enhancement of natural flows of carbon; *dashed box*, carbon store.

can severely compromise the health of marine habitats. Nevertheless, proponents of CDR strategies argue that these risks can be managed if properly planned.

Proponents of solar radiation management strategies argue that, theoretically, geoengineering can be achieved in different ways at different atmospheric levels. One cluster of technologies focuses on restricting the amount of solar energy that gets into the earth's atmosphere by employing technology that is meant to mimic sun shades. This includes the erection of solar shields or the construction of space-based reflecting mirrors.[33] Another cluster of technologies focuses on reducing incoming solar energy by injecting sulfate particles into the stratosphere. A third cluster of technologies would reflect incoming solar radiation back into space by increasing the albedo (reflectiveness) of cirrus clouds. One proposal even involves building floating machines on the ocean—like the one in figure 10.2—that can blow water vapor into the atmosphere to increase the cover of reflective bright clouds.[34] At the earth's surface, there are additional strategies for solar radiation management that include increasing the albedo of the earth's surface, such as by altering vegetation cover (e.g., by replacing dark forests with more reflective grasslands) and by constructing buildings that reflect rather than absorb solar energy (e.g., constructing white roofs and white roads).

Figure 10.2. A Rebuilt Searunner 34 Configured as an "Albedo Removal Boat." *Source:* Brian Launder and J. Michael T. Thompson, eds., *Geo-Engineering Climate Change: Environmental Necessity or Pandora's Box?* (Cambridge: Cambridge University Press, 2010), 170.

In sum, advocates argue that CDR and solar radiation management tech-niques need not threaten the entire planet. One interdisciplinary team of re-spected researchers holds that geoengineering, though seeming horrific to some, may become an inevitable human response to runaway climate change. Therefore, it needs to be publicly and systematically experimented with, the idea being that any risks can be managed through best practices of transpar-ency and proper process controls.[35] Small-scale experiments, the research team assures us, would probably have little to no impact on the environment. As the researchers conclude, a "laissez-faire approach is risky and imprudent . . . Our ignorance of the benefits and problems could become dangerous."[36] Two highly regarded reports from the US National Research Council echo such sentiments, calling for research programs to fill the gaps in our knowledge about geoengineering.[37]

Geoengineering Has Tangible Economic and Political Benefits
Advocates are quick to point out that geoengineering strategies, in aggregate, offer discernible financial benefits. They claim that geoengineering is magni-tudes of order cheaper than mitigation. Some estimates purport that a global geoengineering program can be implemented for only $100 billion, whereas single mitigation efforts can exceed $1 trillion. Carnegie Mellon engineering professor Granger Morgan suggests that geoengineering approaches are likely to cost only a fraction of a percent of GDP, whereas the cost of mitigation will be considerably higher—at least a few percent of GDP, maybe more.[38] A side benefit is that geoengineering research could deliver future technological in-novations that enhance the economic competitiveness of supporting nations. As congressman and former presidential candidate Newt Gingrich said in 2008, "Instead of penalizing ordinary Americans, we would have an option to address global warming by rewarding scientific invention [through geoengineering]," adding: "Bring on the American ingenuity."[39]

Advocates also point out that geoengineering is more acceptable among powerful stakeholder groups than are mitigation or adaptation efforts. The promise of geoengineering initiatives turns the politics of climate change up-side down because these initiatives offer prompt, massive benefits to powerful special interest groups at seemingly minor investment costs.[40] This is in sharp contrast to mitigation or adaptation strategies that are supported by disen-franchised stakeholders and contested by powerful special interest groups. In-deed, even Nobel Laureate Paul Crutzen stated that the lack of progress on mitigation justifies sustained research on geoengineering.[41] Political scientist D. M. Kahan and colleagues also found, after surveying the perceptions of 3,000 participants in a study of the United Kingdom and United States, that geoengineering had the potential to counteract the political polarization over climate change, noting that people expressed more support for and open-mindedness about geoengineering than mitigation or adaptation.[42]

Unlike mitigation options that require cooperation from all major GHG-emitting nations, geoengineering can be undertaken unilaterally by nation-states, corporations, or even wealthy individuals. As Victor et al. state, "It is cheap, easy, and takes only one government with sufficient hubris or desperation to set it [geoengineering practices] in motion."[43] The United States could choose to fund such a solution to save Greenland's ice cap in order to prevent flooding in Florida, or China could choose to fund such a solution to preserve Himalayan glaciers in Bhutan, Nepal, and Tibet to secure freshwater supplies.

Lastly, proponents of geoengineering argue that these technologies involve a measure of symmetry and elegance. Human beings have engineered the problem of climate change through the misuse of technology, which caused excessive GHG concentrations in the atmosphere. If we got into this mess through technology, we should be able to get out of this mess with technology. As the National Academy of Sciences put it, "It is important to recognize that we are at present involved in a large project of inadvertent 'geo-engineering' by altering atmospheric chemistry, and it does not seem inappropriate to inquire if there are countermeasures that might be implemented to address the adverse impacts . . . Our current inadvertent project in 'geo-engineering' involves great uncertainty and great risk. Engineered countermeasures need to be evaluated."[44] A medical corollary would be treating a disease with engineered pharmaceuticals. At the very least, as Michael Oppenheimer explained, having geoengineering available is akin to an "emergency backup" in case mitigation or adaptation efforts falter.[45]

The Other Side: Geoengineering Is an Unnecessary Risk

Opponents of geoengineering respond in kind with their own compelling set of objections. A host of scholars and analysts have identified what they see as serious, ineluctable drawbacks to the deployment of geoengineering in practice. First and foremost on the list of objections is that we do not possess a sufficient understanding of the complex adaptive system that underpins biosphere health. Furthermore, geoengineering presents a set of unavoidable and sobering challenges and risks.

Human-Made Systems Are a Poor Replacement for Natural Ones

Our biosphere is a complex adaptive system: its capacity to successfully self-regulate is predicated on balancing a huge number of influential factors that are interrelated in ways that science is still far too immature to understand. Poor technology was not the source of the failure of Biosphere II. Rather, the Biosphere II project failed because the humans who created the technology were not capable of comprehensively modeling all the critical interconnections necessary for success. Accordingly, opponents of geoengineering would be quick to point out that if we are not even capable of keeping eight people alive in a simulated biosphere for more than two years, is it not the height of folly to invest in technological fixes for climate change—a problem of far greater scale and scope?

Critics argue that geoengineering is a linear solution when systematic thinking is required. David Keith and Mike Hulme write that "deliberately adding one pollutant to temporarily counter another is a brutally ugly technical fix,"[46] and even then, there is no guarantee that the fix will work. Nature's complex cycles—hydrological, carbon, phosphorus, and nitrogen cycles, to name a few—have evolved over billions of years to create a series of interconnected functions that keep our biosphere in precarious balance. Humanity has now upset this balance, and geoengineering advocates propose technological corrections that could all but eviscerate it. Clive Hamilton highlights the arrogance of this approach, writing that "for sheer audacity, no plan by humans exceeds the one now being hatched to take control of the Earth's climate."[47]

Geoengineering Presents Large-Scale, Possibly Irreversible, Technical Challenges

Opponents contend that geoengineering and, in particular, solar radiation management strategies present a set of unique, potentially catastrophic risks. For starters, geoengineering could interfere with our ability to fully exploit renewable forms of energy. Solar radiation management is predicated on reducing the amount of incoming solar energy—energy that could be harvested for generating renewable power. Even the diffusion of light by injecting sulfur particles into the stratosphere would severely undermine solar power production.[48] Changes in incoming solar radiation patterns would also affect wind patterns and thus adversely affect the potential for producing wind power and wave power—to say nothing of possible deleterious effects on forests and the growth of plants.

Another risk associated with solar radiation management is that the evaporative cycle will be adversely affected by less sunlight reaching the earth's surface. Less evaporation means less rain, and less rain means less water on a planet that is already plagued by water scarcity. Such methods also do little to manage atmospheric CO_2 and therefore fail to address other climate change risks such as progressive ocean acidification.

Carbon dioxide removal techniques have their own challenges as well. They have extremely long timeframes and operate on large, geographical scales, and they would require almost perfect operation. To work effectively, any carbon removed from the atmosphere must remain sequestered from the sea or atmosphere for centuries and, ideally, for millennia. Moreover, such projects would have to be global in scale. As one group of international scholars in favor of ocean fertilization, a type of geoengineering, mused, "Local or short-term sequestration is irrelevant."[49]

The irreversibility of a CDR solution—once begun, it will have to be perpetually supported to keep working—may be a significant weakness, since it will require continual and costly maintenance.[50] As Victor argues, "Once the process of geoengineering begins—whether unilateral or collective—it is likely the world will be unable to stop."[51] It is, sort of, the technological cousin of opening a jar of macadamia nuts—expensive and difficult to stop consuming.

Since so little about geoengineering is currently known, we could find ourselves cast out of the frying pan and into the fire. It is an option where the unknown perils from tinkering with the climatic system could be even worse than the more predictable effects of climate change.[52] This challenge is further compounded when complicated geoengineering systems, with entirely different sets of components, interact in serial. As environmentalist James Lovelock cautions, "Before we start geo-engineering we have to raise the following question: are we sufficiently talented to take on what might become the onerous permanent task of keeping the Earth in homeostasis? Consider what might happen if we start by using a stratospheric aerosol to ameliorate global heating; even if it succeeds, it would not be long before we face the additional problem of ocean acidification. This would need another medicine, and so on. We could find ourselves enslaved in a Kafka-like world from which there is no Escape."[53] Geoengineering, under this logic, could create conditions wherein humanity spends generations inventing and implementing risky technological solutions only to maintain the artificial balance that the initial solution engendered.

And what if geoengineering efforts completely misfire? John Cullen, an oceanographer at Dalhousie University, notes that "history is full of examples of ecological manipulations that backfired."[54] One study noted that failure to sustain a partially implemented geoengineering project could lead to climate change at a pace 20 times more severe.[55] For perhaps this reason, the signatories to the 2010 UN Convention on Biodiversity agreed, for all intents and purposes, to "outlaw" geoengineering until "there is an adequate scientific basis on which to justify such activities and appropriate consideration of the associated risks."[56] Consider the significance of this international agreement: more or less the same group of nations that have been locked in ideological contestation over how to mitigate climate change have found common ground over the need to be wary of geoengineering.

Geoengineering Carries Immense Political and Economic Risks

Geoengineering technology, critics argue, is not ready for commercialization, and initiatives have the potential for distracting the international community from enacting lasting, sustainable change and could result in international conflict or even war. Some have argued that merely talking about geoengineering options erodes collaborations aimed at adaptation and mitigation.

Many experts harbor extreme doubts about the current state of geoengineering technology. The US National Research Council warned in 2015 that CDR strategies are "currently limited by cost and at present cannot achieve the desired result of removing climatically important amounts."[57] It also cautioned that after reviewing the risks and rewards of state-of-the-art solar radiation management techniques, it remained opposed to deploying them.[58]

Although we asserted earlier that countries could, in theory, unilaterally implement geoengineering systems, this does not absolve such nations from geopolitical reproach. Gwynne Dwyer, in *Climate Wars*, describes a scenario where poor countries begin an aerosol seeding option and then a war stops the application.[59] The result is cataclysmic. The range of threats associated with geoengineering includes human error (if a system fails, rapid warming and other climate changes are possible) and weaponization (in the hands of terrorists, geoengineering systems could be powerful weapons). As such, geoengineering conjures up the specter of ecological imperialism because of the potential to undertake unilateral actions that could have repercussions across the globe. Failed geoengineering initiatives could, in the extreme, "lead to death on a scale that makes all previous wars, famines and disasters small."[60]

It need not be nation-states or terrorists that launch geoengineering "attacks"; even individuals could do so. For example, to date, the world's largest geoengineering experiment was done in violation of the UN moratorium. In 2012, an American businessmen known for tomfoolery—his previous efforts to create large-scale commercial dumps in the Galápagos and Canary Islands led to his vessels being banned from Spanish and Ecuadorian waters—hoped to profit from the generation of carbon credits associated with iron fertilization, a type of CDR. He controversially dumped about 100 tons of iron sulfate into the Pacific Ocean off the western coast of Canada, hoping to create an artificial plankton bloom that would eventually sink carbon to the ocean floor.[61] The bloom quickly grew out of control and spread across 10,000 square kilometers, as detected by satellites. The businessman failed both to sequester substantial amounts of carbon and to generate any revenue from credits—but his actions underscore the ability of individuals to conduct large-scale uncontrolled experiments with little to no oversight.

In summary, opponents argue that geoengineering is unneeded and unnecessarily risky. To use an apt analogy, climate change is a bit like having somebody point a loaded gun at the metaphorical head of our civilization. Mitigation would be removing the bullets from the gun before it could do any harm. Adaptation would be moving the head so that when the gun fired, it wouldn't do as much damage and would merely nick a cheek. Geoengineering would be trying to build a helmet strong enough to withstand a continuous onslaught of bullets. Even if this were possible, opponents muse, one would need a lifetime supply of headache medicine.

Common Ground: We Need to Study Geoengineering but Restrain from Full Deployment

None of the large-scale geoengineering initiatives are economically viable at the present time. Most that could have a sizable impact on climate change are still at conceptual stages, and the economic challenges of implementing any of these initiatives on a global scale are still far too great for us to seriously contemplate. Research will continue, and some of the more attractive solutions will begin to receive greater political attention; however, actual implementation—at least, a coordinated effort by nation-states—is decades off, and by the time these solutions become economically viable, there may be no need for them.

This does not mean that research into geoengineering should be shelved. Our synthesis, shown in box 10.1, is to study but not to deploy, and to develop public dialogue about responsible use. Although there may be a limited number of scenarios where geoengineering is seen as acceptable, there is a growing consensus that geoengineering research needs to be initiated immediately. Great value can come from knowing earlier rather than later what works and what doesn't. Over the past decade, a handful of scientists have been patiently developing geoengineering options and considering the global policy issues surrounding their use. We may come to deploy some smaller-scale options in the near term, such as white roofs or biological carbon dioxide removal, which have minimal impact and less risk. By simultaneously pursuing research on possibilities, we can better understand the dynamics and implications of larger-scale deployments at a later date, without committing to them.

As our synthesis indicates, there is a strong rationale to continue to study the risks involved in geoengineering. Stakeholders of all stripes and colors should continue to discuss geoengineering, fleshing out the risks and debating when, if ever, initiatives should be deployed. This approach has been utilized with other types of potentially hazardous materials and practices, such as genetic research (the Human Genome Project), standards for genetically modified organisms (World Trade Organization), nuclear weapons and power (the Nuclear Non-Proliferation Treaty), and the continuing study of dangerous diseases such as smallpox (International Task Force for Disease Eradication and the World Health Organization). All of these efforts have encouraged "responsible use" rather than prohibiting use.

The type of transparency that would be necessary to implement our synthesis could help socialize a norm of accountability in geoengineering research

BOX 10.1

Geoengineering and the Common Ground

Should geoengineering be outlawed?

ONE SIDE: Engineers can design technological systems that will enable us to engineer a better climate.

OTHER SIDE: Natural ecosystems and the global climate are irreplaceable assets that are under threat of being irreversibly altered by thoughtless human interventions such as geoengineering.

COMMON GROUND: We should study the geoengineering options but keep all but the least disruptive techniques on the shelf.

and ensure that if humanity does decide to solve the climate problem with technology, then the tools deployed are the ones that have had the most stakeholder scrutiny. The knowledge gleaned might also be useful for guiding solutions to future problems in other environmental areas. As climate expert David Victor warned, making geoengineering "taboo" does not create a lasting deterrent.[62] An outright ban might ensure that only the least responsible governments or groups, or those with the weakest safety standards or governance ideals, control the fate of the technology. Geoengineering shouldn't necessarily be outlawed per se, but it should be placed on probation and carefully monitored.

NOTES

1. Amory B. Lovins, L. Hunter Lovins, and Paul Hawken, "A Road Map for Natural Capitalism," *Harvard Business Review*, May/June 1999, 145.

2. David G. Victor, "On the Regulation of Geoengineering," in *The Economics and Politics of Climate Change*, ed. Dieter Helm and Cameron Hepburn (Oxford: Oxford University Press, 2009), 325.

3. Thomas A. Heberlein, *Navigating Environmental Attitudes* (Oxford: Oxford University Press, 2012), 1–2.

4. Action Group on Erosion, Technology and Concentration (etc Group), "The Geoengineering Moratorium under the UN Convention on Biological Diversity" (November 10, 2010), available at www.etcgroup.org/sites/www.etcgroup.org/files/publication/pdf _file/ETCMoratorium_note101110.pdf.

5. David Keith and Mike Hulme, "Climate Science: Can Geoengineering Save the World?" *Guardian Sustainable Business Blog*, November 29, 2013, available at www.the guardian.com/sustainable-business/blog/climate-science-geoengineering-save-world.

6. M. A. Brown and B. K. Sovacool, *Climate Change and Global Energy Security: Technology and Policy Options* (Cambridge, MA: MIT Press, 2011).

7. W. M. Washington et al., "How Much Climate Change Can Be Avoided by Mitigation?" *Geophysical Research Letters* 36 (2009): L08703.

`8. Intergovernmental Panel on Climate Change (IPCC), "Summary for Policymakers," in *Climate Change 2013: The Physical Science Basis*, Contribution of Working Group I to the Fifth Assessment Report of the Intergovernmental Panel on Climate Change, ed. T. F. Stocker et al. (Cambridge: Cambridge University Press).

9. IPCC, *Greenhouse Gas Emissions Accelerate Despite Reduction Efforts: Many Pathways to Substantial Emissions Reductions Are Available* (Geneva: IPCC, April 13, 2014), available at www.ipcc.ch/pdf/ar5/pr_wg3/20140413_pr_pc_wg3_en.pdfIPCC 2014.

10. Data from table SPM3 in IPCC, "Summary for Policymakers."

11. See J. Hansen et al., "Target Atmospheric CO_2: Where Should Humanity Aim?" *Atmospheric Science Journal* 2 (2008): 217–231; D. Archer, "Fate of Fossil Fuel CO_2 in Geologic Time," *Journal of Geophysical Research* 110 (2005): 26–31.

12. David Victor et al., "The Geoengineering Option: A Last Resort against Global Warming?" *Foreign Affairs* 88 (2009): 65.

13. David Archer, *The Long Thaw* (Princeton: Princeton University Press, 2009).

14. Asian Development Bank (ADB), *The Economics of Climate Change in Southeast Asia: A Regional Review* (Manila, Philippines: ADB, April 2009).

15. Arief Anshory Yusuf and Herminia A. Francisco, *Climate Change Vulnerability Mapping for Southeast Asia* (Singapore: IDRC, January 2009).

16. Gordon McGranahan, Deborah Balk, and Bridget Anderson, "The Rising Tide: Assessing the Risks of Climate Change and Human Settlements in Low Elevation Coastal Zones," *Environment and Urbanization* 19, no. 1 (2007): 17–37.

17. Alex Smith, "Climate Refugees in Maldives Buy Land" (Tree Hugger, press release, November 16, 2008).

18. Michael E. Mann, "Defining Dangerous Anthropogenic Interference," *Proceedings of the National Academy of Sciences USA* 106, no. 11 (2009): 4065–4066.

19. Quoted in B. K. Sovacool, "Perceptions of Climate Change Risks and Resilient Island Planning in the Maldives," *Mitigation and Adaptation of Strategies for Global Change* 17, no. 7 (September 2012): 745.

20. Quoted in A. Rawlani and B. K. Sovacool, "Building Responsiveness to Climate Change through Community Based Adaptation in Bangladesh," *Mitigation and Adaptation Strategies for Global Change* 16, no. 8 (December 2011): 860.

21. "The Melting North: Special Report on the Arctic," *Economist*, June 16, 2012, 3–4.

22. James Hansen et al., *Climate Sensitivity, Sea Level, and Atmospheric CO_2* (New York: NASA Goddard Institute for Space Studies and Columbia University Earth Institute, 2013).

23. Anderson quoted in Paddy Manning, "Too Hot To Handle: Can We Afford a 4 Degree Rise?" *Sydney Morning Herald*, July 9, 2011.

24. Rachel Warren, "The Role of Interactions in a World Implementing Adaptation and Mitigation Solutions to Climate Change," *Philosophical Transactions of the Royal Society A* 369, no. 1934 (January 2011): 217–241.

25. See Thomas Prugh et al., *Natural Capital and Human Economic Survival* (New York; CRC Press, 1995), 19–47. See also Ann P. Kinzig, "Ecosystem Services," in *The Princeton*

Guide to Ecology, ed. Simon A. Levin et al. (Princeton: Princeton University Press, 2009), 573–578.

26. Robert Costanza et al., "The Value of the World's Ecosystem Services and Natural Capital," *Nature* 387 (May 15, 1997): 253–260.

27. Robert Costanza et al., "Changes in the Global Value of Ecosystem Services," *Global Environmental Change* 26 (May 2014): 152.

28. Indur M. Goklany, "Technological Substitution and Augmentation of Ecosystem Services," in Levin et al., *Princeton Guide to Ecology*.

29. Royal Society, *Geoengineering the Climate: Science, Governance, and Uncertainty* (London: Royal Society, September 2009).

30. Ottmar Edenhofer et al., eds., *IPCC Expert Meeting on Geoengineering*, Meeting Report (Geneva: IPCC, 2011).

31. D. W. Keith, "Geoengineering the Climate: History and Prospect," *Annual Review of Energy and Environment* 25 (2000): 245–284.

32. National Academy of Sciences, *Limiting the Magnitude of Future Climate Change* (Washington, DC: National Academies Press, 2010).

33. On solar shields, see Roger Angel, "Feasibility of Cooling the Earth with a Cloud of Small Spacecraft near the Inner Lagrange Point," *Proceedings of the National Academy of Sciences USA* 103, no. 46 (2006): 17,184–17,189.

34. Brian Launder and J. Michael T. Thompson, eds., *Geo-Engineering Climate Change: Environmental Necessity or Pandora's Box?* (Cambridge: Cambridge University Press, 2010), 149–180.

35. Jane C. S. Long, Frank Loy, and M. Granger Morgan, "Start Research on Climate Engineering," *Nature* 518 (February 5, 2015): 29–31.

36. Ibid., 30.

37. The two reports, both authored by the Committee on Geoengineering Climate: Technical Evaluation and Discussion of Impacts, the Board on Atmospheric Sciences and Climate, the Ocean Studies Board, and the Division on Earth and Life Studies, are *Climate Intervention: Carbon Dioxide Removal and Reliable Sequestration* and *Climate Intervention: Reflecting Sunlight to Cool Earth* (Washington, DC: National Research Council, 2015).

38. Granger Morgan, "Governance and Geoengineering: Who Decides and How" (panel discussion at the National Academies Workshop on Geoengineering Options to Respond to Climate Change: Steps to Establish a Research Agenda, Washington, DC, June 15, 2009).

39. Gingrich quoted in Clive Hamilton, "The Risks of Climate Engineering," *New York Times*, February 12, 2015.

40. Victor, "Regulation of Geoengineering," 325–339.

41. P. J. Crutzen, "Albedo Enhancement by Stratospheric Sulfur Injections: A Contribution to Resolve a Policy Dilemma?" *Climatic Change* 77 (2006): 211–219.

42. D. M. Kahan et al., "Geoengineering and Climate Change Polarization: Testing a Two-Channel Model of Science Communication," *Annals of the American Academy of Political and Social Science* 658, no. 1 (2015): 192.

43. Victor et al., "Geoengineering Option."

44. National Academy of Sciences, *Policy Implications of Greenhouse Warming: Mitigation, Adaptation, and the Science Base*, Panel on Policy Implications of Greenhouse Warming, Committee on Science, Engineering, and Public Policy (Washington, DC: National Academies Press, 1992), 433–464.

45. Michael Oppenheimer, "Concluding Remarks" (paper presented at National Academies Workshop on Geoengineering Options to Respond to Climate Change: Steps to Establish a Research Agenda, June 15, 2009).

46. Keith and Hulme, "Climate Science."

47. Clive Hamilton, *Earthmasters: The Dawn of the Age of Climate Engineering* (New Haven: Yale University Press, 2014), 1.

48. Dan Schrag, "Physical Science: Important Questions, State of Knowledge, and Major Uncertainties Related to Selected Geoengineering Options" (paper presented at National Academies Workshop on Geoengineering Options to Respond to Climate Change: Steps to Establish a Research Agenda, Washington, DC, June 15, 2009).

49. R. S. Lampitt et al., "Ocean Fertilization: A Potential Means of Geo-engineering?" in Launder and Thompson, *Geo-Engineering Climate Change*, 149–180.

50. Victor et al., "Geoengineering Option," 64–76.

51. Victor, "Regulation of Geoengineering," 339.

52. Ibid., 325–339.

53. James Lovelock, "A Geophysiologist's Thoughts on Geo-engineering," in Launder and Thompson, *Geo-Engineering Climate Change*, 85.

54. Cullen quoted in Martin Lukacs, "World's Biggest Geoengineering Experiment 'Violates' UN rules," *Guardian* (London), October 15, 2012.

55. H. D. Matthews and K. Caldeira, "Transient Climate-Carbon Simulations of Planetary Geoengineering," *Proceedings of the National Academy of Sciences USA* 104, no. 24 (2007): 9949–9954.

56. Fred Pearce, "What the UN Ban on Geoengineering Really Means," *New Scientist*, November 1, 2010.

57. Committee on Geoengineering Climate et al., *Climate Intervention: Carbon Dioxide Removal*.

58. Committee on Geoengineering Climate et al., *Climate Intervention: Reflecting Sunlight*.

59. Gwynne Dwyer, *Climate Wars: The Fight for Survival as the World Overheats* (New York: Oneworld Publications, 2010).

60. Lovelock, "Geophysiologist's Thoughts," 91.

61. Lukacs, "World's Biggest Geoengineering Experiment."

62. Victor, "Regulation of Geoengineering," 325–339.

Is Clean Coal an Oxymoron?

Concerned about coal's negative image, during the holiday season in 2008, the American Coalition for Clean Coal Electricity sponsored a multimillion-dollar advertising campaign designed to coopt Christmas in the name of commerce. The campaign featured singing lumps of coal called "The Clean Coal Carolers" merrily crooning the benefits of clean coal.[1] One of their songs, to the tune of "Frosty the Snowman," went like this:

> Frosty the Coalman is a jolly happy soul.
> He's abundant here in America, and he helps our economy roll.
>
> Frosty the Coalman's getting cleaner every day.
> He's affordable and adorable and helps workers keep their pay.
>
> There must have been some magic in clean coal technology,
> For when they looked for pollutants there were nearly none to see.

Not to be outdone, the Natural Resources Defense Council, an environmental action group opposed to coal, was quick to react with its own song, to the tune of "Deck the Halls":

> Liquid coal is a big folly,
> Fa la la la la, la la la la.
> Gives a gift to old Peabody,
> Fa la la la la, la la la la.
>
> Some in Congress like to coddle,
> They must know it's a boondoggle.
>
> Coal for fuel is such a bummer,
> Might as well just buy a Hummer.
>
> Driving cars fueled by coal fire,
> Will make the ocean levels higher.
>
> Liquid coal is no solution,
> Doubles global warming pollution.

So we urge our Senators,
Don't give in to polluters.[2]

While this was cute and creative, one must ask why both sides of the "clean coal" debate decided to resort to operatic anthracite rather than enlightened discussion. Is there something hard to understand about making coal clean?

"Clean coal" is an umbrella term for processes and approaches that seek to minimize emissions of greenhouse gases (GHGs) and other pollutants from the use of coal for electricity and industry.[3] As defined by the US Environmental Protection Agency, clean coal "means any technology, including technologies applied at the pre-combustion, combustion or post-combustion stage, at a new or existing facility which will achieve significant reductions in air emissions . . . associated with the utilization of coal in the generation of electricity, process steam, or industrial products."[4] Advocates of this strategy point out that it is relatively easy for carbon dioxide to be sequestered from power plants and other emitting industries. Moreover, they highlight that clean coal offers a major technological "fix" for the problem of climate change, since it has the potential to stop atmospheric GHG concentrations from increasing.

Another side holds that "clean coal" is an oxymoron—that capture or storage technologies are not viable and that coal, no matter how you package it, is a scourge on the environment. These opponents argue that little research has been done to verify the capacity of clean coal solutions to resolve energy security problems in a safe and economically viable manner. Misplaced confidence in clean coal only exacerbates the problem: as the debate continues, so too do the unsustainable consumption practices that have given rise to climate change. Proponents of this view argue that the rule of thumb should be "keep coal in the ground" and avoid diverting funds to these fanciful schemes.

One Side: Clean Coal Is a Necessity

The arguments in favor of clean coal may be more alluring than readers realize. We present four arguments: (1) all energy systems cause some environmental damage, and best practices can minimize the damage coming from the coal sector; (2) use of fossil fuels and particularly coal will increase over the next few decades; (3) carbon capture and storage (CCS) technology is a viable way to sink and store the (inevitable) emissions that come from this increased use; and (4) clean coal is therefore an economic part of the process of global decarbonization.

There Is No Free Lunch in Energy Systems, and Coal Mining Can Be Done Sustainably

Advocates of clean coal are quick to point out that every kilowatt-hour of conventional electricity generated, barrel of oil produced, ton of uranium mined, and cubic foot of natural gas manufactured produces a laundry list of environmental woes. Depending on the technology, these problems may include radioactive waste, toxins associated with abandoned uranium mines and mills, acid rain (and the ensuing damage to fisheries and crops), water pollution, groundwater depletion, pollution by particulate matter, and degradation of ecosystems (with threats to biodiversity due to species loss and habitat destruction). Energy scientists Robert J. Budnitz and John Holdren pointed out almost 40 years ago that "no existing or proposed energy technology is so free of environmental liabilities as to resolve satisfactorily the central dilemma between energy's role in creating and enhancing prosperity and its role in undermining it through environmental and social impacts."[5] As table 11.1 indicates, coal does have its environmental pros and cons, but so do oil, natural gas, hydroelectric dams, and even wind turbines and solar panels. So if environmental damage is inevitable, the argument runs, why not try to mitigate the damages associated with an energy system such as coal that is already cheap and well-established?

Table 11.1. Environmental Sustainability Attributes of Major Energy Systems

Technology	Pro	Con
Oil	Higher energy density, lower emissions of some pollutants compared with natural gas or biofuel	Major source of greenhouse gas (GHG) emissions; risk of damaging spills
Natural gas	Burns more cleanly than coal or oil; can be a bridge technology to renewables	Methane more potent than carbon dioxide; prone to leaks and accidents
Coal	More efficient than other feedstocks such as natural gas or electricity in heavy manufacturing and steel making	Key threat to global warming; huge source of health problems
Hydroelectric dams	No direct combustion; base-load power option; minimal impact on air pollution	GHG emissions from reservoirs and deforestation; significant impact on rivers and fisheries
Wind energy	No direct emissions; minimal amounts of water needed to generate electricity	Avian mortality; land clearing for remote wind farms
Solar energy	No direct emissions; can be integrated into buildings and harnessed in a decentralized way	Manufacturing involves toxic components; needs large areas of land
Nuclear power	Negligible direct GHG emissions; can provide base-load power in large increments with minimal land	Generates radioactive waste; can cause severe damage during accidents

Source: Modified from B. K. Sovacool and H. Saunders. "Competing Policy Packages and the Complexity of Energy Security," *Energy* 67 (April 2014): 641–651.

Clean coal sponsors also mention that best practices can greatly mitigate any damage that might occur during the mining, transport, and combustion of coal. For instance, control methods exist to minimize coal miners' exposure to mine dust, the primary contributor to coal workers' pneumoconiosis, commonly known as black lung disease. These methods include applying water to lower the amount of dust from roads, using brushes to clean conveyor belts, and improving air ventilation and circulation.[6] Furthermore, methane control and prediction systems can be implemented to monitor fugitive emissions and enable the recovery or capture of coalbed methane—which enhances safety, reduces leakage, and provides a revenue source from the relatively clean natural gas by-product.[7] When mines are closed, "integrated closure planning" schemes can rehabilitate and remediate degraded landscapes so that they have productive, second uses.[8] In Australia, such "alternative landform design" practices have already been shown to result in improved hydrology, physical stability, geochemistry, and vegetation patterns sufficient to restore ecosystems to functional health.[9] Downstream at power plants, the installation of flue gas scrubbing systems, baghouse filters, electrostatic precipitators, flue gas desulfurization systems, carbon injection systems to absorb mercury, and selective catalytic reduction equipment can mitigate as much as 90% of the emission of harmful air pollutants. In addition, potentially hazardous releases of chemical pollutants into the air and water can be tracked, with adverse information released to local communities so that they can minimize their exposure.[10]

Fossil Fuels Are Firmly Embedded in Economies and in Future Emissions Trajectories

A second compelling argument in favor of clean coal is that all projections suggest that coal will continue to play a dominant role in electricity generation in the next few decades, regardless of international efforts to mitigate climate change. Therefore, clean coal technology is a prudent response to an economic reality. A study in 2014 estimated that the existing worldwide infrastructure of fossil fuel power plants will emit more than 300 billion tons of carbon dioxide over their expected lifetimes. These "committed" emissions are not shrinking; they are growing by about 4% per year.[11] Similarly, a country-by-country analysis conducted by the International Energy Agency (IEA) indicated that 80% of the total CO_2 emissions projected to result in atmospheric GHG concentrations rising to 450 ppm is "locked in" by existing investment in capital stock.[12] Coal and nuclear power plants typically operate for more than 50 years, while natural gas plants, oil refineries, and chemical and cement plants operate for decades.[13] If climate change is a runaway train unlikely to be stopped by renewables or other mitigation options, then clean coal is the only brake that those in control are willing to pull.

As an indication of technological lock-in in practice, the global investment community continues to channel more money to new fossil fuel infrastructure development each year—today and over the past decade—than to clean renewable energy and low-carbon infrastructure. As figure 11.1 shows, in 2013, investors directed about $250 billion to all forms of renewable energy but sank $950 billion into new coal, oil, and gas infrastructure investments. In the United States alone, annual oil and gas investment soared to $200 billion in 2013, amounting to 20% of total private fixed investment. This matched the volume of investment in home building—a "first" in the country's history.[14] The IEA reports that global investment in fossil fuel initiatives doubled in real terms from 2000 to 2008.[15] These represent investments that are now in place and aren't going anywhere in the near future.

Coal is a firmly entrenched part of the existing and future energy system. In 2013, the world was home to about 24,000 coal mines, and the coal sector provided approximately seven million full-time jobs.[16] Moreover, the use of coal is expected to grow faster than that of other fossil fuels in the near term. For instance, some major economies such as South Africa, the United Kingdom, Poland, and Australia rely on coal to generate 90% or more of their electricity, and the rapidly developing economies of China and India depend on coal for 50% or more of electricity.[17] In South Africa, planners put such high stock in coal-fired electricity plants that they decided to name a 4,800 MW facility under construction "Medupi," which literally means "rain that soaks parched lands, giving prosperity."[18]

Despite global alarm over GHG emissions and climate change, the IEA forecasts that coal could surpass oil as the world's primary source of energy in only a few years.[19] As Maria van der Hoeven, executive director of the IEA, put it when interpreting these data, "Thanks to abundant supplies and insatiable demand for power from emerging markets, coal met nearly half of the rise in global energy demand during the first decade of the 21st Century . . . In fact, the world will burn around 1.2 billion more tonnes of coal per year by 2017 compared to today—equivalent to the current coal consumption of Russia and the United States combined. Coal's share of the global energy mix continues to grow each year, and if no changes are made to current policies, coal will catch oil within a decade."[20] For instance, there are approximately 1,200 coal-fired power plants, constituting 1,400 GW of capacity, in various stages of planning and construction around the world, with significant growth expected in China, India, Russia, and the United States.[21] Coal, in other words, is still flourishing—fa la la la la, la la la la.

Coal use could even surpass expectations if it starts to make inroads as a substitute for oil in the transportation sector. The Fischer-Tropsch process is a gas-to-liquid conversion technique that uses a catalyzed chemical reaction to

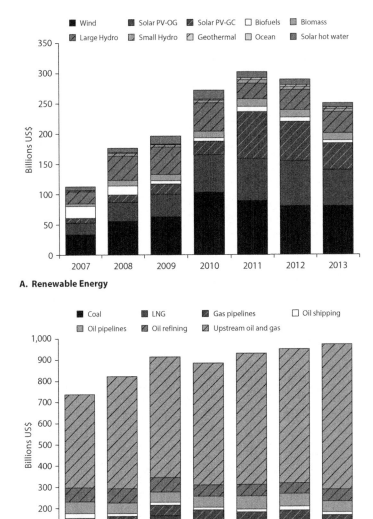

A. Renewable Energy

B. Fossil Fuels

Figure 11.1. Annual Global Investment Rates in Selected Energy Systems, 2007–2013. *Sources: A*, Renewable Energy Policy Network for the 21st Century (REN21), *Renewables 2011: Global Status Report; Renewables 2012: Global Status Report; Renewables Global Status Report 2013;* and *Renewables Global Status Report 2014* (Paris: REN21 Secretariat, 2011, 2012, 2013, and 2014). *B*, Ambrose Evans-Pritchard, "Oil and Gas Investment in the US Has Soared to $200b," *Daily Telegraph* (London), July 10, 2014. *Note:* Biomass includes waste-to-energy, excludes off-grid fuelwood consumption. GC = grid-connected; LNG = liquefied natural gas; OG = off-grid; PV = photovoltaic.

heat coal and transform the resulting carbon monoxide and hydrogen into a colorless, odorless fuel that is interchangeable with diesel fuels. Syntroleum Corporation has used the process to produce more than 400,000 gallons of diesel and jet fuel at a demonstration plant near Tulsa, Oklahoma. One assessment found that Fischer-Tropsch fuel facilities, if expanded, could easily reach production levels of three million barrels per day in the United States, if proper investments were made.[22] Another study noted that "coal-to-liquid (CTL) fuels are attractive due to the vast reserves (over 200 billion short tons of the demonstrated reserve coal base) available for mining"; "the prices for coal are more stable than the prices for natural gas or oil"; and "a transition to a coal economy is desirable to reduce the dependence on foreign oil supplies."[23]

CCS Is Technically Viable

To make the transition to a clean coal economy possible, large volumes of carbon will have to be captured and sequestered. Proponents of clean coal argue that the technical capacity for this currently exists, using the commercially available or technically feasible options summarized in table 11.2. CO_2 is already routinely captured as a by-product of ammonia and hydrogen production and from limestone calcinations. Moreover, engineers have been experimenting with pre-combustion capture, post-combustion capture, and oxyfuel combustion for decades.[24] Four large-scale CCS projects operated around the world in 2008 (in Algeria, in Norway, and (two) in the United States), each capturing approximately one million tons of CO_2 annually.[25] The first three of these projects involved the extraction of CO_2 from natural gas, which often has a higher CO_2 content than is allowed to enter natural gas distribution networks. The fourth project (in North Dakota) involved CO_2 capture from the production of synthetic natural gas from coal. As of late 2012, the number of large-scale industrial facilities operational or under construction exceeded 75 in disparate locations including Australia, China, Canada, Germany, the Netherlands, Poland, and the United Kingdom. Sixteen of these projects are big enough to capture 36 million tons of CO_2 per year.[26]

Three prominent methods for sequestering captured CO_2 are receiving the bulk of attention: geological storage, marine storage, and terrestrial/biological storage. Geological storage strategies attempt to take advantage of natural geological storage chambers to safely sequester CO_2. The technology already exists and has been proven in practice because, for decades, oil companies have been pumping CO_2 back into oil wells to enhance recovery of oil. Promising geological storage strategies include trapping and storing CO_2 in depleted oil and gas reservoirs, saline formations, and commercially unviable coal seams. Marine storage involves injecting CO_2 into ocean depths where it liquefies under pressure or becomes sequestered in mineral carbonates. Advocates of

Table 11.2. Commercially Available and Technically Feasible Carbon Capture
and Storage Options

Process	Commercially available on the market today	Technically feasible but not yet commercially available
Carbon capture	Chilled ammonia capture process Amine scrubbing	Post-combustion capture Oxyfuel combustion Pre-combustion capture
Geological storage	CO_2 injection with oil or methane recovery Geological monitoring and modeling methods for CO_2 transport	Saline formations Deep-seam coal beds
Terrestrial sequestration	Cropland, forestland, and grazing management with advanced information technologies	Genetic engineering to enhance biological carbon uptake

Source: Modified from M. A. Brown and B. K. Sovacool, *Climate Change and Global Energy Security: Technology and Policy Options* (Cambridge, MA: MIT Press, 2011).

these approaches contend that technologies for processing, compressing, and transporting CO_2, as well as techniques for subsurface reservoir engineering, can be readily adapted from the petroleum industry.[27]

The final sequestration strategy—terrestrial or biological sequestration—involves the conversion of atmospheric CO_2 to carbon and oxygen through absorption and photosynthesis.[28] Compared with geological or marine storage, terrestrial sequestration techniques are far less environmentally invasive and tend to carry far less risk of release. Many technologies and practices that sequester carbon in this manner, such as improved soil conservation or sustainable forest management, have already been widely practiced and proven.

Clean Coal Enables Decarbonization without Economic Collapse or Behavioral Change

Advocates of clean coal save perhaps their most compelling argument for last. They claim that given the link between fossil fuel combustion and climate change and the global projections that fossil fuel consumption will not end soon, if society does not develop this technology, humanity will simply be unable to sustain standards of life as we know it in industrialized countries. According to this stream of logic, society must undergo a process of decarbonization.[29] As researchers noted in 2014, to avert economic, social, and environmental disaster, global society will need to achieve 95% reductions based on year 2000 GHG emissions by 2050, thereafter entering the phases of "carbon neutrality" and then "carbon dioxide removal," as illustrated in figure 11.2. That

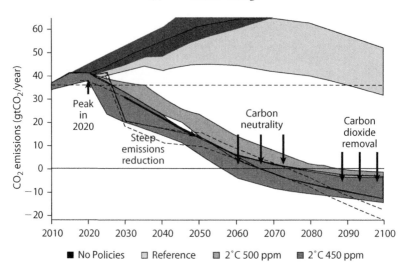

Figure 11.2. Phases of Global Decarbonization, 2010–2100. *Sources:* Modified from Elmar Kriegler et al., "What Does the 2C Target Imply for a Global Climate Agreement in 2020? The Limits Study on Durban Platform Scenarios," *Climate Change Economics* 4, no. 4 (2013): 1340008; and M. Tavoni et al., LIMITS Consortium, "Limiting Global Warming to 2°C: Policy Findings from Durban Platform Scenario Analyses" (LIMITS Synthesis Report), available at www.feem-project.net/limits. Figure courtesy of Elmar Kriegler.

is, even after very deep cuts in emissions, we will need to reach net-negative emissions—that is, a state where society stores or sinks more carbon than it emits. CCS is a key technology to enable this final phase of decarbonization.

This finding in support of the necessity of CCS even holds true for societies already undertaking low-carbon energy transitions. The IEA notes that in the Nordic region, a part of Europe with some of the world's most proactive climate and energy policies (Denmark, for example, is aiming to be fossil fuel–free by 2050), achieving GHG emission targets will be impossible without CCS. If these Nordic countries expect to retain any of their heavy industries, they will need to successfully equip 50% of cement plants and 30% of iron, steel, and chemical plants with CCS by 2050. As the IEA and Nordic Energy Research report puts it, "CCS represents the most important option among new technologies for reducing industrial CO_2 emissions after 2030."[30]

The Other Side: Coal Can Never Be Clean

Opponents contend that clean coal is so fraught with risks and liabilities that it should be avoided entirely. They point out that in a world where some industries exist solely for the purpose of cleaning up after failed technology, there is little likelihood of pioneering systems that could ever safely capture and store carbon for thousands or hundreds of thousands of years. As biologist E. O. Wilson once calculated, if every person in the world were to reach present levels of consumption in the United States, with the best available, most efficient present-day technology, we would need four more planet Earths to handle their waste streams.[31] In short, we are expecting a lot from our technologies. We have had hundreds of thousands of years to perfect human tools and socio-technical systems, but nature has had billions of years in which to foster the evolution of its interconnected ecosystems, which exist in precarious balance. The phrase "clean coal" is an oxymoron because it attempts to reengineer this delicate balance at multiple points and, at best, presents an unproven tool that faces huge technical and social implementation challenges. As Page et al. conclude, "CCS is not presently a near-term measure for mitigating greenhouse gas emissions . . . In light of the tension between the current status of CCS and the need for rapid and deep emissions contractions . . . the value of further investment in CCS must be seriously questioned."[32] We consider here three arguments that cast doubt on the viability of CCS.

CCS Does Not Negate Coal's Massive Social and Environmental Impacts

Perhaps the most convincing rebuke of clean coal and CCS is that the technologies do little to minimize the social and environmental damages from coal during other stages of its life cycle. These include coal mining pollution, occupational hazards to coal miners, and air pollution (other than CO_2).

One of the most destructive aspects of coal's life cycle occurs during mining, especially mountaintop removal. Of the more than one billion tons of coal mined in the United States annually, roughly 30% comes from mountaintop removal. The overall process has destroyed ecosystems, blighted landscapes, and diminished the water quality of rural communities. Tragically, one recent study noted that "there is, to date, no evidence to suggest that the extensive chemical and hydrologic alterations of streams by mountaintop mining with valley fill (MTVF) can be offset or reversed by currently required reclamation and mitigation practices."[33] Another survey of 78 MTVF streams in one region found that 73 of them had water pollution levels far exceeding the threshold for toxic bioaccumulation.[34] As one of the authors of that study stated, "The scientific evidence of the severe environmental and human impacts from mountaintop mining is strong and irrefutable. Its impacts are pervasive and long lasting and there is no evidence that any mitigation practices successfully reverse the damage it causes."[35] A separate assessment from a team of economists estimated that the life cycle impacts of coal and the waste stream it generates cost the US public "a third to over one half a trillion dollars annually," meaning that "accounting for the damages conservatively doubles to triples the price of electricity from coal per kWh generated."[36]

Notwithstanding more rigid safety regulations and improved medical treatment, underground coal mining is still a very dangerous activity. Miners frequently encounter pockets of highly volatile underground methane gas, which is produced in significant quantities when coal is removed. Shifting and unpredictable geological conditions destabilize the roofs of mines. And miners face the ever-present risks of flooding and fire, hazards that have increased in recent years as miners cultivate deeper seams.[37]

Statistics highlight the perils of coal mining. More than 3,300 accidents in Chinese coal mines led to 5,938 deaths in 2005, followed by 4,746 mining deaths in 2006, with an additional 163 deaths per year resulting from coal-related black lung disease—currently suffered by an estimated 800,000 Chinese coal miners.[38] This is not just an issue in the developing world. Though the numbers are not nearly as large, more than 700 coal miners died of work-related accidents in the United States over the past few decades,[39] and fatal accidents occur every few years in large coal-producing countries such as Australia.[40]

Even though the occupational hazards associated with mining are well known, and mitigation strategies for coalbed methane leaks have been understood for two centuries now, studies confirm that as many as 12% of all coal miners develop one of several possibly fatal diseases over the course of their work, ranging from pneumoconiosis, progressive massive fibrosis, and emphysema to chronic bronchitis and accelerated loss of lung function.[41] Advocates for the industry strive to ensure that these deaths are treated as a "cost of doing business" rather than as a stimulus for social change, even in the face of major tragedies.[42]

Mining's social impacts can be almost as grave. One assessment linked coal mine sites to communities plagued by corruption, civil unrest, and lack of participation in civil society.[43] The study warned that nearly one-quarter of active mines and exploration sites are located in countries with the weakest governance structures. These regimes also have the least stringent environmental controls and rarely regulate or prohibit dumping and disposing of mine wastes. Overall, more than one-quarter of the world's active mines overlap with or are within a 10-kilometer radius of strictly protected areas or ecosystems of high conservation value, about one-third of all active mines are located in stressed watersheds, and one-fifth of active mines and exploration sites are in areas of high or very high seismic hazard.

Indeed, when one attempts to internalize these externalities into energy prices, some economists calculate that coal is by far the dirtiest—the most polluting, the most damaging—of all energy sources, even compared with oil or nuclear power. Thomas Sundqvist and Patrik Soderholm found that when av-

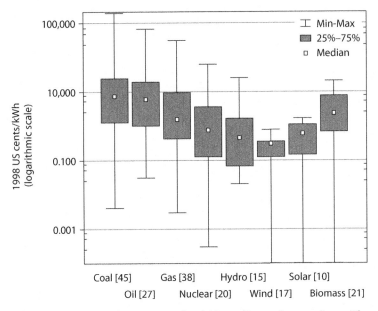

Figure 11.3. Negative Externalities Associated with Various Energy Systems. *Source:* Thomas Sundqvist, "What Causes the Disparity of Electricity Externality Estimates?" *Energy Policy* 32 (2004): 1753–1766. *Note:* Numbers in brackets on the *x*-axis refer to number of estimates included for each energy system.

eraged across studies and converted into a price per unit of electricity, externalities for wind energy would add a meager 0.29 US cents/kWh to the price of electricity, but coal's true price would be 14.87 US cents/kWh higher.[44] Figure 11.3 provides a full breakdown of Sundqvist and Soderholm's estimates. Other respected studies have verified that coal is king of the externalities.[45] Admittedly, clean coal technologies could help to mitigate these external costs, but the technology comes at a price and, as we will see, is still commercially unproven.

Best practices and pollution controls do already exist, but these are frequently ignored or rejected on the grounds that they are too expensive. In Russia, electrostatic precipitators, which reduce particulate matter emissions, are installed at only 39% of thermal units; Venturi scrubbers are installed at less than half of units; and multi-cyclones are installed at 10% of units.[46] In the United States, the "grandfathering" of old coal-fired power plants under the Clean Air Act amendments means that even in 2014, 80% of coal-fired power plants are projected to fall short of minimum pollution requirements for nitrogen oxide and/or sulfur dioxide.[47] In sum, opponents hold that the added cost of clean coal would encourage coal users to adopt shortcuts and obfuscate rules designed to keep coal clean. After all, coal is a dirty business.

Carbon Capture Is Not Yet Commercially Viable, and for Good Reason

Opponents also point out a key technological flaw associated with carbon capture that renders it a less than appealing solution: it makes an already inefficient process even less efficient. The IEA notes that carbon capture requires a substantial amount of heat and involves complex processes such as amine solvent regeneration and flue gas pretreatments, all of which, when combined with the need for auxiliary power, blowers, pumps, and compressors, reduces the operating efficiency of a coal power plant by 8% to 10%.[48] According to the Intergovernmental Panel on Climate Change, widespread adoption of CCS could erase all energy efficiency gains made in coal technology over the past 50 years and increase coal consumption by one-third.[49] To compound the problem, with current technologies, carbon capture does not capture all CO_2, with up to 15% estimated to escape into the environment.[50]

The energy used in running a CCS system has been dubbed an energy penalty. The actual estimates are contentious. For example, one global study found that these energy penalties tend to be higher when calculated by independent researchers. In the study, industry projections estimated penalties to be in the range of 8% to 15.4%. In contrast, independent studies estimated energy penalties to range from a best theoretical case of 11% to an upper level of 40%.[51] There is less contention over coal-to-liquid conversion energy penalties. Even industry experts calculate the thermal efficiency of direct liquefaction of coal to be between 50% and 54%.[52]

Many of the efficiency challenges to CCS have been confirmed by considering operational performance at existing plants. While defending one energy utility's plan to avoid pursuing clean coal in the United States, an executive stated that "IGCC [integrated gasification combined cycle] is a promising technology, but is not yet viable on a large-scale commercial basis . . . There are only two IGCC units in operation today in the U.S.—both are small, were heavily subsidized, and actually have dirtier emissions profiles than the supercritical plants we have proposed. Further, both these plants continue to operate at low reliability levels more than five years after coming on line."[53]

Clean coal technology has been challenged on the basis of similar operating experiences. An assessment of the 15 IGCC power plants operating around the world since 1984 (without carbon sequestration) in Germany, Italy, Japan, the Netherlands, Singapore, Spain, and the United States found that all were prone to a host of reliability issues, including difficulties with maintenance, high sensitivity to the types of coal used, and frequent malfunctions in auxiliary systems. Given these problems, facilities, in aggregate, operated only 28.5% of the time, or 35,000 hours out of a potential 122,640 hours.[54] Consider what such delays would do to the financial profile of such a plant.

The implication here is that in the absence of substantive technological progress, if coal-fired power plants equipped with CCS use more energy, then there will be a corresponding increase in coal mining operations and all of the associated negative externalities discussed above. The Department of Trade and Industry in the United Kingdom echoed this concern in noting that power plants with CCS will "increase [GHG] emissions and air pollutants per unit of net delivered power and will increase all ecological, land-use, air-pollution, and water-pollution impacts from coal mining, transport, and processing, because the CCS system requires 25% more energy, thus 25% more coal combustion, than does a system without CCS."[55]

These technical and environmental problems could become even worse, not better, over time. Two experts looked at future projections of CCS costs and found that a common assumption that expenses would decrease over time due to learning and experience were wrong.[56] After surveying a large number of experts in the field, they found that costs were likely to go up due to "negative learning" and the risk of leakage. By their calculations, even an annual global leakage rate of 0.5%, excellent in terms of operational efficiency, would undermine climatic stability. To successfully address climate change, these imperfect systems would need to achieve near perfection.

Carbon Storage Has Its Own Serious Drawbacks

Lastly, opponents suggest that carbon sequestration is plagued by an assortment of technical, economic, and political difficulties. Foremost on the list of disadvantages is that, so far, geological storage has been successfully implemented only on a very small scale. The sites operating in Norway, Algeria, and the United States may sound as if they are capable of processing large amounts of carbon (one million tons per year), but, in reality, each project is capable of managing only the CO_2 produced by a small (150 MW) coal-fired power plant.[57] Based on these limited results, accepting an argument that CCS is capable of managing CO_2 on a global scale is akin to a person declaring himself capable of flying because he has successfully leapt one foot into the air.

One significant obstacle lies in transporting carbon after it has been captured. Opponents point to the logistical Mount Everest facing a global network for distributing or transporting carbon. If all of the CO_2 emitted from power plants in the United States were transported for sequestration, the quantity would be three times the weight and one-third the volume of natural gas transported by the nation's entire natural gas pipeline system annually.[58] If buildable, a suitable CCS pipeline network could cost as much as $71.4 trillion.[59] This estimate may even be conservative: CCS would also need to be deployed at biomass and natural gas power plants, in the fuel transformation and gas-processing sectors, and in emissions-intensive sectors such as cement, iron, steel, chemicals, and pulp and paper manufacturing.[60]

There is also the question of storage space. If just 60% of the CO_2 emissions were to be captured and compressed in liquid form for geological sequestration, the volume of liquid carbon that would have to be managed each day would be equivalent to all of the oil currently consumed in the United States (about 20 million barrels per day). Even if this were manageable, over time, storage space would become an issue. Experts in Europe, after examining the storage capability of all large potential storage sites—including deep saline aquifers, depleted hydrocarbon fields, and unmineable coal beds—predicted that CCS systems had enough capacity to store only 36 years' worth of emissions.[61] Terrestrial and biological sequestration initiatives face similar dilemmas in that they must confront the ever-present risk of having to counter deforestation pressures exerted by the logging industry, expansion of agriculture, and natural disasters (e.g., fires and floods).[62]

Considering all these issues, opponents point out that the costs of managing a global network of sequestered carbon would be enormous. An interdisciplinary team at the Massachusetts Institute of Technology calculated that CCS technologies would almost double the cost of coal-fired power.[63] In a survey of state-of-the-art CCS technologies, Davison and Thambimuthu determined that implementation would result in an increased gross cost of electricity generation of 30% to 80%.[64]

Lastly, and perhaps just as significant, even if global CCS were to become technically and economically viable, opponents argue, it would give rise to a future governance problem. One international research team likened CCS to nuclear power in that both create "Faustian bargains."[65] As in managing high-level nuclear waste, CCS systems would need to safely sequester CO_2 for staggeringly long periods of time. They would have to operate within narrow margins of error to achieve climate change goals and to protect human health and the environment. Imagine the difficulty the International Atomic Energy Agency and Nuclear Regulatory Commission already have in managing relatively small volumes of nuclear waste, then consider the enormity of the management challenge of global CCS, given that the amount of CO_2 emitted into the atmosphere each year is in the billions of tons. Better yet, given that the amount of effluent associated with CCS has been equated to the volume of oil, one might reasonably speculate that the probability of problem-free management of the process is similar to the probability of eliminating all oil spills.

Common Ground: We Should Contract and Converge Emissions and Capture and Store the Rest

When one weighs the pros and cons of both perspectives, it becomes possible to synthesize these two starkly opposing theses. As box 11.1 indicates, our synthesis contains three prongs: (1) to reduce emissions by cutting coal consumption so there is less carbon to capture, transport, and sequester; (2) to better account for and minimize coal's environmental damages upstream; and (3) to continue to research and improve CCS and deploy it in certain industrial niches—in other words, CCS might have a role to play during a phase-out of coal, but it should not be a platform for ramping up coal dependence.

The first prong of this strategy is to encourage universal contraction of GHG emissions while acknowledging that the pace of contraction cannot be the same in industrialized and developing nations—a process known as contraction and convergence. Under this ideology, industrialized nations must accept the responsibility to facilitate deep reductions in GHG emissions (70% to 80%) over the next three decades. Meanwhile, leaders of developing economies need to commit to slower GHG emission reduction trajectories. In doing so, the worst effects of climate change can be avoided or at least attenuated, and gradually, developed and developing nations will converge to a low-carbon economy. This low-carbon convergence strategy ascribes to the principle of "common but differentiated responsibility" by acknowledging that all of us are responsible for the problem, but to different degrees, and all are responsible for the solution, but to different degrees.[66] The contraction and convergence strategy also ensures there will be considerably less carbon that needs to be integrated into a CCS system, thereby averting the trials of trying to manage a mammoth CCS network.

The second prong is to make coal as clean as it can be. It is hard to disagree with the contention that coal can be made cleaner—though some see coal's polluting properties as a strong justification for condemnation, it also means there is ample room for improvement. Better safety standards and occupational training can lessen the risk of coal mining accidents. Particulate matter control can lower the risks for black lung disease among miners. Methane monitoring and capture systems can minimize fugitive emissions, and integrated closure planning programs can protect the environmental and socioeconomic health of mining communities. Degraded mining sites can be remediated, and pollution control equipment can be established at coal-fired power plants.

Our third prong would see deployment of CCS on smaller scales where risks are more manageable and continuing research and improvement of the technology. We are all but certain that CCS will remain a critical technology for sinking carbon, though not necessarily at large-volume coal-fired power plants. It may be

BOX 11.1

Clean Coal and the Common Ground

Is clean coal an oxymoron?

ONE SIDE: Clean coal systems are an essential part of our economic future and play a key part in the process of global decarbonization.

OTHER SIDE: Capture and storage technology does not mitigate coal's upstream damage, and it faces serious geological, economic, and political barriers.

COMMON GROUND: We should contract and converge emissions, make mining more suitable, and capture and store what we can.

---

more viable at factories, especially those in the cement, steel, and chemical industries. Moreover, once perfected, clean coal technologies could sequester CO_2 associated with natural gas—a technology that we highlighted as a possible bridge to renewable electricity (see chapter 6)—and emissions associated with electric vehicles (chapter 7), biofuel (chapter 8), and nuclear power (chapter 12). The promise of CCS therefore justifies a certain amount of research and development related to its eventual use, even if its ultimate application ends up being confined to strategic niches rather than a worldwide system.

NOTES

1. Dan Radmacher, "Effort to Clean Coal's Image Won't Work," *Roanoke Times* December 21, 2008, 7.

2. Omitting here the "Fa la la la la, la la la la" after each line in the second and subsequent verses. See Lloyd Alter, "Clean Coal Carolers from an Industry Run by Morons" (Tree Hugger, December 11, 2008).

3. In this chapter we understand "clean coal" to refer to a collection of four technologies and processes: (1) supercritical pulverized coal plants that boost thermal efficiency by operating at higher temperatures, (2) integrated gasification combined cycle (IGCC) plants that use chemical processes to gasify coal and remove sulfur and mercury, (3) pressurized fluid bed combustion plants that use elevated pressure to capture sulfur dioxide and nitrogen oxides, and (4) carbon capture and storage (CCS) techniques such as deep underground geological formations that are engineered to capture and store excess CO_2.

4. US Environmental Protection Agency (EPA), "Clean Air Markets, Section 415: Clean Coal Technology Regulatory Incentives" (July 25, 2012), available at www.epa.gov /airmarkets/progsregs/arp/sec415.html.

5. Robert J. Budnitz and John Holdren, "Social and Environmental Costs of Energy Systems," *Annual Review of Energy* 1 (1976): 579.

6. Jay Colinet et al., *Best Practices for Dust Control in Coal Mining* (Atlanta, GA: National Institute for Occupational Safety and Health, Office of Mine Safety and Health Research, 2010).

7. C. Özgen Karacan et al., "Coal Mine Methane: A Review of Capture and Utilization Practices with Benefits to Mining Safety and to Greenhouse Gas Reduction," *International Journal of Coal Geology* 86, no. 2–3 (May 1, 2011): 121–156.

8. D. Limpitlaw et al., *Post-Mining Rehabilitation, Land Use, and Pollution at Collieries in South Africa* (Johannesburg: South African Institute of Mining and Metallurgy, 2005).

9. L. C. Bell, "Establishment of Native Ecosystems after Mining—Australian Experience across Diverse Biogeographic Zones," *Ecological Engineering* 17, no. 2–3 (July 1, 2001): 179–186.

10. Daniel M. Franks, David Brereton, and Chris J. Moran, "Managing the Cumulative Impacts of Coal Mining on Regional Communities and Environments in Australia," *Impact Assessment and Project Appraisal* 28, no. 4 (2010): 299–312.

11. Steven J. Davis and Robert H. Socolow, "Commitment Accounting of CO_2 Emissions," *Environmental Research Letters* 9 (2014): 084018

12. International Energy Agency (IEA), *World Energy Outlook 2011* (Paris: OECD, 2011).

13. Cedric Philibert, *Technology Penetration and Capital Stock Turnover: Lessons from IEA Scenario Analysis* (Parise: IEA, 2008); IEA, *World Energy Outlook 2008* (Paris: OECD, 2008), 73–75; M. A. Brown and B. K. Sovacool, *Climate Change and Global Energy Security: Technology and Policy Options* (Cambridge, MA: MIT Press, 2011), fig. 3.2

14. Ambrose Evans-Pritchard, "Oil and Gas Investment in the US Has Soared to $200b," *Daily Telegraph* (London), July 10, 2014.

15. Ibid.

16. World Coal Mining, "Industry Overview" (2013), available at www.mbendi.com /indy/ming/coal/p0005.htm.

17. IEA, *World Energy Outlook 2011*.

18. W. Rafey and B. K. Sovacool, "Competing Discourses of Energy Development: The Implications of the Medupi Coal-Fired Power Plant in South Africa," *Global Environmental Change* 21, no. 3 (August 2011): 1141–1151.

19. IEA, *World Energy Outlook 2014* (Paris: OECD, 2014).

20. van der Hoeven quoted in IEA, "Coal's Share of Global Energy Mix to Continue Rising, with Coal Closing in on Oil as World's Top Energy Source by 2017" (December 17, 2012), available at www.iea.org/newsroomandevents/pressreleases/2012/december/name,34441, en.html.

21. Ailun Yang and Yiyun Cui, *Global Coal Risk Assessment: Data Analysis and Market Research* (Washington, DC: World Resources Institute, November 2012).

22. EPA, "Clean Alternative Fuels" (2002), available at www.epa.gov/OMS/consumer /fuels/altfuels/420f00036.pdf.

23. J. Brathwaite, S. Horst, and J. Iacobucci, "Maximizing Efficiency in the Transition to a Coal-Based Economy," *Energy Policy* 38 (2010): 6086.

24. US Department of Energy, Office of Fossil Energy, "Carbon Sequestration Technology Roadmap and Program Plan 2006" (2006), 11–12, available at http://fossil .energy.gov/sequestration/publications/programplans/2006/2006_sequestration _roadmap.pdf.

25. IEA, *World Energy Outlook 2008*, 74.

26. Global Carbon Capture and Storage Institute, *The Global Status of CCS: 2012* (Docklands, Australia: Global CCS Institute, October 2012).

27. Climate Change Technology Program, *Technology Options for the Near and Long Term* (Washington, DC: US Department of Energy, November 2003).

28. US Department of Energy (DOE), "Terrestrial Sequestration Research," available at www.fossil.energy.gov/programs/sequestration/terrestrial/index.html.

29. A. Grubler and N. Nakicenovic, "Decarbonizing the Global Energy System," *Technological Forecasting and Social Change* 53 (1996): 97–100.

30. IEA and Nordic Energy Research, *Nordic Energy Technology Perspectives: Pathways to a Carbon Neutral Energy Future* (Paris: OECD, 2013).

31. E. O. Wilson, *The Future of Life* (New York: Random House, 2002), 23.

32. S. C. Page, A. G. Williamson, and I. G. Mason, "Carbon Capture and Storage: Fundamental Thermodynamics and Current Technology," *Energy Policy* 37, no. 9 (September 2009): 3314–3324.

33. Emily S. Bernhardt and Margaret A. Palmer, "The Environmental Costs of Mountaintop Mining Valley Fill Operations for Aquatic Ecosystems of the Central Appalachians," *Annals of the New York Academy of Sciences* 1223 (2001): 57.

34. M. A. Palmer et al., "Mountaintop Mining Consequences," *Science* 327 (January 8, 2010): 148–149.

35. Quoted in Natural Resources Defense Council, "Mountain Top Removal Coal Mining" (June 23, 2011), available at www.nrdc.org/energy/coal/mtr.

36. Nicholas Z. Muller, Robert Mendelsohn, and William Nordhaus, "Environmental Accounting for Pollution in the United States Economy," *American Economic Review* 101, no. 5 (August 2011): 1674.

37. US Government Accountability Office (GAO), *Additional Guidance and Oversight of Mines' Emergency Response Plans Would Improve the Safety of Underground Coal Miners*, GAO-08-424 (Washington, DC: GAO, April 2008).

38. World Wildlife Fund (WWF), *Coming Clean: The Truth and Future of Coal in the Asia Pacific* (Washington, DC: WWF, 2007).

39. United States Mine Rescue Association, "Historical Data on Mine Disasters in the United States" (July 2013), available at www.usmra.com/saxsewell/historical.htm.

40. Andrew Fraser, "Coalmining's Invisible Killer," *Australian*, November 22, 2010.

41. Anil Markandya and Paul Wilkinson, "Electricity Generation and Health," *Lancet* 370 (2007): 979–990.

42. For instance, the New Zealand Royal Commission on the Pike River Coal Mine Tragedy's final report (October 30, 2012) notes: "But, We declare that you are not, under this Our Commission, to inquire into and report upon the wider social, economic, or environmental issues, such as the following . . . the merits of coal mining, or any other mining, and related operations in New Zealand."

43. Marta Miranda, *Mining and Critical Ecosystems: Mapping the Risks* (Washington, DC: World Resources Institute, 2004).

44. Thomas Sundqvist and Patrik Soderholm, "Valuing the Environmental Impacts of Electricity Generation: A Critical Survey," *Journal of Energy Literature* 8, no. 2 (2002): 1–18; Thomas Sundqvist, "What Causes the Disparity of Electricity Externality Estimates?" *Energy Policy* 32 (2004): 1753–1766.

45. Ian F. Roth and Lawrence L. Ambs, "Incorporating Externalities into a Full Cost Approach to Electric Power Generation Life-Cycle Costing," *Energy* 29 (2004): 2125–2144; Daniel Kammen and Sergio Pacca, "Assessing the Costs of Electricity," *Annual Review of Environment and Resources* 29 (2004): 301–344.

46. Alexander Romanov, Lesley Sloss, and Wojciech Jozewicz, "Mercury Emissions from the Coal-Fired Energy Generation Sector of the Russian Federation," *Energy Fuels* 26, no. 8 (2012): 4647–4654.

47. Sarah K. Adair, David C. Hoppock, and Jonas J. Monast, "New Source Review and Coal Plant Efficiency Gains: How New and Forthcoming Air Regulations Affect Outcomes," *Energy Policy* 70 (2014): 183–192.

48. IEA, *Global Gaps in Clean Energy Research, Development, and Demonstration* (Paris: OECD, 2009).

49. Intergovernmental Panel on Climate Change (IPPC), *Carbon Dioxide Capture and Storage* (Geneva: IPCC, 2005).

50. IEA, *Global Gaps*, 13.

51. Kurt Zenz House et al., "The Energy Penalty of Post-combustion CO_2 Capture & Storage and Its Implications for Retrofitting the U.S. Installed Base," *Energy and Environmental Science* 2 (2009): 193–205.

52. D. Bellman et al., "Coal to Liquids" (Working Document No. 18 of the NPC Global Oil and Gas Study, July 18, 2007).

53. Brian Tulloh, "Letters from Readers," *EnergyBiz Insider*, February 8, 2007, 16.

54. Alessandro Franco and Ana R. Diaz, "The Future Challenges for Clean Coal Technologies: Joining Efficiency Increase and Pollutant Emission Control," *Energy* 34 (2009): 348–354.

55. UK Department of Trade and Industry, "The Energy Challenge" (2010), available at http://webarchive.nationalarchives.gov.uk/20101209210643/webarchive.nationalarchives .gov.uk/+/http://www.berr.gov.uk/files/file32014.pdf.

56. Anders Hansson and Marten Bryngelsson, "Expert Opinions on Carbon Dioxide Capture and Storage—A Framing of Uncertainties and Possibilities," *Energy Policy* 37 (2009): 2273–2282.

57. DOE, Office of Fossil Energy, "Geologic Sequestration Research" (June 20, 2010), available at http://fossil.energy.gov/sequestration/geologic/index.html.

58. James Katzer et al., *The Future of Coal: Options for a Carbon-Constrained World*, Interdisciplinary MIT Study (Cambridge, MA: MIT, 2007).

59. B. K. Sovacool, C. Cooper, and P. Parenteau, "From a Hard Place to a Rock: Questioning the Energy Security of a Coal-Based Economy," *Energy Policy* 39, no. 8 (August 2011): 4664–4670.

60. IEA, *Technology Roadmap: Carbon Capture and Storage* (Paris: OECD, 2010).

61. EU Geocapacity Project, "Assessing European Capacity for Geological Storage of Carbon Dioxide" (2009), S21, available at www.geology.cz/geocapacity.

62. B. Metz et al., eds., *Climate Change 2007: Mitigation of Climate Change: Contribution of Working Group III to the Fourth Assessment Report of the Intergovernmental Panel on Climate Change, 2007* (Cambridge: Cambridge University Press, 2007).

63. Katzer et al., *Future of Coal.*

64. J. Davison and K. Thambimuthu, "An Overview of Technologies and Costs of Carbon Dioxide Capture in Power Generation," *Proceedings of the Institute of Mechanical Engineers* 223 (2009): 201–212

65. Daniel Spreng, Gregg Marland, and Alvin M. Weinberg, "CO_2 Capture and Storage: Another Faustian Bargain?" *Energy Policy* 35 (2007): 850–854.

66. B. K. Sovacool and M. H. Dworkin, *Global Energy Justice: Problems, Principles, and Practices* (Cambridge: Cambridge University Press, 2014).

IV. ENERGY SECURITY AND ENERGY TRANSITIONS

Is Nuclear Energy Worth the Risk?

The era of atomic energy began with a whimper, not a bang, on December 7, 1942. On the polished wood floors of a World War II–appropriated squash court at the University of Chicago, physicists Enrico Fermi and Leó Szilárd inserted about 50 tons of uranium oxide into 400 carefully constructed graphite blocks. A small puff of heat heralded the first self-sustaining nuclear reaction, a bottle of Chianti was shared among the scientists present, and nuclear power was born.[1]

Since then, many people have harbored grand visions of nuclear possibilities. Early advocates in the United States promised a future of electricity too cheap to meter, an age of peace and plenty without high prices and shortages, with atomic energy providing the power needed to desalinate water for the thirsty, irrigate deserts for the hungry, fuel interstellar travel, and, farcically, enable the development of atomic golf balls that could always be found thanks to their radioactive signature.[2] Soviet engineers speculated that the atomic age would allow gamma-ray prospecting of minerals and oil, alleviate all energy shortages, and eliminate hunger through the creation of atomically enhanced fertilizers and the irradiation of food to prolong shelf-life.[3] Charles de Gaulle justified the French nuclear program on the grounds that it would liberate the republic from the "yoke of double hegemony" by setting France on its own, independent technological pathway, making it an international superpower similar to the Soviet Union or United States.[4] Jawaharlal Nehru praised the Indian program for its potential to catalyze an industrial revolution that would eradicate poverty and prevent the country from being recolonized.[5]

Advocates of nuclear power have recently framed it as an important part of any solution aimed at fighting climate change. This narrative makes intuitive sense—nothing is burned inside a nuclear reactor because it relies on fission, meaning that reactors produce no smoke and no direct greenhouse gas (GHG) emissions. Nuclear reactors can produce massive amounts of energy with facility lifetimes often exceeding 40 to 60 years. Nuclear power has been depicted as a godsend for the environment and for national economies, especially in the developing world.

Opponents argue that nuclear power is "too cheap to meter" only if one's meter is broken. Moreover, the wonder of creating energy from small amounts of materials is not the only thing that matters. Nuclear power presents severe economic risks, such as liability for accidents, construction cost overruns, and dependence on subsidies. When one takes into account the carbon-equivalent emissions associated with the entire nuclear life cycle, nuclear reactors do contribute to climate change. Those who hold this contrarian view conclude that when the safety, waste storage, and environmental risks become visible, nuclear energy is unjustifiable.

The common ground tends, as always, to be heavily influenced by the strength of the arguments on each side. So let's deal out the cards on the pros and cons of nuclear power and see which side has the strongest hand.

One Side: Nuclear Is an Inexpensive, Reliable, Low-Carbon Favorite

Today's nuclear plant operators rely on a mix of technologies and suppliers—mostly light-water reactors, pressurized-water reactors, and heavy-water reactors—to produce electricity. Light-water reactors constitute more than 80% of the world's nuclear power fleet. Light-water reactors use ordinary water, called "light" in the early days of the industry to distinguish it from "heavy" water, in which one hydrogen atom is replaced with deuterium. Light-water reactors are of two basic types: pressurized-water and boiling-water reactors. Nuclear engineers demarcate four generations of plant design stemming from development of the technology in the early 1940s (see table 12.1).

Proponents of nuclear power often describe three main driving factors in favor of the technology: (1) the need to keep costs low—both to ratepayers and to governments, (2) the need to provide basic energy services to the world's poor, and (3) the need to find sources of energy that are less GHG-intensive. Proponents claim that nuclear power is the only technology that can satisfy all three of these critical needs.

Nuclear Power Is Cheap and Reliable

Unlike electricity-generating resources such as wind turbines and solar panels—which operate in decentralized configurations, in small increments, and dependent on intermittent fuel—nuclear reactors generate electricity almost continuously. Once a nuclear reaction starts, each reactor will produce energy for an extended period of time (usually about a year) until it needs to be refueled. This makes nuclear reactors well suited to providing base-load power, electricity that is "always on" and is the backbone of national electricity grids.[6] Put in a global context, some countries such as China and India need large amounts of base-load power to drive industrialization and grow their economies, making nuclear energy an attractive option.[7]

Table 12.1. Generations of Nuclear Power Technologies

Generation	Years	Description	Specific designs/examples
I	1948–1965	Early prototype reactors	Shippingport, Dresden, Magnox
II	1966–1995	Commercial power reactors	Pressurized Water Reactors, Boiling Water Reactors, Canada Deuterium Uranium (CANDU)
III	1996–2012	Advanced light-water reactors	CANDU6, System 80+, Westinghouse AP600, Pebble Bed Modular Reactors
III+	2012–2030	Evolutionary designs	Advanced Boiling Water Reactor, Advanced CANDU Reactor 1000, Westinghouse AP1000, Mitsubishi Advanced Pressurized Water Reactor, European Pressurized Reactor, Economic Simplified Boiling Water Reactor
IV	2030+	Revolutionary designs	Gas-cooled Fast Reactor, Very-high-temperature Reactor, Supercritical Water Reactor, Sodium-cooled Fast Reactor, Lead-cooled Fast Reactor, Molten Salt Reactor

Source: Modified from Trevor Findlay, *Nuclear Energy and Global Governance: Ensuring Safety, Security, and Non-Proliferation* (London: Routledge, 2011), 20.

Data suggest that nuclear power plants do have very low historical production costs and, recently, have enjoyed improved performance due to better capacity factors and operating procedures. Although particular production costs differ with design, site requirements, and rate of capital depreciation, existing light-water reactors, which make up a majority of the world's nuclear power fleet, produce electricity at costs between 2.5 and 7 cents/kWh. This makes them cost-competitive with or cheaper than many other sources of electricity on the global market today (if externalities are not factored into costs).[8] Another advantage is improvements in operational performance and maintenance, due largely to decades of research and development and government subsidies. The Energy Information Administration tells us that existing plants in the United States, for example, improved their capacity factor by 40% from 1990 to 2008, with the average plant operating 90.1% of the time in 2013.[9]

Also, existing nuclear reactors have become safer over time as fail-safes have improved and new safety regulations have tightened oversight. The industry has incorporated research findings on human factors and safety culture provided by groups and organizations such as the International Atomic Energy Agency (IAEA) and World Nuclear Association of Nuclear Operators.[10]

These efforts have produced dividends, as evidenced by the finding in a meta-survey of nuclear power plant performance that existing plants were getting safer. The study found that industrywide, 1.04 accidents occurred per

200,000 worker hours in 1990, but only 0.28 per 200,000 worker hours in 2003.[11] The study noted that, overall, better occupational safety and health regulations in Europe and North America, improved medical knowledge, and better emergency care and first-aid techniques have reduced the likelihood of accidents and lessened their impact when they do occur. The Fukushima accident, nuclear advocates tell us, was an anomaly never to be repeated—a simultaneous earthquake and tsunami interacting with unique design flaws in 40-year old reactors at a particular site—and one in which no directly caused deaths occurred, at least so far.[12]

Nuclear Energy Can Fight Energy Poverty

Denying electricity and the services it supports to those in need promotes discrimination in the vein of what is sometimes called "energy poverty."[13] At least one billion people—roughly one-sixth of the global population—have little or no access to electricity. Without electricity, millions of women and children are particularly disadvantaged, forced to spend significant amounts of time searching for firewood and then burning wood and charcoal indoors to heat their homes and prepare meals. Just 64% of the population in developing countries has access to electricity, and in Asia and Africa the numbers are even lower: about 40% for South and Southeast Asia, 30% for Africa, and 22% for sub-Saharan Africa.[14]

As we noted in chapter 1, the pollution and health consequences of cooking with coal and wood in poorly ventilated rooms are significant: upwards of four million deaths per year, or 7.6 deaths per minute. Most of these are deaths of children under the age of five, who must suffer their final months of life dealing with debilitating respiratory infections, chronic obstructive pulmonary disease, and lung cancer.[15] Yet, as one researcher at the IAEA explains, "The role of [nuclear power] could be to increase the availability of clean energy in a variety of usable forms for all regions of the world, to broaden the access to clean and affordable and diverse energy products and, in this way, to contribute to the eradication of poverty and, subsequently, to peace and stability in the world."[16] In essence, nuclear power is seen as an option that can prevent a form of "energy apartheid" in which people in the Western world use large amounts of energy and have higher standards of living and longer life expectancies, while those in the less-developed nations have no access to energy, suffer from impoverished conditions, and die earlier.[17]

Nuclear Power Produces Large Amounts of Energy with Minimal Waste

A final argument in favor of nuclear power is that it creates negligible waste streams compared with fossil fuels. It should therefore be considered a logical technology for supplying energy in a carbon-constrained world.

Advocates point out that nuclear fission involves no combustion. A reactor is a device that manages controlled nuclear reactions over many months. The ongoing reaction heats fuel rods, which in turn boil water and generate steam to drive an electrical turbine. From this perspective, the claim that the process is emissions-free does have some merit. Moreover, fission can produce scads of energy in this emissions-free manner. The energy released by fission of one gram of uranium-235 is 2.5 million times the energy released in burning one gram of fossil fuel. To illustrate this point, one group of environmental scientists argued that "a golf-ball-sized lump of uranium would supply the lifetime's energy needs of a typical person, equivalent to 56 tanker trucks of natural gas, 800 elephant-sized bags of coal or a renewable battery as tall as 16 'super' skyscraper buildings placed one on top of the other."[18] Talk about a bang for one's buck.

Writers attempting to familiarize the public with atomic energy back in the 1940s used to calculate the untapped atomic power potential in the molecules of everyday objects: one pound of water had enough energy to heat 100 million tons of water; a handful of snow could power an entire city; an airplane or car could run its entire lifetime on a pellet of atomic energy the size of a vitamin pill.[19] Environmentalist Stuart Brand notes that, in modern times, the nuclear waste produced by creating enough power to supply one person with a lifetime of electricity would fit in a Coca Cola can.[20] The waste generated by a large nuclear plant each year is 2 million times smaller by weight and a billion times smaller by volume than waste from a coal-burning plant.[21]

Nuclear power advocates argue that a substantial portion of the fossil fuel waste stream is released into the environment (in the form of stack gases and particulates), whereas the waste from nuclear power is easier to manage and control. Hans Blix, former director of the IAEA, explained it this way: "A 1,000 megawatt equivalent (MWe) coal plant with optimal pollution abatement equipment will annually emit into the atmosphere 900 tons of sulfur dioxide, 4,500 tons of nitrous oxide, 1,300 tons of particulates, and 6.5 million tons of carbon dioxide . . . By contrast, a nuclear plant of 1,000 MWe capacity produces annually some 35 tons of highly radioactive spent fuel."[22] The direct impact on human health from fossil fuel combustion, Blix concluded, was far more severe than that from nuclear power: nuclear waste becomes less toxic with time as radioactive materials decay, whereas the chemicals emitted from coal combustion are ingested by humans and other organisms, giving rise to progressive problems.

Indeed, some advocates claim that nuclear power has saved lives and continues to do so. Using historical production data, two climate experts estimated that from 1971 to 2009, commercial nuclear reactors around the world prevented 1.84 million air pollution–related deaths (mainly by offsetting particulate matter).[23] They argued that these saved lives eclipse the 4,900 deaths directly attributable to nuclear power accidents such as those at Chernobyl. They concluded that nuclear power could save an additional 420,000 to 7 million deaths by 2050 if nuclear power were to expand to substitute for fossil fuels.

Another environmental advantage is that nuclear power plants do not emit carbon dioxide. The Nuclear Energy Institute promotes nuclear power as a "carbon-free electricity source" and expounds on the importance of building "emission-free sources of energy like nuclear."[24] The two climate experts cited above argued that "because nuclear power is an abundant, low-carbon source of base-load power, it could make a large contribution to mitigation of global climate change."[25] When US President George W. Bush signed the Energy Policy Act in August 2005, he remarked that "only nuclear power plants can generate massive amounts of electricity without emitting an ounce of air pollution or greenhouse gases."[26] Nicholas Ridley, former secretary of state for the environment in the United Kingdom, was even more unequivocal in his support for nuclear power, stating on BBC television that "there is absolutely no doubt that if you want to arrest the Greenhouse Effect you should concentrate on a massive increase in nuclear generating capacity. Nuclear power stations give out no sulfur and carbon dioxide, so they are the cleanest form of power generation."[27]

Perhaps for these reasons, nuclear power technology has featured prominently in the energy projections undertaken by groups such as the International Energy Agency (IEA), as well as in some academic studies. The IEA includes nuclear power as one of its select "low-carbon technologies" and argues that if the world is to see a 50% drop in energy-related CO_2 emissions, something they call their "Blue Scenario," then nuclear energy must expand rapidly.[28] It projects that nuclear power needs to expand on a massive scale to provide about one-quarter of electricity by 2050, a requisite jump from less than 400 GW today (generating less than 2,500 TWh of electricity) to 1,200 GW in 2050 (with annual generation of almost 10,000 TWh). This would make nuclear energy the "the single largest source of electricity in 2050, with a correspondingly significant contribution to cutting greenhouse gas emissions,"[29] as depicted in figure 12.1. Achieving this level of nuclear capacity would require about $4 trillion of additional investment, more than any other equivalent source of electricity would require.[30] Similar projections reflecting a massive increase in nuclear generation also appear in the IEA's *Energy Technology Perspectives 2012*: nuclear power is expected, without any breakthroughs in technology, to triple in aggregate capacity by 2050.[31] As the IAEA puts it in one of its flagship publications, "Nuclear power can make an important contribution to reducing greenhouse gas emissions while delivering energy in the increasingly large quantities needed for global socioeconomic development."[32]

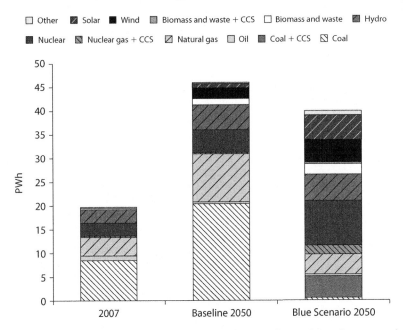

Figure 12.1. Global Electricity Production by Source, 2007 and 2050. *Source:* International Energy Agency, *Technology Roadmap: Nuclear Energy* (Paris: OECD, June 2010). *Note:* CCS = carbon capture and storage; PWh = petawatt-hour.

As noted earlier, even for some prominent figures in the environmental community, nuclear power, properly designed, goes hand-in-hand with climate protection. NASA climate expert James Hansen has become so passionate about the topic that, with other influential environmental researchers, he wrote a public, open letter addressed to "those influencing environmental policy but opposed to nuclear power." In the letter, he wrote that "in the real world there is no credible path to climate stabilization that does not include a substantial role for nuclear power" and "with the planet warming and carbon dioxide emissions rising faster than ever, we cannot afford to turn away from any technology that has the potential to displace a large fraction of our carbon emissions."[33] Patrick Moore, cofounder of Greenpeace, has publicly stated that "nuclear energy is the only non-greenhouse gas emitting energy source that can effectively replace fossil fuels and satisfy global demand."[34] Environmentalist James Lovelock has gone so far as to say that "we have no time to experiment with visionary energy sources; civilization is in imminent danger and has to use nuclear—the one safe, available, energy source—now or suffer the pain soon to be inflicted by our outraged planet."[35] Barry W. Brook and Corey J. A. Bradshaw argue that "although the environmental movement has historically rejected the nuclear energy option . . . society cannot afford to risk wholesale failure to address energy-related biodiversity impacts because of preconceived notions and ideals."[36] Their article convinced 69 conservation scientists to openly endorse nuclear power for reasons of biodiversity protection.[37]

The Other Side: Nuclear Is a Costly, Dangerous Foible

The articulators of the antithesis suggest that the situation is a bit more complicated than "embrace nuclear power or else suffer climate change." Environmentalists, groups of scientists, and others against nuclear power are quick to point out three serious challenges: (1) high future costs and dependence on subsidies, (2) the threats posed by nuclear waste and weapons proliferation, and (3) environmental damage from other stages of the nuclear fuel cycle. When these factors are considered, geographer Vaclav Smil concludes, "because of hasty commercialization, safety concerns, and unresolved long-term storage of its wastes, the first nuclear era has been a peculiarly successful failure."[38]

Nuclear Power Has High Future Costs, Made More Expensive by Accidents
Opponents begin by pointing out that the key word in any stated advantage of nuclear power from an economic standpoint is *historical*: only at plants that have been operating for years, where capital costs have been greatly subsidized and are all but paid off, does the technology make economic sense. The expense of building new nuclear plants is immense. Independent assessments suggest that costs run in the range of $5,500/kW to $8,100/kW, translating into an elephantine $6 to $9 billion price tag for each 1,100 MW plant.[39] Researchers at the Keystone Center, a nonpartisan think tank, consulted with representatives from 27 nuclear power companies and contractors and concluded that projected operating costs for these plants would be shockingly high: 30 cents/kWh for the first 13 years until construction costs were recovered, followed by 18 cents/kWh over the remaining lifetime of the plant.[40] These capital and operating costs are so high that they deter most investors and utilities from considering nuclear power. Costs upward of $8 billion to build a new reactor are equivalent to 75% to 100% of the capitalization of assets for an entire electric utility. The costs of small modular reactors, which have been touted as the next big thing to "save" the industry, are expected to be much higher than those for commercially available wind and solar systems of the same capacity,[41] as figure 12.2 shows.

Due to long construction timetables, nuclear plants are also exceptionally prone to cost overruns. The independent US Congressional Budget Office estimates that actual construction costs for American reactors average twice as much as projected costs and that the risk of default on loan guarantees exceeds 50%.[42] Another survey of global construction times and costs for 180 reactors

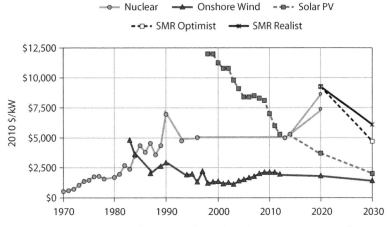

Figure 12.2. Future Cost Trends for Wind, Solar, and Nuclear Power. *Source:* Mark Cooper, "Small Modular Reactors and the Future of Nuclear Power in the United States," *Energy Research and Social Science* 3 (September 2014): 161–177. *Note:* PV = photovoltaic; SMR = small modular reactor.

across seven countries found that cost overruns afflicted more than 97% of nuclear projects and led to a mean cost escalation of 117% per project—so, on average, each plant cost more than twice what its sponsors said it would.[43] And these overruns exclude the costs of major accidents such as Three Mile Island in 1979, Chernobyl in 1986, and Fukushima in 2011, which resulted in upwards of $170 billion in combined property or human health damage.[44]

Opponents caution that it is a matter of "when" not "if" future accidents will occur. Using some of the most advanced probabilistic risk assessment (PRA) tools available at the time, an interdisciplinary team at the Massachusetts Institute of Technology identified possible reactor failures and predicted that the best estimate of core damage frequency was around one incident every 10,000 reactor years.[45] In terms of the expected global growth scenario for nuclear power from 2005 to 2055, the MIT team estimated that at least four serious core damage accidents (with a high probability of offsite releases) would occur worldwide in that time period. The study was done in 2003, so if Fukushima counts as one of these accidents and if the projections are accurate, we have three left. Such immense safety risks convinced the MIT authors to conclude that "both the historical and the PRA data show an unacceptable accident frequency."[46] The point here is not that systems fail—all energy technologies have their imperfections—but that nuclear power systems are so catastrophically damaging when they do fail: a billion-dollar asset can become a trillion-dollar liability in a matter of seconds.

Nuclear power therefore requires and, indeed, has received a flood of hidden subsidies just to stop prospective investors fleeing at breakneck speed. These subsidies include limited liability for accidents, loan guarantees, funds for decommissioning, and government-backed waste storage. Advocates contend that these gifts are essential: without them, the nuclear industry would not be able to compete in the electricity marketplace. Douglas Koplow looked at five decades' worth of subsidy data and concluded that "subsidies to the nuclear fuel cycle have often exceeded the value of the power produced. This means that buying power on the open market and giving it away for free would have been less costly than subsidizing the construction and operation of nuclear power plants."[47] This dependence on subsidies caused Peter Bradford, a former commissioner for the US Nuclear Regulatory Commission, to joke that the best way to stop the nuclear renaissance would be to simply "do nothing."[48]

Opponents conclude that the large price tag for nuclear plants and the dependence on subsidies render it a nonstarter for most emerging economies and developing countries—the very nations where advocates contend that nuclear power has a role to play in eliminating energy poverty. Small-island developing states such as Fiji and the Maldives and least-developed countries such as Bhutan and Mali have entire electricity sectors with only a few hundred million dollars of investment and small amounts of installed capacity. How are they to afford the billions needed for a nuclear reactor? Furthermore, electricity generation does not usually eliminate dependence on solid fuels and the hazards of energy poverty. The IEA notes that "there is a widespread misconception that electricity substitutes for biomass. Poor families use electricity selectively, mostly for lighting and communications. They often continue to cook and heat with wood or dung, or with fossil-based fuels like LPG [liquefied petroleum gas] and kerosene."[49] The best energy option for these countries is to expand access to improved cookstoves, micro-hydro dams, solar home systems, and micro-grids,[50] rather than to install nuclear technology. For instance, in India, $2 billion could be spent on a new midsize nuclear reactor (giving rise to additional connectivity expenses) or could provide 114 million households at the "bottom of the pyramid" with solar lanterns, cookstoves, and small hydropower systems.[51]

The Industry Still Doesn't Have a Solution to Its Waste Problem

A second quandary for the nuclear power industry is how to manage radioactive waste. Because nothing is burned or oxidized during the fission process, nuclear plants convert almost all fuel to waste with little reduction in mass. Typically, a single nuclear reactor consumes an average of 32,000 fuel rods over the course of its lifetime and produces 20 to 30 tons of spent nuclear fuel per year—an average of about 2,200 metric tons annually for the entire US nuclear fleet.[52] The global nuclear fleet creates almost 10,000 metric tons of high-level spent nuclear fuel each year. About 85% of this waste is not reprocessed, and most of it is stored onsite in special facilities because no community wishes to host long-term nuclear storage facilities.

Determining a final resting site for this waste is a pernicious problem in search of a solution. In the United States, plans to build the nation's first permanent underground repository at Yucca Mountain were indefinitely suspended.[53] This is not an exception; it is the norm. As one study concluded, "The management and disposal of irradiated fuel from nuclear power reactors is an issue that burdens all nations that have nuclear power programs. None has implemented a permanent solution to the problem of disposing of high-level nuclear waste, and many are wrestling with solutions to the short-term problem of where to put the spent, or irradiated, fuel as their cooling pools fill."[54]

The sheer longevity of the radioactive threat from this waste—some of it will persist in hazardous form for upwards of 100,000 years—could be why physicist Alvin M. Weinberg called nuclear power a "Faustian bargain." Nuclear power trades off short-term gain for long-term pain: society receives electricity in exchange for the burden of storing long-lasting nuclear waste for millennia. Weinberg commented that this makes nuclear power "immortal" as an energy system.[55] Unlike Faust, who was ultimately able to renege on his bargain with the Devil, with nuclear energy, society binds itself in perpetuity "to the remarkable belief that it can devise social institutions that are stable for periods equivalent to geologic ages."[56]

A final concern about waste is the security of handling it—and the threat that it or nuclear materials could fall into the hands of rogue nations or terrorists. To date, several countries have tried to or managed to develop nuclear weapons under the guise of civilian nuclear energy programs.[57] The Nobel Prize–winning nuclear physicist Hannes Alfven once said that "atoms for peace and atoms for war are Siamese twins."[58] The four countries with the largest reprocessing capacity—Belgium, France, Germany, and the United Kingdom—have acknowledged that they possess at least 190 tons of separated plutonium, mostly stored as plutonium dioxide powder at aboveground sites and fuel manufacturing complexes, a volume that is enough to manufacture 20,900 nuclear weapons.[59] If we doubled the number of nuclear reactors worldwide, we would double the possibility that countries without weapons might obtain them. No other energy system has such an acute link to weapons of mass destruction.

The Nuclear Life Cycle Has Catastrophic Effects on the Environment
The nuclear fuel cycle involves some of the most hazardous elements known to humankind, including more than 100 dangerous radionuclides and carcinogens, such as strontium-90, iodine-131, and cesium-137. These are the same toxins found in the fallout from nuclear weapons. Opponents contend that there are

three environmental issues that are exacerbated by nuclear power expansion: mining pollution, water degradation, and climate change.

In the uranium mining industry, to produce the 25 tons of uranium needed to keep a typical reactor working for one year, 500,000 tons of waste rock and 100,000 tons of mill tailings (toxic for hundreds of thousands of years) are created, along with an extra 144 tons of solid waste and 1,343 cubic meters of liquid waste.[60] Open-pit mining contributes to the degradation of land, as kilometer-wide craters are formed around uranium deposits, which then interfere with the flow of groundwater as far as 10 kilometers away.[61] These pits are also prone to accumulations of radioactive gases. In Australia, a major producer of uranium, mines frequently discharge acidic liquid waste directly into freshwater and groundwater supplies and contaminate thousands of hectares of land with radium-226.[62] It may thus come as no surprise that the nonpartisan Senate References and Legislation Committee, part of the Australian federal government, documented a practice norm at uranium mines in which "short-term considerations have been given greater weight than the potential for permanent damage to the environment."[63]

A second concern is water. Like all thermoelectric plants, nuclear reactors draw on vast quantities of water both for cooling cycles and for dissipating waste streams. According to a study conducted by the Electric Power Research Institute, nuclear plants are the most water-intensive of any source of commercial electricity. They utilize 25% to 50% more water per unit of electricity generated than a fossil fuel–powered plant with an equivalent cooling system.[64] Given that many regions around the world are grappling with water scarcity problems, many opponents of nuclear power question the logic behind supporting a transition to yet another water-depleting technology.

This water requirement is not just of environmental concern; it is also of strategic concern. Droughts and extended periods of high temperatures can cripple nuclear power generation. It is often during heat waves that electricity demand is highest because of increased air conditioning and refrigeration loads and diminished hydroelectric capacity. This mismatch was poignantly demonstrated in Europe after a series of heat waves in 2003 forced France to cut back 6,000 MW of capacity. Several German reactors were also forced to operate at diminished capacity.[65]

The third concern is the link between nuclear power and climate change. Opponents point out that the claim that nuclear reactors produce no carbon dioxide when in operation is disingenuous. It is akin to saying that an atomic bomb is safe because no one has ever been hit by it as it falls. Opponents of nuclear power argue that each of the nuclear power life-cycle stages depicted in figure 12.3 emits GHG and other undesirable pollutants.

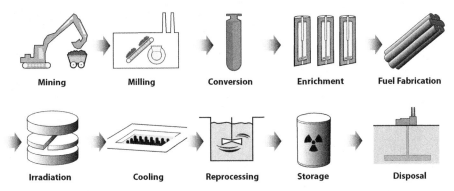

Figure 12.3. Depiction of Nuclear Power's Complete Life Cycle. *Source:* International Institute for Strategic Studies (IISS), *Preventing Nuclear Dangers in Southeast Asia and Australasia* (London: IISS, September 2009).

In aggregate, emissions from these stages of the life cycle can be sizable, such that nuclear power is of dubious merit, especially when compared with the carbon footprint of renewables.[66] Opponents charge that nuclear power is much more carbon-intensive across its life cycle than the representatives of the industry (and its advocates) claim and is a worse option than numerous alternatives (see table 12.2).

Opponents caution that GHG emissions associated with nuclear power are also destined to escalate. High-quality uranium ore is increasingly difficult (and more energy-intensive) to find, which means the carbon footprint of nuclear power is likely to rise over time, not fall.[67] Carbon-equivalent emissions of the nuclear life cycle will also worsen as reprocessed fuel is depleted, necessitating a shift to fresh ore. The Oxford Research Group projects that if the percentage of world nuclear capacity remains what it is today, and given the inevitable shift to lower-quality uranium ore as higher-quality ores diminish, by 2050 nuclear power will generate as much CO_2 per kilowatt-hour as comparable natural gas–fired power stations.[68] With very low ore grades in use, some nuclear power plants currently emit the equivalent of 337 gCO_2/kWh, making them already as environmentally damaging as equivalent-sized gas-fired power plants.[69]

Scholarship challenging the low-carbon promise of nuclear power goes on to point out other alternatives that cost less and abate more GHG emissions. For example, each dollar invested in energy efficiency displaces nearly seven times as much CO_2 as a dollar invested in nuclear power, since plants for the latter take so long to build compared with practices for the former.[70] McKinsey

Table 12.2. Comparative Life-Cycle Greenhouse Gas Emissions Estimates for Sources of Electricity

Technology	Capacity/configuration/fuel	Mean estimate (gCO_2e/kWh)
Hydroelectric	3.1 MW, reservoir	10
Biogas	Anaerobic digestion	11
Hydroelectric	300 kW, run-of-river	13
Solar thermal	80 MW, parabolic trough	13
Biomass	Forest wood steam turbine	22
Wind	Various sizes and configurations	34
Geothermal	80 MW, hot dry rock	38
Solar photovoltaic	Various sizes and configurations	50
Nuclear	Various reactor types	66
Natural gas (conventional)	Various combined cycle turbines	443
Natural gas (fracking)	Combined cycle turbines using fuel from hydraulic fracturing	492
Natural gas (liquefied, LNG)	Combined cycle turbines using LNG	611
Fuel cell	Hydrogen from gas re-forming	664
Diesel and oil	Various generator and turbine types	778
Coal	Various generator types with scrubbing	960
Coal	Various generator types without scrubbing	1,050

Source: Modified from D. Nugent and B. K. Sovacool. "Assessing the Lifecycle Greenhouse Gas Emissions from Solar PV and Wind Energy: A Critical Meta-survey," *Energy Policy* 64 (February 2014): 229–244.

& Company's cost abatement curves have repeatedly affirmed this finding, concluding that nuclear power is significantly more expensive at displacing carbon than are well-placed investments in efficiency, waste recycling, geothermal energy, and small hydropower, among others.[71] Indeed, if one factors in levelized cost and capacity, hydroelectricity is nine times more effective than nuclear power in terms of dollar per ton of CO_2 avoided, wind energy is about five times more effective, and concentrating solar power is almost twice as effective.[72] As two of us (Sovacool and Valentine), along with colleagues, have written elsewhere, "Renewables and efficiency . . . get you faster climate protection as well as more carbon displaced per dollar expended. The urgency of world hunger does not require us to fight it with caviar, no matter how nourishing fish eggs might be. In the end, buying the most expensive remedies first will only diminish what we can, and must, spend on more promising approaches. Given the opportunity costs involved, nuclear power could reduce and retard the climate protection [that environmentalists] so rightly seek."[73]

A 2009 assessment of nuclear energy in the United States noted that the cost of building 100 new reactors rather than pursuing cheaper methods such as energy efficiency would require an extra investment of $1.9 to $4.4 trillion over the life of the reactors.[74] Nuclear energy, according to this analysis, is an unnecessary and perhaps even counterproductive detour on the already treacherous path to climate stability.

Common Ground: Only Safe, Reliable, and Low-Carbon Reactors Should Be Endorsed

Before presenting our synthesis of these opposing views, we offer three key observations about the nuclear debate. First, nuclear power is a technology that brings with it both risks and rewards. On the one hand, for nations that have already committed the research and development investments to commercialize nuclear power and build plants, daily operating costs are low. This is important because for many countries that already have nuclear power programs, the low operating-cost profile engenders a degree of support that is hard to argue with. Additionally, compared with coal-fired power, which is the other major source of base-load electricity, nuclear power does have lower GHG emissions even when the entire life cycle is factored in. In other words, for nations with established nuclear plants, nuclear energy can be a valid technology for climate change mitigation. On the other hand, these benefits come with other costs. For nations that are considering adoption of nuclear power programs, the expense of designing and building reactors radically alters the economic profile of nuclear-generated electricity for the worse. When waste management and decommissioning are factored in, the competitive allure of nuclear power is eroded even further. Last but not least, the risk of release of radioactivity at a nuclear power plant, whether due to equipment failure, human error, or natural disaster, will exist as long as nuclear power plants are operating. In sum, nuclear power enhances some dimensions of sustainable development or energy security but only by creating other environmental challenges and elevated risks.

The second observation is that the climatic impact of nuclear power depends on how it is designed, managed, and governed. A CANDU reactor in Canada that has a load factor above 90% and a 40-year lifetime, that is built with skilled labor and advanced construction techniques, and that utilizes uranium that is produced domestically, relatively close to reactor sites, and is enriched with advanced technologies in a regulatory environment with rigorous environmental controls, can produce a carbon footprint as low as 15 $gCO_2e/$ kWh.[75] A Chinese light-water reactor that is built using low-tech construction techniques, that has a load factor below 70% and a 20-year lifetime, that utilizes uranium imported thousands of miles from Australia, with enrichment supported by coal-fired power plants with less stringent environmental and air-quality controls, can see emissions increase tenfold to 150 $gCO_2e/kWh.$[76] Rather than detailing the complexities inherent in calculating the CO_2 footprint of the nuclear life cycle and carefully and openly weighing its risks, most studies obscure them, especially those attempting to make nuclear energy look cleaner or dirtier than it really is.

Finally, what appears to be true on the surface might not represent the whole story. Nuclear power plants seem, to many, to pose no threat to the climate because reactors do not combust fossil fuels. They seem to hum along merrily, obscured from public sightlines, producing emission-free energy. The unversed observer misses the myriad of problems emerging at other stages—at the uranium mine and mill, during construction, when storing waste, while being decommissioned, and so forth—where associated process emissions contribute to climate change. Perhaps because these sources of CO_2 are not concentrated in one place and instead are spread across a complex fuel cycle involving hidden processes in geographically dispersed locations, they remain unaccounted for, even by some energy analysts, politicians, and environmentalists. This out-of-sight, out-of-mind mentality convinces many smart people to underestimate the environmental risks associated with nuclear power in the long run.

Many opponents of nuclear power concede that, compared with the climate change perils posed by continued reliance on fossil fuel energy sources, nuclear energy might indeed be a better option. However, they further contend that such a comparison is akin to an argument over who would be a more desirable leader, Adolf Hitler or Pol Pot. The answer, of course, is neither, thank you very much. The point is that comparatively speaking, nuclear power gives rise to a host of environmental problems and social threats that make it a poor contender for replacing fossil fuel technologies.

Our synthesis, shown in box 12.1, advances a set of three interrelated streams of logic. First, accept that there is a degree of logic underpinning initiatives to support nuclear power in situations where its total cost (including externalities and risk) is less than that of alternatives, including renewable sources of electricity and energy efficiency. On the carbon front, while nuclear energy is in no way carbon-free or emissions-free, it is less carbon-intensive than coal, oil, and natural gas electricity generators (meaning it can substitute for these systems in some situations), but it is generally more carbon-intensive than renewable technologies.

Second, when the risks of the entire nuclear power life cycle are accounted for by ratepayers or insurers, and when communities are well informed about and accept externalities, nuclear power becomes a social choice. When all factors are included, this choice might still be heavily swayed by adverse economics— as one nuclear engineer noted wryly, "Nuclear [energy] can be safe, or it can be cheap, but it can't be both."[77] Even if nuclear power is adopted, its diffusion may still present compelling justice, moral, or geopolitical challenges when accidents can affect neighboring countries or when the upstream damages from uranium

BOX 12.1

Nuclear Energy and the Common Ground

Is nuclear energy worth the risk?

ONE SIDE: Nuclear reactors produce no direct carbon emissions, create minimal waste and pollution, and should play a key role in the global energy system.

OTHER SIDE: Nuclear reactors suffer from high costs, are dependent on subsidies, produce radioactive waste, and are an environmental blight across the nuclear energy life cycle.

COMMON GROUND: Nuclear energy could be promoted where it has a low carbon emissions profile, where its safety record is sound, and where risks can be made transparent and fully subsumed by consumers.

---

mining or downstream damages from spent waste are passed on to remote communities or future generations.

Third, given that nuclear technology already exists at an advanced stage of development and historical investment has already been committed, the prodigious amounts invested in the research and development process should not go entirely to waste, especially given the theoretical promise of nuclear technology. We should continue to invest in better safety and security standards and procedures to ensure that nations that choose to adopt nuclear power do so in the safest manner. Policymakers and communities may come to decide (as they have in Japan) that new facilities should be avoided but that reactors already built (and largely paid off), with high capacity factors, low-emissions profiles, and a strong track record of safety, are worth the risk. We should also ensure adequate funding for developing effective long-term nuclear waste storage solutions, since even a complete cessation of nuclear electricity generation would still leave us with thousands of tons of spent fuel.

Our synthesis incorporates these insights by acknowledging that if investment in new electricity generation is to be made, it should prioritize replacing nuclear power plants with renewable electricity or energy efficiency. Scarce investment funds ought to focus first on replacing fossil fuel–fired power plants with less risky technologies that mitigate emissions faster, sooner, and less dangerously. Therefore, nuclear power, like shale gas or clean coal, represents a possible bridge. This does not mean that humanity should turn its back on nuclear power entirely. Research should continue on extending the nuclear fuel cycle and designing advanced applications of nuclear technology. Fifty years from now, when wind turbines have become a regular sight in rural landscapes, we might find that advanced nuclear technology represents a less invasive solution. But for the time being, it is an imperfect technology that places current generations at risk in the event of an accident and saddles future generations with massive waste management burdens. As social critic Slavoj Zizek noted, "It is indeed true that we live in a society of risky choices, but it is one in which only some do the choosing, while others do the risking."[78] Our synthesis attempts to ensure that those who do the choosing of nuclear power or the consuming of its energy are those who subsume its risks.

NOTES

1. Norman Metzger, *Energy: The Continuing Crisis* (New York: Thomas Y. Crowell Company, 1984).

2. Richard Munson, *From Edison to Enron: The Business of Power and What It Means for the Future of Electricity* (London: Praeger, 2005); Otis Dudley Duncan, "Sociologists Should Reconsider Nuclear Energy," *Social Forces* 57, no. 1 (September 1978): 1–22.

3. Paul R. Josephson, *Red Atom: Russia's Nuclear Power Program from Stalin to Today* (New York: W. H. Freeman, 1999); Paul R. Josephson, "'Projects of the Century' in Soviet History: Large-Scale Technologies from Lenin to Gorbachev," *Technology and Culture* 36, no. 3 (1995): 519–559.

4. Albert Donnay and Martin Kuster, "France," in *Nuclear Wastelands: A Global Guide to Nuclear Weapons Production and Its Health and Environmental Effects*, ed. Arjun Makhijani, Howard Hu, and Katherine Yih (Cambridge, MA: MIT Press, 1995), 439.

5. B. K. Sovacool and S. V. Valentine, *The National Politics of Nuclear Power: Economics, Security, and Governance* (London: Routledge, 2012).

6. Quirin Schiermeier et al., "Electricity without Carbon," *Nature* 454 (August 2008): 816–823.

7. L. Echavarri, "Is Nuclear Energy at a Turning Point?" *Electricity Journal* 20, no. 9 (2008): 89–97.

8. Schiermeier et al., "Electricity without Carbon."

9. US Energy Information Administration (EIA), *Monthly Energy Review*, DOE/EIA-0035(2014/12) (Washington, DC: EIA, December 2014).

10. M. V. Ramana, "Nuclear Power: Economic, Safety, Health, and Environmental Issues of Near-Term Technologies," *Annual Review of Environment and Resources* 34 (2009): 127–152.

11. L. C. Cadwallader, *Occupational Safety Review of High Technology Facilities*, INEEL/EXT-05-02616 (Idaho Falls: Idaho National Engineering and Environmental Laboratory, January 2005).

12. World Nuclear Association, "Fukushima Accident" (November 2014), available at www.world-nuclear.org/info/Safety-and-Security/Safety-of-Plants/Fukushima-Accident.

The report states: "There have been no deaths or cases of radiation sickness from the nuclear accident."

13. B. K. Sovacool, "The Political Economy of Energy Poverty: A Review of Key Challenges," *Energy for Sustainable Development* 16, no. 3 (September 2012): 272–282.

14. See International Energy Agency (IEA), *World Energy Outlook 2009* (Paris: OECD, 2009).

15. John P. Holdren and Kirk R. Smith, "Energy, the Environment, and Health," in *World Energy Assessment: Energy and the Challenge of Sustainability*, ed. Tord Kjellstrom, David Streets, and Xiaodong Wang (New York: United Nations Development Programme, 2000), 61–110.

16. V. Kuznetsov, "Design and Technology Development Status and Design Considerations for Innovative Small and Medium Sized Reactors" (paper presented at Sixteenth International Conference on Nuclear Engineering, Orlando, FL, 2008).

17. Denis E. Beller, "Atomic Time Machines: Back to the Nuclear Future," *Journal of Land Resources and Environmental Law* 24 (2004): 41–61.

18. Steve Connor, "Nuclear Power Is the Greenest Option, Say Top Scientists," *Independent* (London), January 4, 2015.

19. See Editors of Pocket Books, *The Atomic Age Opens* (New York: Pocket Books, 1945), 202–203; Arjun Makhijani and Scott Saleska, *The Nuclear Power Deception: U.S. Nuclear Mythology from Electricity Too Cheap to Meter to Inherently Safe Reactors* (New York: Apex Press, 1999), 17.

20. Stuart Brand, *Whole Earth Discipline: An Ecopragmatist Manifesto* (New York: Viking Press, 2009).

21. B. L. Cohen, "Perspectives on the High Level Waste Disposal Problem," *Interdisciplinary Science Reviews* 23 (1998): 193–203.

22. H. Blix, "Nuclear Energy in the 21st Century," *Nuclear News*, September 1997, 34–48.

23. P. A. Kharecha and J. E. Hansen, "Prevented Mortality and Greenhouse Gas Emissions from Historical and Projected Nuclear Power," *Environmental Science and Technology* 47 (2013): 4889–4895.

24. Nuclear Energy Institute quoted in B. K. Sovacool, "Valuing the Greenhouse Gas Emissions from Nuclear Power: A Critical Survey," *Energy Policy* 36, no. 8 (August 2008): 2940–2953.

25. Kharecha and Hansen, "Prevented Mortality," 4890.

26. Bush quoted in Benjamin K. Sovacool and Christopher Cooper, "Nuclear Nonsense: Why Nuclear Power Is No Answer to Climate Change and the World's Post-Kyoto Energy Challenges," *William and Mary Environmental Law and Policy Review* 33, no. 1 (Fall 2008): 22.

27. Ridley quoted in ibid., 23.

28. IEA, *Technology Roadmap: Nuclear Energy* (Paris: OECD, June 2010).

29. Ibid., v.

30. Ibid., table 6, which lists $3,974 billion in funds needed from 2010 to 2050 in constant 2008 dollars.

31. IEA, *Energy Technology Perspectives 2012* (Paris: OECD, 2012).

32. International Atomic Energy Agency (IAEA), *Climate Change and Nuclear Power* (Vienna: IAEA, 2013), 5.

33. Letter quoted in John Upton, "More Nukes: James Hansen Leads Call for 'Safer Nuclear' Power to Save Climate," *Grist*, November 4, 2013, available at http://grist.org /news/more-nukes-james-hansen-leads-call-for-safer-nuclear-power-to-save-climate.

34. Moore quoted in "Greenpeace Co-Founder Says Nuclear Energy Is 'Only Option' " (Environmental News Service, July 2005), available at www.euronuclear.org/e-news/e -news-9/greenpeace.htm.

35. James Lovelock, "Nuclear Power Is the Only Green Solution," *Independent* (London), May 24, 2004.

36. Barry W. Brook and Corey J. A. Bradshaw, "Key Role for Nuclear Energy in Global Biodiversity Conservation," *Conservation Biology* 29, no. 3 (June 2015): 702–712.

37. "An Open Letter to Environmentalists on Nuclear Energy" (December 15, 2014), available at http://bravenewclimate.com/2014/12/15/an-open-letter-to-environmentalists -on-nuclear-energy.

38. Vaclav Smil, "Energy in the Twentieth Century: Resources, Conversions, Costs, Uses, and Consequences," *Annual Review of Energy and Environment* 25 (2000): 46.

39. P. R. Russell, "Prices Are Rising: Nuclear Cost Estimates under Pressure," *Energy-Biz Insider*, May–June 2008, 22.

40. Keystone Center, *Nuclear Power Joint Fact-Finding* (Keystone, CO: Keystone Center, June 2007).

41. Mark Cooper, "From Climate Deniers to Cost Deceivers: The Importance of Reality Based, Forward Looking Analysis in Low Carbon Resources Acquisition" (Institute for Energy and the Environment, Vermont Law School, 2014).

42. US Congressional Budget Office (CBO), *Nuclear Power's Role in Generating Electricity* (Washington, DC: CBO, May 2008).

43. B. K. Sovacool, D. Nugent, and A. Gilbert. "An International Comparative Assessment of Construction Cost Overruns for Electricity Infrastructure," *Energy Research and Social Science* 3 (September 2014): 152–160.

44. B. K. Sovacool, "Questioning the Safety and Reliability of Nuclear Power: An Assessment of Nuclear Incidents and Accidents," *Gaia* 20, no. 2 (June 2011): 95–103.

45. Massachusetts Institute of Technology, "The Future of Nuclear Power" (Interdisciplinary MIT Study, 2003), available at http://web.mit.edu/nuclearpower/pdf/nuclearpower -summary.pdf.

46. Ibid., 68.

47. Douglas Koplow, *Nuclear Power: Still Not Viable without Subsidies* (Washington, DC: Union of Concerned Scientists, February 2011), 2–3.

48. P. A. Bradford, "How to Close the US Nuclear Industry: Do Nothing," *Bulletin of the Atomic Scientists* 69, no. 2 (2013): 12–21.

49. IEA, *World Energy Outlook 2002* (Paris: OECD, 2002).

50. B. K. Sovacool, "Deploying Off-grid Technology to Eradicate Energy Poverty," *Science* 338 (October 5, 2012): 47–48.

51. Sreyamsa Bairiganjan et al., *Power to the People: Investing in Clean Energy for the Base of the Pyramid in India* (Washington, DC: World Resources Institute, 2010).

52. See A. Funk and B. K. Sovacool, "Wasted Opportunities: Resolving the Impasse in United States Nuclear Waste Policy," *Energy Law Journal* 34, no. 1 (May 2013): 113–147; B. K. Sovacool and A. Funk. "Wrestling with the Hydra of Nuclear Waste Storage in the United States," *Electricity Journal* 26, no. 2 (March 2013): 67–78.

53. Blue Ribbon Commission on America's Nuclear Future, *Report to the Secretary of Energy* (Washington, DC: US Department of Energy, January 2012), 14.

54. Allison Macfarlane, "Interim Storage of Spent Fuel in the United States," *Annual Review of Energy and Environment* 26 (2001): 222.

55. Alvin M. Weinberg, "Immortal Energy Systems and Intergenerational Justice," *Energy Policy*, February 1985, 51–59.

56. Alvin M. Weinberg, "Social Institutions and Nuclear Energy," *Science* 177, no. 4043 (1972): 27–34.

57. T. B. Taylor, "Nuclear Power and Nuclear Weapons." *Science and Global Security* 13 (2005): 117–128.

58. Alfven quoted in Alexander Shlyakhter, Klaus Stadie, and Richard Wilson, *Constraints Limiting the Expansion of Nuclear Energy* (Washington, DC: Global Strategy Council, 1995).

59. Sovacool, *Contesting the Future*, chap. 6.

60. David Thorpe, "Extracting Disaster," *Guardian* (London), December 5, 2008.

61. V. V. Shatalov et al., "Ecological Safety of Underground Leaching of Uranium," *Atomic Energy* 91, no. 6 (2001): 1009–1015.

62. Gavin M. Mudd and Mark Disendorf, "Sustainability of Uranium Mining and Milling: Toward Quantifying Resources and Eco-Efficiency," *Environmental Science and Technology* 42 (2008): 2624–2629.

63. Thorpe, "Extracting Disaster."

64. Electric Power Research Institute (EPRI), *Water & Sustainability, Volumes 1–4: U.S. Electricity Consumption for Water Supply & Treatment—The Next Half Century*, Topical Report 1006787 (Concord, CA: EPRI, March 2002).

65. Paul Faeth et al., *A Clash of Competing Necessities: Water Adequacy and Electric Reliability in China, India, France, and Texas* (Alexandria, VA: CNA, July 2014).

66. B. K. Sovacool, "Valuing the Greenhouse Gas Emissions from Nuclear Power: A Critical Survey," *Energy Policy* 36, no. 8 (August 2008): 2940–2953.

67. Mudd and Disendorf, "Sustainability of Uranium Mining."

68. Oxford Research Group quoted in Sovacool and Cooper, "Nuclear Nonsense," 98.

69. Jef Beerten et al., "Greenhouse Gas Emissions in the Nuclear Life Cycle: A Balanced Appraisal," *Energy Policy* 37, no. 12 (December 2009): 5056–5068.

70. Bill Keepin and Gregory Kats, "Greenhouse Warming: Comparative Analysis of Nuclear and Efficiency Abatement Strategies," *Energy Policy* 16, no. 6 (December 1988): 538–561.

71. McKinsey & Company, "Impact of the Financial Crisis on Carbon Economics: Version 2.1 of the Global Greenhouse Gas Abatement Cost Curve" (2010), 8, available at www.mckinsey.com/client_service/sustainability/latest_thinking/greenhouse_gas _abatement_cost_curves.

72. B. K. Sovacool, "Exposing the Paradoxes of Climate and Energy Governance," *International Studies Review* 16, no. 2 (June 2014): 294–297.

73. B. K. Sovacool et al., "Comment on 'Prevented Mortality and Greenhouse Gas Emissions from Historical and Projected Nuclear Power,'" *Environmental Science and Technology* 47, no. 12 (May 2013): 6717.

74. Mark Cooper, *The Economics of Nuclear Reactors: Renaissance or Relapse?* (Montpellier, VT: Institute for Energy and the Environment, June 2009).

75. Sovacool, "Valuing Greenhouse Gas Emissions."

76. Ibid.

77. Comment by Peter Bradford in "Nuclear Power Is Unsafe Because the Operators Are Pinching Pennies and Cutting Corners," *Washington Post*, December 21, 2011.

78. Slavoj Žižek, *First as Tragedy, Then as Farce* (New York: Verso, 2009), 13.

Is National Energy Independence Feasible and Desirable?

US President George W. Bush did not cite jobs, climate change, or even technological competitiveness as justification for launching the "Hydrogen Fuel Initiative" in 2003, even though the initiative would contribute to these areas. He simply explained it in this way: "We need energy independence for this Nation. It's important for our country to understand—I think most Americans do—that we import over half of our crude oil stocks from abroad. And sometimes we import that oil from countries that don't particularly like us. It puts us at a—it jeopardizes our national security to be dependent on sources of energy from countries that don't care for America, what we stand for, what we love. It's also a matter of economic security, to be dependent on energy from volatile regions of the world. Our economy becomes subject to price shocks or shortages or disruptions or, one time in our history, cartels."[1]

Prince Turki al-Faisal, director of intelligence of Saudi Arabia (the nation that boasts the world's largest oil reserves) and ambassador to the United States, harbors a competing perspective on the importance of energy independence. His view is that calls for "energy independence" represent "political posturing at its worst—a concept that is unrealistic, misguided, and ultimately harmful to energy-producing and -consuming countries."[2]

Neither perspective should come as a surprise. In 2003, the United States was still recovering from September 11, and its military was mired in operations designed to track down al-Qaida terrorists. It made some sense for leaders of a country like the United States to minimize dependence on imported oil, given that such imports result in transfer of an enormous amount of wealth overseas, which in some circumstances has been channeled to terrorist groups. It also makes sense for leaders of a country like Saudi Arabia to encourage oil interdependence. The country derives 85% of its total revenue from the oil industry. Every dollar increase in the price of a barrel of oil translated into a gain of about $3 billion for the Saudi treasury.[3]

We explore here the desirability of energy independence in terms of enhancing energy security for a given nation. Many national leaders have outlined aspirations for their country to be energy-independent—free from the need to import energy fuels or systems—or at least to be less import-dependent. In the

United States, for example, every administration since Richard Nixon was president has announced a goal of increasing US energy independence. The premise appears to be that domestic energy production and self-sufficiency should be goals that every country should aspire to. Indeed, a number of countries around the world—including Russia, Saudi Arabia, Australia, and Denmark—have achieved a degree of independence over the past few decades.

There is an alternate perspective. In today's globalized world, no country can ever be truly energy-independent, nor should they try to be, because it is a suboptimal strategy. Daniel Yergin wrote that "the U.S. must face the uncomfortable fact that its goal of energy independence . . . is increasingly at odds with reality"[4] And C. Fred Bergsten stated that the idea of oil independence is "ridiculous," since it implies that a country would be committed to decreasing its reliance on imports at any cost.[5] Both perspectives are underpinned by some strong arguments.

One Side: Energy Independence Is A Compelling Goal

We review three of the more prominent arguments in favor of enhanced energy self-sufficiency and independence: (1) dependence exacerbates economic risk, (2) dependence is a geopolitical quagmire, and (3) many countries can achieve a high degree of self-sufficiency. To proponents of this view, it seems illogical not to want control over a resource as important as energy.

Energy Dependence Increases Economic Risk

In the United States, the world's largest importer of crude oil (by volume), reliance on imports has transferred immense wealth to petroleum producers. Researchers at Oak Ridge National Laboratory (ORNL) estimated that from 1970 to 2004, US dependence on imported oil cost the country $5.6 to $14.6 trillion in lost wealth. This exceeds the financial losses from all wars fought by the United States going back to the Revolutionary War.[6] The leak of wealth continues in the new millennium. As figure 13.1 indicates, from 2005 to 2010, import dependence resulted in another $2.1 trillion in lost wealth, when factoring in the indirect economic costs of macroeconomic shocks, economic dislocation, and transfers of wealth.[7]

Although ORNL's numbers are limited to the United States, all major economies face similar risks. The emerging projections of oil production and consumption over the next two decades suggest that oil dependence in Europe, China, India, and other Asian countries will grow rapidly, each nation or region relying on imports to meet at least 75% of oil demand by 2030. Consequently, when it comes to oil, most major economies will face much the same security risks as the United States.[8]

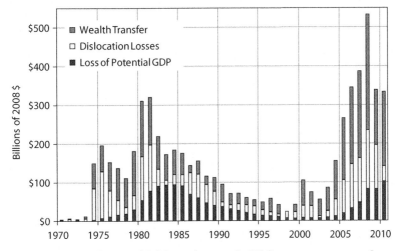

Figure 13.1. Estimated Costs of Oil Dependence to the US Economy, 1970–2010. *Source:* David L. Greene et al., *OPEC and the Costs to the U.S. Economy of Oil Dependence: 1970–2010*, White Paper 1-13 (Knoxville, TN: Howard Baker Center for Public Policy, 2013).

The risks of oil dependence may be even more precarious for many developing countries. Increases in the costs of crude oil and gasoline suggest that the foreign exchange expended to procure oil imports will adversely affect the balance of trade in many developing countries. This will result in a massive transfer of wealth that could otherwise go to bolstering domestic industrial capacity. Although developed countries spend just 2% or 3% of gross domestic product (GDP) on imported oil, nations in the developing world spend an average of 4.5% to 9% of GDP on oil imports. One study looked at the economic impact of oil price inflation in 161 countries from 1996 to 2006, when prices increased by a factor of seven. The authors concluded that lower-middle-income countries were the most vulnerable, followed by low-income countries, even though they consumed less oil per capita than industrialized or high-income countries.[9]

Energy Dependence Is a Geopolitical Nightmare

In addition to the economic costs associated with keeping oil supply channels running smoothly, concern over geopolitical costs has begun to take center stage in the argument for self-sufficiency. There is evidence that some of the oil revenues of certain nations are being channeled to groups that sponsor terrorism. As Pulitzer Prize–winning journalist Thomas Friedman put it, "Through our energy purchases, we are funding both sides of the war on terror."[10] Friedman went on to draw links between the $165 to $180 billion that Saudi Arabia earned each year between 2006 and 2008 and the financing of terrorist groups. Aside from support for terrorists, energy revenues in nations such as Russia and Venezuela are emboldening leaders of these nations to exploit their energy riches to alter geopolitics. For example, in 2007, Russia shut down a pipeline carrying oil exports to Europe, on the heels of a dispute with Belarus and Ukraine. Venezuela has increased oil exports to Iran and China in a strategy that some analysts believe is intended to undermine US interests.[11]

With oil profits funding terrorist activities and supporting the antics of authoritarian leaders, some analysts are now arguing that the best way to enhance global stability is to curtail access to oil revenues by reducing import dependency. Former CIA director James Woolsey pointed out that ending US addiction to oil imported from the Middle East would go a long way toward defanging al-Qaida. In fact, until the execution of Osama Bin Laden, Woolsey drove an electric vehicle sporting a bumper sticker that read "Bin Laden Hates This Car."[12]

To be sure, there is a degree of interconnectedness between economic security and geopolitical energy risks. For the United States and other global powers, dependence on imported oil has engendered significant costs in preserving supply channels. These costs are seldom factored into the cost of oil because they are buried within defense budgets—but they are readily acknowledged and accepted by global powers as the necessary facet of oil dependency. This has been true as far back as the OPEC embargo of 1973. Then, President Gerald Ford asked the Congressional Research Service to draft a report assessing the consequences of "the possible use of U.S. military force to occupy foreign oil fields in exigency."[13] Similarly, at the height of the Iran-Iraq war in the 1980s, when Iranian forces began attacking Kuwaiti oil tankers traveling through the Persian Gulf in an attempt to discourage Kuwait from supplying loans to Iraq for arms procurement, President Roald Reagan authorized the reflagging of Kuwaiti tankers with the US ensign to afford them naval protection. The Clinton administration and both Bush administrations funneled billions of dollars into protecting the Persian Gulf and other oil assets. The US Southern Command now promotes security cooperation activities to expand US influence in oil-producing regions of South America. These activities include training, equipping, and developing security forces to protect refineries and offshore oil and gas platforms. In central Asia, the United States still operates programs to train and equip Georgian and Uzbeki security forces with the capability to maintain the free flow of oil essential to the US economy. In western Africa, military aid and training have been funneled to Nigeria (the third-largest supplier of oil to the United States) to help bolster the security of its oil infrastructure.[14]

The cost—both economic and political—of such activities quickly adds up. One assessment calculated that the costs of the military operations in the Persian Gulf for protecting oil assets and infrastructure range from $50 to $100 billion per year.[15] Another study put the figure at between $29 and $80 billion per year.[16] Summing up the situation, energy analyst Gal Luft concluded that "because the American way of life is one of the most energy-intensive in the world, U.S. oil dependence is a source of great national security threats."[17] Indeed, it seems that if the United States weaned itself from oil imports, there would be a huge number of military personnel looking for work.

Many Countries Can Achieve a High Degree of Self-Sufficiency
One of the biggest objections to self-sufficiency in energy is that for many nations, self-sufficiency simply is not possible. However, data suggest that prospects are rosier than critics contend. Of the 137 countries on which the International Energy Agency (IEA) compiled data in 2011, 47 were self-sufficient when measured in terms of energy production divided by total primary energy consumption. Although this list contained many small economies, it also included Russia (180% self-sufficiency), Saudi Arabia (322%), and Australia (241%). Many South American nations were energy self-sufficient, as was the entire continent of energy-poor Africa. The United States was 81% self-sufficient based on this metric.[18]

One thing becomes clear when it comes to energy self-sufficiency: holding on to energy revenues enhances domestic financial clout. Take Russia, for instance, which is "resource and energy-product rich."[19] It possesses the world's largest natural gas reserves (second in production, behind the United States), second-largest coal reserves (sixth in production), and seventh-largest oil reserves (third in production, behind Saudi Arabia and the United States).[20] Consequently, oil and gas contribute to roughly half of Russia's GDP and two-thirds of its exports. As a recent assessment surmised, "Russia's status as a current and future energy producer is close to unrivaled."[21] Russia's energy resources are so large that some have speculated that the country can use its "energy weapon" to leverage higher prices with European customers and to further its national and strategic goals.[22]

Saudi Arabia serves as another example. It is the world's largest producer and exporter of crude oil, has the most spare crude oil production capacity, and is the second-largest refiner of crude oil and petroleum products. Not only does the nation sit atop one-quarter of the earth's proven oil reserves, but it also enjoys extremely low production costs due to the shallow depth at which oil is found. A barrel of oil costs less than $5 to produce. In 2009, Saudi Arabia earned a staggering $153 billion from oil exports. The same year, it was responsible for 17% of the world's oil exports, the single largest share of any country.[23]

Australia is "generally self-sufficient in energy resources particularly coal, gas and oil."[24] Black coal deposits are found in every state, and the country also has close to 40% of the world's uranium reserves. Since 1986, Australia has been the largest exporter of coal, and in 1989, it became the largest exporter of uranium.[25] It also possesses the largest reserves of natural gas in the Asia Pacific region, enabling it to export significant quantities of liquefied natural gas (LNG). Hence, as one energy security scholar put it, "in a region where the general trend is clearly towards lower levels of energy self-sufficiency, Australia is already a net energy exporter of considerable significance—and the degree of that significance looks likely to increase."[26]

Even in the United States, a nation renowned for problems caused by dependency on energy imports, there is a drive to improve energy self-sufficiency, and evidence indicates that the nation has sufficient resource potential to achieve this goal. A comprehensive study undertaken by the US Department of Energy in the late 1980s calculated that the "total resource base of energy" for the country amounted to more than 657,000 billion barrels of oil equivalent, or more than 46,800 times the annual rate of national energy consumption at that time. Of this amount, 115,000 billion barrels of oil equivalent were estimated to be commercially viable, still amounting to thousands of times the national energy use.[27] This study was validated by a broad group of researchers at ORNL, US Geological Survey, Pacific Northwest National Laboratory, Sandia National Laboratory, National Renewable Energy Laboratory, Colorado School of Mines, and Pennsylvania State University. Perhaps the most remarkable feature of this estimate is that it was undertaken before the so-called shale gas boom (see chapter 5). As one energy analyst put it, "Not only could the U.S. be energy independent, but [it] could be a major exporter to other nations."[28]

The Other Side: Interdependence Better Reflects Energy Realities

The opposing view argues that energy interdependence, rather than independence, is what nations should aspire to. All energy producers and exporters are bound to consumers and importers in mutual interdependencies through interlocked markets. Critics argue that for multiple reasons—the interconnectivity of energy systems, established global supply networks, consortiums of investors, and governing institutions—it is both impossible and undesirable to be truly independent. For many of these reasons, Jason Grumet told US senators that oil independence "is an emotionally compelling concept, but it's a vestige of a world that no longer exists."[29] Philip J. Deutch argues that "it may be a noble statement of ultimate intentions, but as a practical matter energy independence is absurd."[30]

Even Self-Sufficient Countries Are Not Necessarily Energy-Secure

Advocates of interdependence point out that energy self-sufficient countries may not necessarily be more energy-secure. Moreover, energy independence can be a deceptive concept. Many of the nations identified above as being energy self-sufficient (in terms of total primary energy production divided by demand) are not self-sufficient in all forms of energy, and all are affected by global energy market dynamics and dependent on outside cooperation.

Russia is not an energy island unto itself. It exports gas and oil but must import uranium, energy machinery and equipment, and even electricity at times. From 1999 to 2011, for example, Russia depended on the European Union for two-thirds of its imports of industrial machinery, computers, and transport equipment. Many of these technologies were then utilized by the energy industry to extract, convert, and export hydrocarbon-laden fuels back to the EU countries.[31] In 2012, Russia imported billions of kilowatt-hours of electricity from countries such as Finland, Kazakhstan, and Mongolia.[32]

Saudi Arabia also exhibits gaps in energy independence. It is a net exporter of petroleum products, but it must import refined gasoline. During the summer of 2013, the nation imported approximately 4.5 million barrels of gasoline and 8.9 million barrels of diesel fuel per month.[33] Moreover, its extractive industries sector heavily depends on imported equipment and foreign workers: 7 out of 10 jobs in the country (and close to 9 out of every 10 private sector jobs) are filled by foreigners.[34] Saudi Arabia also depends on the international economy to supply computer energy software systems, SCADA (supervisory control and data acquisition) devices, and even drill bits for oil and gas wells.[35] If all nations decided to build fences around their energy support industries, Saudi Arabia's economy would collapse.

Australia's energy independence is also far from universal. It is one of the world's largest exporters of coal and uranium, but it is only 50% self-sufficient in oil and gas, and it is a significant importer of refined diesel fuel. Furthermore, its energy riches have not made it immune to disruptions in supply. A September 1988 explosion and fire at an Esso gas plant caused $1.3 billion in damage and disrupted service for more than four million customers. In 2003, severe storms interrupted the delivery of aviation fuel to Sydney's international airport, resulting in the need to ration fuel for aircraft and cancel flights. In 2008, an explosion at a natural gas facility on Varanus Island caused a 30% reduction in supply, with adverse consequences for industrial customers, households, and farmers.[36]

No Nation Can Escape the Influence of Global Energy Markets

Many critics argue that energy independence is not even feasible in a globalized marketplace. This premise rests on four interconnected claims related to (1) the global trade in energy fuels, (2) the transnational impact of major national energy events, (3) global investment consortiums entrenched in energy infrastructure ownership, and (4) the political clout of institutions that govern energy production and use.

First, fossil fuel and uranium resources are fungible goods, and there is a global market for these goods. Changes in global price affect any nation that utilizes these commodities. If the cost of a barrel of oil skyrockets, nations that use a lot of oil will lose competitive advantage to nations that do not. If the price of oil drops, the reverse applies, and nations that use a lot of oil will gain competitive advantage. Since changes in the cost of energy commodities alter the comparative attractiveness of energy technologies, price inflation for any given energy commodity could cause some nations that are dependent on that resource to consider a shift to another. The global energy market is a butterfly of behemoth proportions, and when it flaps its wings, domestic energy markets feel the draft.

Second, because of the interrelated nature of the global energy market, actions taken by individual countries or accidents such as Fukushima can have ramifications well beyond the national borders where the event occurred. No country is immune to the actions of others. For example, the shale gas boom in the United States has undermined the US market for coal and has motivated many US coal suppliers to turn to export markets. Consequently, US coal has recently inundated global markets, depressing coal prices.[37] The abundance of cheap shale gas in the United States is also having significant repercussions in global LNG markets. The flood of US shale gas has forced LNG producers in Qatar, Trinidad and Tobago, Oman, and other areas to rein in production. In 2013, analysts determined that North American output of low-priced shale gas had placed $160 billion worth of natural gas investments by major companies such as Chevron and ExxonMobil at risk in Australia.[38]

The impact of the nuclear accident in Fukushima in March 2011 extended far beyond Japanese borders. The global price of uranium promptly dropped 25% (as supply exceeded demand due to the unexpected shutdown of dozens of Japanese reactors). Conversely, there was a sudden increase in the cost of energy substitutes such as oil, coal, and natural gas. In particular, the global price for LNG rose considerably in response to news that Japan would switch to natural gas power to make up for its shortfall in electricity.[39] Within a week of the disaster, gas and coal prices increased 13.4% and 10.8%, respectively, underscoring the global ramifications of a single energy-related event.[40]

Third, energy independence is seldom feasible because national generation capacity is often privately owned, and these private entities tend to operate across borders and prefer to do so in order to diversify risk. In 2012, roughly 40% of energy infrastructure projects around the world had capital value of $10 billion or more, meaning that they required large international consortiums of financiers across numerous countries to back them.[41] The $4.6 billion Baku-Tbilisi-Ceyhan oil pipeline, for instance, was financed by a consortium led by British Petroleum and including nine American, Turkish, Norwegian, Italian, French, and Japanese oil companies, the World Bank Group's International Finance Corporation, the European Bank for Reconstruction and Development, the export credit agencies of seven countries, and a syndicate of 15 commercial banks. The $40 billion Trans-ASEAN Natural Gas Pipeline Network in Asia featured a list of 37 formal stakeholders.[42] Even a single exploratory oil rig such as *Deepwater Horizon* (which exploded in the Gulf of Mexico in 2010) was built by Hyundai Heavy Industries in South Korea, flagged by the Marshall Islands, operated by the Swiss conglomerate Transocean, and leased to a British firm (BP) to explore for oil in the United States.[43] In short, even if nations wanted to secure domestic energy independence, chances are that the companies that own the infrastructure would still be exposed to external market risk.

Finally, any major nation that aspired to go it alone in terms of energy self-sufficiency would have to extricate itself from global alliances that currently provide a significant amount of energy security. There are a number of such entities. Some are intergovernmental organizations (IGOs), created and funded by national governments, which have secretariats that answer to a governing body, such as the IEA. Some are entities that have arisen from summit processes that offer a sort of halfway house between formal IGOs and the normal practices of diplomacy between national governments. These entities typically have no charter, fixed membership, or secretariat, but they offer a flexible way to address pressing multilateral problems. Others are international nongovernmental organizations, which usually have boards and receive funding from both the public and private sectors. Other alliances include multilateral financial institutions, regional organizations that involve two or more countries as members, and even hybrid entities including everything from transnational networks of advocacy to quasi-regulatory private bodies, global policy networks, and public-private partnerships. In 2012, no less than 50 of these global actors operated in the energy landscape, with some having done so for decades (see table 13.1).[44] These alliances provide a high degree of economic and supply-chain security that arguably outweighs any benefits associated with severing ties and going all out for independence.

Table 13.1. Fifty Major Global Institutions Involved in the Governance of Energy

Institution	Form	Year of creation	Central location(s)
World Energy Council	INGO	1923	London, UK
World Bank Group	MFI	1944	Washington, DC, USA
United Nations System	IGO	1945	New York, USA; Vienna, Austria; Geneva, Switzerland
The Inter-American Development Bank	MFI	1959	Washington, DC, USA
Organization of Petroleum Exporting Countries	IGO	1960	Vienna, Austria
Organization of Economic Cooperation and Development	IGO	1961	Paris, France
African Development Bank	MFI	1964	Abidjan, Côte d'Ivoire
Asian Development Bank	MFI	1966	Manila, Philippines
Association of Southeast Asian Nations	RO	1967	Jakarta, Indonesia
Organización Latinoamericana de Energía	IGO	1973	Quito, Ecuador
International Energy Agency	IGO	1974	Paris, France
Group of Eight	SP	1975	Rotating meetings hosted in turn by member states
International Fund for Agricultural Development	MFI	1977	Rome, Italy
International Institute for Energy Conservation	Hybrid	1984	Vienna, VA, USA
South Asian Association for Regional Cooperation	RO	1985	Katmandu, Nepal
Asia-Pacific Economic Cooperation	RO	1989	Singapore
Solar Electric Light Fund	Hybrid	1990	Washington, DC, USA
Global Environment Facility	IGO	1991	Washington, DC, USA
Energy Charter Conference Treaty	IGO	1991	Brussels, Belgium
International Energy Forum	IGO	1991	Riyadh, Saudi Arabia
Global Energy Network Institute	INGO	1991	San Diego, CA, USA
European Bank for Reconstruction and Development	MFI	1991	London, UK
Organization of the Black Sea Economic Cooperation	RO	1992	Istanbul, Turkey
Southern African Development Community	RO	1992	Gaborone, Botswana
International Network for Sustainable Energy	Hybrid	1992	Hjortshøj, Denmark
European Union	RO	1993	Brussels, Belgium
Summit of the Americas	SP	1994	Washington, DC, USA

Name	Type	Year	Location
World Business Council on Sustainable Development	Hybrid	1995	Geneva, Switzerland
International Network on Gender and Sustainable Energy	Hybrid	1996	Leusden, Netherlands
Central Asia Regional Economic Cooperation	Hybrid	1997	Manila, Philippines
Energy Through Enterprise	Hybrid	1997	Bloomfield, NJ, USA
Collaborative Labeling and Appliance Standards Program	Hybrid	1999	Washington, DC, USA
Generation IV International Forum	IGO	2001	Paris, France
Gas Exporting Countries Forum	IGO	2001	Doha, Qatar
World Council for Renewable Energies	INGO	2001	Bonn, Germany
Shanghai Cooperation Organization	RO	2001	Beijing, China.
Acumen Fund	Hybrid	2001	New York, NY, USA
Renewable Energy and Energy Efficiency Partnership	Hybrid	2002	Vienna, Austria
Global Network on Energy for Sustainable Development	Hybrid	2002	Roskilde, Denmark
Efficient Energy for Sustainable Development Partnership	Hybrid	2002	Washington, DC, USA
Partnership for Clean Fuels and Vehicles	Hybrid	2002	Nairobi, Kenya
Small-Scale Sustainable Infrastructure Development Fund	Hybrid	2002	Cambridge, MA, USA
International Partnership for the Hydrogen Economy	IGO	2003	Berlin, Germany
Global Energy Efficiency and Renewable Energy Fund	Hybrid	2004	European Investment Bank, Luxembourg
Appropriate Infrastructure Development Group	Hybrid	2005	Boston, USA
Global Village Energy Partnership	Hybrid	2005	London, UK
Clinton Climate Initiative	Hybrid	2006	New York, NY, USA
International Renewable Energy Agency	IGO	2009	Abu Dhabi, United Arab Emirates
Global Alliance for Clean Cookstoves	Hybrid	2010	Washington, DC, USA
Green Climate Fund	Hybrid	2013	Songdo, South Korea

Source: B. K. Sovacool and A. E. Florini, "Examining the Complications of Climate and Energy Governance," *Journal of Energy and Natural Resources Law* 30, no. 3 (August 2012): 235–263.

Notes: Hybrid = public private partnership or various combinations of other forms; IGO = intergovernmental organization; INGO = international nongovernmental organization; MFI = multilateral financial institution; RO = regional organization; SP = summit process.

Many Energy Externalities Cross National Borders

Another line of reasoning in support of the argument that energy independence is not feasible concerns externalities from the energy sector. Decisions that any given nation makes in terms of energy mix can significantly affect the welfare of other nations. Therefore, as long as there are energy externalities, there will be a cross-border component to energy security.

One example of this involves the strategic decision to adopt "risky" technologies such as nuclear power (see chapter 11). A nation adopting nuclear power places its neighbors at risk—hence Singapore's concerns over a nuclear accident in Indonesia. Such a decision can also alter the strategic decisions being made within any given country. For example, China's rapid buildup of nuclear power capacity doesn't just put South Korea, Japan, and Taiwan at risk in the event of a nuclear disaster; it also enables support for domestic buildup of nuclear power in these nations—as the argument goes, if Japan is at risk due to China's nuclear program, the added risk of continuing with its own domestic program is negligible.

Energy interdependence related to externalities also involves other forms of transboundary pollution stemming from virtually every form of conventional energy, carbon dioxide being the most notable but far from the only important pollutant. Consider mercury emissions from coal-fired power plants. Approximately two-thirds of the anthropogenic mercury emissions generated in the United States, largely from coal, are dispersed to regions outside its borders. Conversely, 40% of the mercury emissions settling on US land come from elsewhere, mostly from Russia and the Asia Pacific.[45] Similarly for sulfur dioxide emissions, also associated with coal-fired power, between 50% and 70% of Canada's acid rain comes from the United States, while 2% to 10% of acid rain in the United States stems from Canadian sulfur emissions.[46] Across the ocean in Europe, roughly half of ambient air pollution there is caused by the global circulation of pollutants from outside the region.[47]

With regard to nuclear power externalities, modelers at the US National Oceanic and Atmospheric Administration's Pacific Marine Environmental Laboratory and Germany's GEOMAR Research Center for Marine Geosciences calculated that, due to ocean currents and the way that cesium-137 concentrates in marine ecosystems, the radioactive plume from the Fukushima accident in Japan would be as much as 10 times more dangerous than it was in coastal Japan when it reached the Pacific Coast of North America in 2015.[48] Scientists determined that the plume would affect a sizable part of the ocean and pass near the borders of almost a dozen countries. Some scenarios, such as the one shown in figure 13.2, anticipate contamination across the entire hemisphere within little more than a decade.

TIME: 19–FEB–2027 10:38   DATA SET: res4–0000

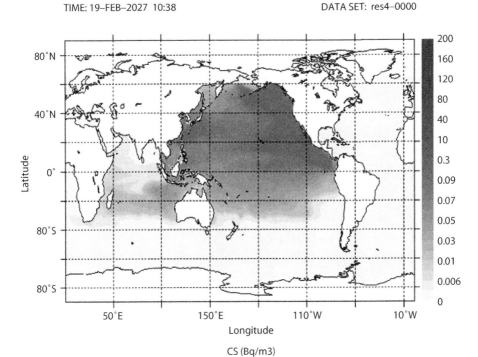

Figure 13.2. Dispersal of Cesium-137 Following the Fukushima Nuclear Accident, 2011–2027. *Source:* China-Korea Joint Ocean Research Center, "Research Activities, Development of Models, Long-Term Dispersion Model," available at www.mrcor.org/Research.aspx?m_id =5. *Note:* Bq = becquerel.

As an example of hydropower exploitation, Turkey, Syria, and Iran have heavily dammed the headwaters that flow into the Tigris and Euphrates Rivers (for electricity and irrigation), impairing the capacity of the Shatt al Arab River to supply water for livestock, crops, and drinking. This has forced tens of thousands of Iraqi farmers to abandon their fields.[49]

With regard to biofuel, forest fires in Indonesia, largely caused by slash-and-burn techniques to clear ground for palm oil plantations (for biodiesel), cause transboundary haze that threatens Malaysia and Singapore every year—amplifying hospital admissions and stifling tourism.[50]

Clearly, given these externalities, no nation can claim to be fully self-reliant in energy without accepting a moral obligation to cooperate on mitigating such external damage. This suggests that energy choices also have to be made with the interests of neighboring states in mind.

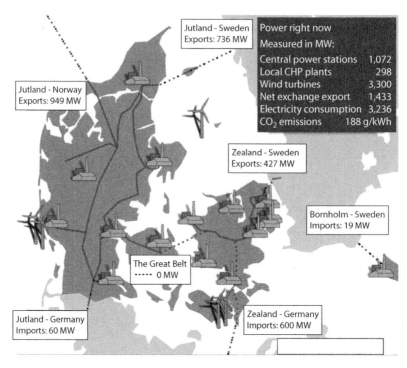

Jutland - Sweden
Exports: 736 MW

Power right now
Measured in MW:
Central power stations   1,072
Local CHP plants   298
Wind turbines   3,300
Net exchange export   1,433
Electricity consumption   3,236
CO_2 emissions   188 g/kWh

Jutland - Norway
Exports: 949 MW

Zealand - Sweden
Exports: 427 MW

Bornholm - Sweden
Imports: 19 MW

The Great Belt
····· 0 MW

Jutland - Germany
Imports: 60 MW

Zealand - Germany
Imports: 600 MW

Figure 13.3. Imports and Exports on the Danish National Electricity Grid, June 2013. *Source*: Energinet.dk (September 2014). *Note:* CHP = combined heat and power.

Interdependence Optimizes Technological Integration

A final strong argument in support of energy interdependence stems from evidence that many energy systems, such as renewable electricity, become more reliable (and less expensive) when they are more interconnected. In other words, grid interconnectivity infuses a degree of stability into a nation's electricity network.

As testament to this, many nations of the European Union are enjoying burgeoning renewable energy markets because the EU grid allows nations with high renewable energy capacity to balance supply. For example, without Denmark's connectivity to Germany and Sweden, it is doubtful that the nation would have achieved the high levels of wind power penetration that it currently enjoys and underpins why some experts deem it feasible for the nation to meet 100% of its electricity needs through renewable energy in the future.[51] As figure 13.3 shows, the country depends heavily on Germany and its neighbors in the Nordic Power Pool to balance electricity loads and to

BOX 13.1
Energy Independence and the Common Ground

Is national energy independence feasible or desirable?

ONE SIDE: Energy independence provides a buffer against energy price volatility, reduces international wealth transfers, and is a compelling and achievable goal for many countries.

OTHER SIDE: No country is truly energy-independent, and energy fuels, investment, governance, and pollution are international in scope.

COMMON GROUND: We should encourage self-sufficiency for all of its benefits, but recognize global dependencies.

---

provide customers for surplus wind-generated electricity. Conversely, the absence of grid interconnectivity has proved to be an obstacle to diffusion of renewable energy. For example, Japan's failure to connect the Japanese grid to the rest of Asia is one of the critical barriers to renewable energy diffusion in that nation.[52]

Common Ground: We Can Pursue Self-Sufficiency while Recognizing Interdependence

These two incompatible views have commensurate elements leading us to the common ground shown in box 13.1: pursue self-sufficiency and all of the benefits it can bring, but recognize global interdependencies. Energy self-sufficient nations, although they may still depend on others for some key fuels or technologies, are able to exert more control over their own destinies. They are more immune to interruptions in global supply, more able to create positive trading relationships that do not transfer wealth abroad, and more resilient to sudden changes in price. These are powerful reasons to seek to enhance prospects for self-sufficiency.

To say that self-sufficiency is an admirable aspiration, however, does not mean that dependencies do not exist. Even the world's major energy superpowers depend on other countries for some forms of fuel, skilled labor, access to technology, critical materials, and demand for their products. Moreover, no country is immune to the impact of global energy market dynamics, market disruptions, and energy mishaps. It's not only the global energy system that is interdependent; so, too, are the associated problems that arise from energy use. Energy-related pollution problems are largely transboundary—such as carbon dioxide's role in climate change, mercury contamination of a river, or acid rain. No nation is an island unto itself when it comes to energy. In terms of national capacities to enhance energy self-sufficiency, we may not all be in the same type of boat, but we are all subject to the same weather. This means that independence, in its purest sense, is probably unachievable.

Independence does not necessarily mean complete cessation of energy imports. ORNL analysts David Greene and Paul Leiby argue that the United States could accomplish a form of oil independence if the meaning of the term is viewed as achieving a state in which the nation's decisions are not subject to influence by other nations because of its need for oil.[53] Conceptualized in this manner, oil independence does not mean eliminating imports or weaning the country completely off oil; instead, it means creating a world where the estimated total economic costs of oil dependence would be less than 1% of US GDP by 2030. Greene and Leiby calculated that the United States could accomplish that task by decreasing demand for oil by 7.22 million barrels per day and increasing supply by 3 million barrels per day by 2030.

As a follow-up, one of us (Sovacool) formulated a "national energy dependence plan" using Green and Leiby's numbers and calculated that a package of concerted efforts would enable the United States to achieve oil independence by 2030.[54] To reduce dependence on foreign sources of oil according to this plan, US policymakers would need to accomplish the following:

- Lower demand for oil by legislating more stringent fuel economy standards for light- and heavy-duty vehicles or lowering the interstate speed limit.
- Promote alternatives in choices of mode of transportation, such as mass transit, light rail, and carpooling.
- Establish telecommuting centers and incentives for commuters to work from home.
- Promote rigorous standards for tire inflation, and reduce oil consumption in other sectors of the economy.
- Increase alternative domestic supplies of oil.
- Mandate the use of advanced oil recovery and extraction techniques.
- Promote alternatives to oil such as ethanol, biodiesel, and Fischer-Tropsch fuels.
- Most importantly, aggressively push plug-in hybrid electric vehicles.

If taken together, such actions would decrease oil demand by between 8.2 and 15.8 million barrels of oil per day by 2030 and increase supply by between 12.4 and 15.6 million barrels of oil per day (well over the target of 3.0). Under this strategy, the United States would meet the targets within Greene and Leiby's definition of oil independence.

Our common ground suggests that energy security is optimized by balancing strategic energy independence with dependence. For example, expanding domestic capacity in commercially viable forms of renewable energy can help insulate a nation from economic disruption caused by unexpected upsets in global energy markets. However, this strategy is optimized by bolstering interconnectivity with neighboring grids—creating a strategic dependency. Analogously, prolonging domestic fossil fuel reserves is a prudent strategy for attaining self-sufficiency, but dependency on imports allows nations to avoid depleting domestic reserves too quickly. To optimize national security through balancing independence and interdependence, it is also necessary to be more strategic about the types of fossil fuel reserves used and the types of nations from which fossil fuel resources are imported. One can be dependent on imports from Iran or imports from Canada. One can switch dependency from oil imports from the Middle East to natural gas imports from the North Sea. Dependency is characterized by a range of risks. When a nation combines strategic independence with strategic interdependence, the result is a balance that acknowledges the complicated tight-wire act of trying to optimize energy security in a dynamic world.

NOTES

1. George W. Bush, "Remarks on Energy Independence" (February 6, 2003), available at www.presidency.ucsb.edu/ws/?pid=63784.

2. Turki al-Faisal, "Don't Be Crude," *Foreign Affairs*, September/October 2009, available at www.foreignpolicy.com/articles/2009/08/17/dont_be_crude.

3. B. K. Sovacool, "The Political Economy of Oil and Gas in Southeast Asia: Heading towards the Natural Resource Curse?" *Pacific Review* 23, no. 2 (May 2010): 225–259.

4. D. Yergin, "Ensuring Energy Security," *Foreign Affairs* 85, no. 2 (March/April 2006): 71.

5. Bergsten quoted in J. J. Fialka, "Energy Independence: A Dry Hole?" *Wall Street Journal*, July 5, 2006, A4.

6. David Greene and Sanjana Ahmad, *Costs of U.S. Oil Dependence: 2005 Update*, Report to the US DOE, ORNL/TM-2005/45 (Washington, DC: US Department of Energy, January 2005).

7. David L. Greene et al., *OPEC and the Costs to the U.S. Economy of Oil Dependence: 1970–2010*, White Paper 1-13 (Knoxville, TN: Howard Baker Center for Public Policy, 2013). See also David L. Greene, "Measuring Energy Security: Can the United States Achieve Oil Independence?" *Energy Policy* 38 (2010): 1614–1621.

8. International Energy Agency (IEA), *World Energy Outlook 2008* (Paris: OECD, 2008), 105, fig. 3.10.

9. R. Bacon and M. Kojima, *Vulnerability to Oil Price Increases: A Decomposition Analysis of 161 Countries*, World Bank Extractive Industries for Development Series (Washington, DC: World Bank, 2008).

10. T. L. Friedman, *Hot, Flat, and Crowded* (New York: Farrar, Strauss and Giroux, 2008), 80.

11. K. M. Campbell and J. Price, "The Global Politics of Energy: An Aspen Strategy Group Workshop," in *The Global Politics of Energy*, ed. K. M. Campbell and J. Price (Washington, DC: Aspen Institute, 2008), 11–23.

12. Jim Acosta and Erika Dimmler, "Another Way to Beat Al Qaeda: Energy Independence?" (CNN, May 7, 2011), available at http://edition.cnn.com/2011/POLITICS/05/06/al.qaeda.energy.

13. Committee on International Relations, *Oil Fields as Military Objectives: A Feasibility Study* (Washington, DC: Congressional Research Service, August 21, 1975), 1.

14. M. T. Klare, "The Futile Pursuit of Energy Security by Military Force," *Brown Journal of World Affairs* 13, no. 2 (2007): 139–153.

15. Michael O'Hanlon, "How Much Does the United States Spend Protecting Persian Gulf Oil?" in *Energy Security: Economics, Politics, Strategies, and Implications*, ed. Carlos Pascual and Jonathan Elkind (Washington, DC: Brookings Institution Press, 2010), 59–72.

16. M. A. Delucchi and J. J. Murphy, "US Military Expenditures to Protect the Use of Persian Gulf Oil for Motor Vehicles," *Energy Policy* 36 (2008): 2253–2264.

17. Gal Luft, "United States: A Shackled Superpower," in *Energy Security Challenges for the 21st Century*, ed. Gal Luft and Anne Korin (Denver, CO: Praeger International/ABC-CLIO, 2009), 66.

18. IEA, *Energy Self-Sufficiency by World Region* (Paris: OECD, 2011).

19. Jae-Young Lee and Alexey Novitskiy, "Russia's Energy Policy and Its Impacts on Northeast Asian Energy Security," *International Area Studies Review* 13, no. 1 (Spring 2010): 41–61.

20. US Energy Information Administration (EIA), *Country Brief: Russia* (Washington, DC: EIA, 2013).

21. Amy Myers Jaffe and Ronald Soligo, "Energy Security: The Russian Connection," in *Energy Security and Global Politics: The Militarization of Resource Management*, ed. Daniel Moran and James A. Russell (Abingdon, UK: Routledge, 2009), 122.

22. Karen Smith Stegen, "Deconstructing the 'Energy Weapon': Russia's Threat to Europe as Case Study," *Energy Policy* 39 (2011): 6505–6513.

23. M. J. Bambawale and B. K. Sovacool, "Sheikhs on Barrels: What Saudi Arabians Think about Energy Security," *Contemporary Arab Affairs* 4, no. 2 (April–June 2011): 208–224.

24. Andrew Forbes, "Australian Energy Security: The Benefits of Self-sufficiency," in *Energy Security: Asia Pacific Perspectives*, ed. Virendra Gupta and Chong Guan Kwa (New Delhi: Manas Publishers, 2010), 120.

25. Ibid., 120–124.

26. Richard Leaver, "Australia and Asia-Pacific Energy Security: The Rhymes of History," in *Energy Security in Asia*, ed. Michael Wesley (Abingdon, UK: Routledge, 2007), 92.

27. US Department of Energy (DOE), *Characterization of U.S. Energy Resources and Reserves*, DOE/CE-0279 (Washington, DC: DOE, 1989).

28. Alan Caruba, "Obama Is Denying Energy Independence to America," *Warning Signs*, August 19, 2013.

29. Grumet quoted in J. J. Fialka, "Energy Independence: A Dry Hole?" *Wall Street Journal*, July 5, 2006, A4.

30. P. J. Deutch, "Think Again: Energy Independence," *Foreign Policy*, November/December 2005, 20.

31. Masaaki Kuboniwa, "Russia's Input-Output Relations with EU in Light of Energy and Machinery Industries" (paper presented at Conference on European and Asian Energy Markets, Tokyo, April 9–10, 2013).

32. Anna Kireeva, "High Electricity Prices Make Russia Switch from Energy Exporter to Importer," *Barents Observer*, April 15, 2013; Vladimir Dzaguto, "While Russia's Electricity Exports Decline, Imports Rise," *Russia Beyond the Headlines*, May 15, 2013.

33. Reuters, "Saudi Arabia to Import Near Record High Diesel This Summer" (May 20, 2013).

34. Robert Baer, "The Fall of the House of Saud," *Atlantic Monthly*, May 2003, 34–48.

35. Council for Australian-Arab Relations (CAAR) and Australia Arab Chamber of Commerce and Industry (AACCI), *Business Guides to the Arab Gulf* (Riyadh: Kingdom of Saudi Arabia, 2013).

36. Forbes, "Australian Energy Security."

37. Patrick Parenteau and Abigail Barnes, "A Bridge Too Far: Building Off-ramps on the Shale Gas Superhighway," *Idaho Law Review* 49 (2013): 325–365.

38. Ross Kelly, "U.S. Shale Boom Threatens Australian Gas Projects," *Wall Street Journal*, July 16, 2013.

39. Guy Chazan, "Japan to Use More LNG," *Wall Street Journal*, March 16, 2011, 25.

40. Philip Stafford, Javier Blas, and Jack Farchy, "Nuclear Problems Put Energy Markets in a Spin," *Financial Times*, March 17, 2011, 15.

41. B. K. Sovacool and C. J. Cooper, *The Governance of Energy Megaprojects: Politics, Hubris, and Energy Security* (Cheltenham, UK: Edward Elgar, 2013).

42. B. K. Sovacool, "A Critical Stakeholder Analysis of the Trans-ASEAN Gas Pipeline (TAGP) Network," *Land Use Policy* 27, no. 3 (July 2010): 788–797.

43. Malcolm Sharples et al., *Post Mortem Failure Assessment of MODUs during Hurricane Ivan*, for Offshore Risk & Technology Consulting Inc., Under Contract to Minerals Management Service, Order No. 0105PO39221 (Washington, DC: US Government Minerals Management Service, April 2006).

44. B. K. Sovacool and A. E. Florini. "Examining the Complications of Climate and Energy Governance," *Journal of Energy and Natural Resources Law* 30, no. 3 (August 2012): 235–263.

45. US Environmental Protection Agency (EPA), *Mercury Study Report to Congress, Volume 1: Executive Summary*, EPA-452/R-97-003 (Washington, DC: EPA, December 1997).

46. Michael Hopkin, "Acid Rain Still hurting Canada," *Nature*, August 10, 2005.

47. Jørgen Brandt et al., "Contributions from the Ten Major Emission Sectors in Europe and Denmark to the Health-Cost Externalities of Air Pollution Using the EVA Model System: An Integrated Modelling Approach," *Atmospheric Chemistry and Physics* 13, no. 5 (2013): 7725–7746.

48. Erik Behrens et al., "Model Simulations on the Long-Term Dispersal of ^{137}Cs Released into the Pacific Ocean off Fukushima," *Environmental Research Letters* 7 (2012): 034004.

49. Steven Lee Myers, "Lament for a Once-Lovely Waterway," *New York Times*, June 12, 2010; Steven Lee Myers, "Vital River Is Withering, and Iraq Has No Answer," *New York Times*, June 12, 2010.

50. J. Jackson Ewing and Elizabeth McRae, "Transboundary Haze in Southeast Asia: Challenges and Pathways Forward," *NTS Alert*, October 2012.

51. H. Lund and B. V. Mathiesen, "Energy System Analysis of 100% Renewable Energy Systems: The Case of Denmark in Years 2030 and 2050," *Energy* 34 (2009): 524–531.

52. D. Englander, "Japan's Wind-Power Problem" (Greentech Media, 2008), available at www.greentechmedia.com/articles/read/japans-wind-power-problem-828.

53. D. L. Greene and P. N. Leiby, *The Oil Security Metrics Model: A Tool for Evaluating the Prospective Oil Security Benefits of DOE's Energy Efficiency and Renewable Energy R&D Programs*, ORNL/TM-2006/505 (Oak Ridge, TN: Oak Ridge National Laboratory, May 2006).

54. B. K. Sovacool, "Energy Security and Mitigating Climate Change: Plug-in Hybrid Electric Vehicles (PHEVs) and Alternatives to Oil in Asia," in *Energy Issues in the Asia-Pacific Region*, ed. Amy Lugg and Mark Hong (Singapore: Institute for Southeast Asian Studies, 2010), 203–216. See also B. K. Sovacool, "Solving the Oil Independence Problem: Is It Possible?" *Energy Policy* 35, no. 11 (November 2007): 5505–5514.

Are We Nearing a Global Energy Crisis?

In 1999, the US National Academy of Engineering was assigned the task of ranking the top engineering achievements of the twentieth century. Consider how difficult this process would be. How would one weigh the merits of sewage technologies that whisk away human waste and disease against agricultural technologies that can now support seven billion people? The academy published its list in 2003: the automobile, the airplane, water supply and distribution, electronics, radio and television, agricultural mechanization, computers, the telephone, and air conditioning and refrigeration were in spots 2 through 10. The one engineering feat judged to trump all others: electrification.[1]

The ranking of electrification as the top engineering feat of the twentieth century speaks to the omnipresence of electricity—it is the largest energy system as measured by capital investment and the largest sector responsible for greenhouse gas (GHG) emissions—and the extent to which it has permeated our daily lives and our economies. In combination with liquid petroleum fuels, electricity has become ubiquitous. Consider this: the nine other engineering feats on the academy's top ten list all require some sort of energy input. Without energy, modern technology goes by a different name—scrap.

In this chapter, we explore the important topic of global energy security and evaluate two competing views: one arguing that energy security is improving, and the other arguing that energy security is eroding and heading toward a global crisis. The argument for improving energy security has its conceptual roots planted firmly in the 1970s. The supply-side oil shocks of the 1970s resulted in creation of the International Energy Agency (IEA), established under the Organization for Economic Cooperation and Development (OECD) to ensure that the global energy network could withstand disruptions in oil supply without catalyzing global economic chaos. As part of the strategy, many OECD nations agreed to boost their energy inventories (to 90 days' supply for oil), began to diversify their energy portfolios to reduce oil dependence, established broader energy trading ties to minimize the coercive power of any one supplier nation, enhanced domestic fossil fuel exploration and development efforts, and started to finance the development of cleaner, more efficient energy systems. By and large, many would argue that this has resulted in a far more secure global energy regime.

The opposing view is that despite these stellar efforts, most countries have regressed in terms of energy security. In many nations, the argument goes, the safety and reliability of conventional energy infrastructure are eroding, the quality and availability of energy fuels and services are declining, and competition for the finite energy resources that remain is increasing. To compound the problem, critics would argue, the environmental harm caused by our energy systems has reached new heights. Therefore, these critics find ample justification for the contention that in many ways, our collective energy security has worsened to the point where a global crisis may be impending.

One Side: Our Global Energy System Is Doing Fine

High-profile support abounds for the contention that global energy security— generally meant to refer to the availability, affordability, efficiency, and stewardship of energy services—has improved over the past century. Looking to the future, Pulitzer Prize–winning author and energy security guru Daniel Yergin claims that "the global supply system will become more resilient, our energy supplies will become more secure, and [nations] will have more flexibility in dealing with crises."[2] We present here some of the key reasons that many experts see energy security in such a positive light, presenting evidence that advances in production coupled with enhancements in regulatory governance and technology have resulted in improved energy security and are likely to do so well into the future.

Major Advances Are Occurring in Energy Production and Technical Development

Those who contend that energy security has improved often cite data to demonstrate that global energy supplies continue to expand. From 1900 to 2000, the annual average supply of energy per capita expanded from 14 GJ (gigajoules) to roughly 60 GJ. This growth rate tracked the growth of the human population, which almost quadrupled from 1.6 billion in 1900 to 6.1 billion in 2000. Over this same period, energy consumption more than tripled in the United States, quadrupled in Japan, and increased thirteenfold in China.[3]

In parallel with increases in global energy supply, technological advances have vastly improved the effectiveness of the energy being employed. The first airliner, Igor Sikorsky's *Ilya Muromets*, featured four 100-horsepower engines. When the airplane made its debut flight in 1913, it was considered a true engineering marvel, equipped with a passenger saloon and an onboard bathroom. Today, an Airbus A380 is also powered by four engines, each certified at

70,000 pounds of thrust, roughly equivalent to 31,800 horsepower. On board, the A380's two decks can accommodate 525 passengers (853 in a single-class configuration), who are treated to individual in-flight entertainment systems and the use of 17 bathrooms. At the beginning of the twentieth century, the typical American farmer harnessed about 5 kW of power from a team of six horses to plow his field. A century later, a typical American farmer sits atop a diesel-powered tractor that delivers 250 kW of power, and she does so in an air-conditioned cabin while enjoying her favorite country-western tunes via satellite radio.[4]

Indeed, if one looks at the energy sector as a whole, there is ample evidence to justify feeling sanguine. As figure 14.1 shows, from 1973—the height of the first oil shock—to 2011, total primary energy supply for the world more than doubled. The production of crude oil, natural gas, and coal substantially increased. Electricity generation tripled, and nuclear energy production grew

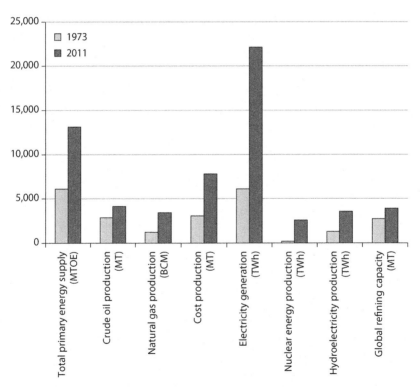

Figure 14.1. Major Global Energy Indicators, 1973 and 2011. *Source:* International Energy Agency, *Key World Energy Statistics 2013* (Paris: OECD, 2013). *Note:* BCM = billion cubic meters; MT = metric ton: MTOE = metric ton of oil equivalent.

twelvefold. Both refining capacity and hydroelectric generation capacity increased notably.[5] The data are hard to refute: we have access to far more energy resources than we did 40 years ago.

Even in the United States, where criticism over the decline of energy security has become a regular topic on the electoral policy agenda, supply-side resilience appears to be high. Testifying in 2013 before the House Natural Resources Subcommittee on Energy and Mineral Resources, Dan Simmons of the Institute for Energy Research argued that both the United States and Mexico are energy-rich countries with total recoverable oil reserves exceeding 1.7 trillion barrels. This is enough fuel for the next 242 years at the current rate of use.[6] Regarding other fossil fuels, Simmons argued that North America has enough natural gas for 176 years at current rates of use, and enough coal for nearly 500 years.[7] In 2012, domestic oil production climbed to its highest level in 15 years, and natural gas production reached an all-time high. Correspondingly, net imports of foreign oil fell to their lowest level in more than two decades, declining from about 60% of national consumption in 2005 to 40% in 2012.[8] In every category, ranging from coal and crude oil to biomass, biofuels, wind power, and solar panels, the United States is both producing more energy and using the energy it produces far more efficiently.[9]

Enhancements in Policy and Governance Are on the Right Track

Those who believe energy security is improving point out that in addition to the expansion of supply and technologically enhanced energy efficiency, substantive improvements in global energy governance have further bolstered energy security. The energy crisis of 1973–74 was a watershed moment. It resulted in establishment of the IEA, creation of strategic petroleum reserves among its members, and diversification of the fuel base for electricity in most countries (transitioning away from oil dependency by seeking alternative ways to produce electricity). In the United States, the crisis forced sweeping energy legislation through Congress, resulted in establishment of the US Department of Energy, and even provoked President Jimmy Carter in 1977 to label the challenge of securing energy reserves "the moral equivalent of war."[10]

This transition away from a dependence on oil was matched by initiatives in improved energy efficiency, demand-side energy reduction, renewable energy, and market restructuring, all of which have enhanced energy mix diversification. In the United States, for example, many states have implemented aggressive renewable portfolio standards, launched emissions trading schemes, and invested heavily in alternative fuels such as hydrogen, ethanol, and biodiesel. Globally, the renewable energy share of final energy consumption in 2011 reached 19%,[11] suggesting a reduced risk of energy price inflation in the face of rising fossil fuel costs. Chapter 3 surveys a variety of tools implemented around the world to improve energy security and/or promote low-carbon energy systems, such as feed-in tariffs, carbon taxes, and renewable fuel standards.

Future Technological Breakthroughs Will Make Us More Secure

A final argument favoring an optimistic perspective on global energy trends is that future innovations and improvements in energy efficiency will enable humanity to capitalize significantly on gains already made. Oil companies appear to be particularly buoyant in this regard. In its *Outlook for Energy*, ExxonMobil spotlighted how energy efficiency practices will enable most North American and European countries to offset increases in energy demand, even as economic output grows significantly (by more than 80%) between 2013 and 2040.[12] The report notes how "technology is enabling the safe development of once hard-to-produce energy resources, significantly expanding available supplies to meet the world's changing energy needs" and that "evolving demand and supply patterns will open the door for increased global trade opportunities." The first page of the executive summary even states, in bold text, that "modern technology is developing new resources and making energy more affordable, all while creating new jobs and expanding trade around the world."[13]

The World Economic Forum, a nonprofit international organization committed to "improving the state of the world,"[14] mirrored this rosy outlook in its 2013 *Global Energy Architecture Performance Index*, claiming that higher levels of gross domestic product (GDP) correlated strongly with "high performing energy systems."[15] It argued that many "top performers" source a large part of their energy mix from "alternative energy sources" such as nuclear power, wind, solar, and biomass. It further noted that "Europe dominates the leader board" due to "concerted regional action on environmental sustainability, better energy efficiency across the value chain and the adoption of clean technologies," as well as significant improvements in terms of lower energy-related GHG and particulate matter emissions. The report concluded that "large natural energy resource endowment is not a critical performance factor" and that even countries with relatively few resources can create energy-secure societies, with Switzerland, Latvia, and France all serving as examples.[16]

So there we have it: our energy outlook has never been brighter. But a competing perspective paints an entirely different picture.

The Other Side: Our Global Energy System Is Heading toward a Crisis

Those who embrace the opposite view that a global energy crisis is imminent can also tap into plenty of evidence to make their case. One assessment of more than 130 countries concluded that all of them had at least one serious energy security vulnerability and that many had multiple vulnerabilities, spread across the dimensions of robustness, sovereignty, and resilience. Threats to robustness included lack of sufficient energy resources, unreliable infrastructure, and rising energy prices. Threats to sovereignty included exposure to terrorist attacks and geopolitical rivalries. Threats to resilience included unstable energy infrastructure that has been severely affected by natural disasters or human disruptions.[17]

Although so many nations face energy vulnerabilities, this does not mean that energy security has decreased: these vulnerabilities might always have been there. Not so, say proponents of the imminent crisis thesis. For example, comparisons of key energy security trends among 22 countries in the OECD found that every country experienced a deterioration of energy security since 1970, and most countries experienced a *net* degradation of energy security. Even the best performers recorded diminished security across a variety of sectors.[18] Another study encompassing Asia's four largest energy users—China, India, Japan, and South Korea—and the developing countries of Southeast Asia reached the same conclusion.[19]

What is behind these deteriorating trends? Crisis proponents point to at least three separate causal factors: (1) declining availability of energy fuels, (2) eroding safety and reliability, and (3) socioenvironmental degradation caused by energy use. This last one, some argue, because it encompasses climate change, could not only lead to crisis but end human civilization as we know it.

High-Quality Energy Resources Are Becoming Less Available

Reserves of the world's four primary energy fuels are concentrated in a just a few countries, creating import dependence that cannot be sustained. Strikingly, 80% of the world's oil is found in 9 countries that have only 5% of the world population and generate 5% of global GDP; 80% of the world's natural gas reserves are found in 13 countries that have 12% of the global population and generate 26% of global GDP; and 80% of the world's coal reserves are in 6 countries, although these countries have 45% of the global population and generate 46% of global GDP.[20] Six other nations (many of which are also coal-rich) control more than 80% of global uranium resources.[21]

On the surface, the physical scope of these conventional energy reserves is eye-popping. For example, Canada's Athabasca tar sands cover an area larger than Greece. However, a blossoming worldwide demand for electricity and liquid fuels threatens to exhaust these reserves in an alarmingly short period of time.

Supply and demand data suggest that transition to a resource-constrained energy supply future is afoot. If levels of production were to remain at current levels, projections indicate, known coal reserves would be exhausted within 137 years, and petroleum and natural gas reserves (excluding unconventional reserves such as tar sands or shale gas) would be exhausted within the next half-century (see table 14.1). However, production cannot remain at current levels

Table 14.1. Life Expectancy of Proven Fossil Fuel and Uranium Resources, 2012

Fuel	Proven reserves	Current production	Life expectancy (years)		
			0% annual production growth rate	1.6% annual production growth rate	2.5% annual production growth rate
Coal	930,400 million short tons	6,807 million short tons	137	74	61
Natural gas	6,189 trillion cubic feet	104.0 trillion cubic feet	60	43	37
Petroleum	1,317 billion barrels	30.560 billion barrels	43	34	30
Uranium	4,743,000 metric tons (at $130/kg of uranium)	40,260 metric tons	118	67	56

Source: Modified from BK Sovacool and SV Valentine, "Sounding the Alarm: Global Energy Security in the 21st Century," in BK Sovacool (ed.) *Energy Security* (London: Sage Library of International Security, 2013), pp. xxxv–lxxviii.

because world energy demand is expected to expand 45% between now and 2030, and more than 300% by the end of the century.[22] If rates of production increased to keep up with growing demand, known fossil fuel reserves would be depleted even more rapidly. As table 14.1 shows, even at a modest 2.5% annual growth rate in demand/production, humanity will face the economic perils associated with scarcity in depressingly short order. Moreover, as finite fuels such as oil, coal, natural gas, and uranium are extracted, it becomes harder (and more energy-intensive) to explore, produce, and deliver them—meaning that the overall carbon footprint of conventional resources is bound to increase.[23]

Not all of the 196 nations on our planet enjoy access to energy reserves. Indeed, due to the "tyranny of terrain," most developing economies lack significant energy deposits, which suggests that a future dominated by conventional energy will be a future where the fortunes of winners and losers are amplified. A survey of the 24 least-developed countries found that 22 each had less than 1% of their region's total energy resources.[24] As an article in *Science* noted, "Large fossil fuel reserves are concentrated in a small number of countries, with half of the low-income countries and more than a third of the middle-income countries having no fossil fuel reserves whatsoever."[25]

Energy Safety and Reliability Are Eroding

There is evidence that in many nations, the energy infrastructure that has emerged to deliver energy fuels and services is eroding precipitously. Many parties that are concerned with global energy security would argue that not only is the continued availability of energy increasingly in doubt, but the ability to turn that energy into usable services is also diminishing.

Much of this problem stems from aging equipment and underinvestment in maintenance. The US transmission and distribution network exemplifies the state of affairs in many developed nations. The US power grid has become so prone to failure that the American Society of Civil Engineers gave it a D+ grade (one grade up from F, for failure) and warned that it was in "urgent need of modernization." Its assessment specifically emphasized that 43% of distribution assets required immediate upgrades (which they were unlikely to receive).[26] In New York, for instance, the aging underground electricity grid is so degraded that it costs local utilities more than $1 billion a year just to replace corroded cable to make the system function at current demand levels. James Gallagher, director of the Office of Electricity and Environment for the New York State Public Service Commission, adds that the billion dollar tab "doesn't even allow us to get proactive to get ahead of the problem. The scale of the challenge is enormous."[27]

The ultimate consequence of aging and inefficient infrastructure is an increased incidence of power outages. The Institute of Electrical and Electronics Engineers estimates that between 1964 and 2004, no less than 17 major blackouts affected more than 195 million residential, commercial, and industrial customers in the United States alone—with 7 of these major blackouts occurring in the past 10 years. Sixty-six smaller blackouts (affecting between 50,000 and 600,000 customers) occurred from 1991 to 1995, and 76 between 1996 and 2000.[28] The costs of these blackouts are monumental. The US Department of Energy estimates that power outages and power quality disturbances cost customers as much as $206 billion annually, or more than the entire nation's electricity bill for 1990. Between 2003 and 2012, an astonishingly high 679 large-scale power outages occurred in the United States due to severe weather.[29]

The problem is not confined to the United States. Fifteen major global blackouts over the past few decades have affected, in aggregate, almost 1.4 billion people and markedly impaired economic activity (see table 14.2). In the developing world, countries tend to lose 1% to 2% of GDP growth potential due to blackouts, overinvestment in backup electricity generators, and inefficient use of resources.[30] Nigerians, for instance, live with such persistent power outages that one government official characterized the power supply as "epileptic."[31] The Nepal Electricity Authority supplies electricity to Nepal's capital, Kathmandu, for less than 8 hours a day, with load shedding occurring over the remaining 16 hours. In the bazaars at night, shops rely more on candles than on electric lightbulbs to exhibit their wares.[32] In July 2012, India suffered an "unprecedented grid failure" that affected 670 million people—more than half

Table 14.2. Major Electricity Blackouts, 1965–2014

Date	Location (event)	Extent of damage / duration of blackout
Nov 2014	Bangladesh	150 million people for 12 hours
Oct 2012	USA (Hurricane Sandy)	8.2 million people in 17 states, DC, and Canada for 2 weeks
July 2012	Northern India	670 million people over half of India for 20 hours
June 2012	USA (wind storm / derecho)	4.2 million customers in 11 states and DC for 10 days
Oct 2011	Northeast USA (storm)	3 million customers in New England for 10 days
Sept 2011	California and Arizona, USA	2.7 million people for 12 hours
Nov 2009	Brazil and Paraguay	60 million people for 24 hours
Feb 2008	Chenzhou, China	4.6 million people for 18 hours
Aug 2005	Indonesia	100 million people across Java and Bali for 14 hours
Sept 2003	Austria, Croatia, Italy, Slovenia, and Switzerland	56 million people for 16 hours
Aug 2003	Northeast USA (blackout)	50 million people in 8 states and Ontario for up to 4 days
Jan 2001	India	230 million people for 20 hours
Mar 1999	Southern Brazil	99 million people for 10 hours
July 1996	West Coast, USA	2 million people in USA, Canada, and Mexico for minutes to hours
Aug 1996	West Coast, USA	7.5 million customers in 7 states/provinces in western USA and Canada for 6 hours
Mar 1989	Quebec, Canada	6 million people for 12 hours
Dec 1982	West Coast, USA	5 million people for various durations
July 1977	New York City, USA	9 million people for up to 26 hours
Mar 1978	Thailand	40 million people for 9 hours
Nov 1965	Northeast USA and Ontario, Canada	30 million people for up to 13 hours

Source: Compiled by the authors from various press reports.

the country's population, or roughly 10% of the world's population—in a single outage.[33]

Energy Is Fueling Mounting Environmental and Social Costs

The final factor is perhaps the most serious. As energy consumption has grown, so, too, have the unwanted externalities associated with energy production and use, such as climate change, air pollution, and social degradation. Put in monetary terms, the social and environmental damages from just one type of energy—worldwide electricity generation—amounted to roughly $2.6 trillion in 2010.[34] When improperly managed, energy systems adversely affect not only natural ecosystems but also human welfare.

The most pronounced social and environmental costs associated with current patterns of energy production and use stem from climate change (see chapters 9, 10, and 11 for further coverage of this). The social and environmental damage from excessive GHG emissions—costs that are not currently factored into most energy markets—could far surpass the cost of previous global threats such as AIDS, the Great Depression, and global terrorism. The Center for Climate and Energy Solutions estimates that "waiting until the future" to address global climate change might bankrupt the US economy.[35] The *Stern Review* in 2006 projected that the overall costs and risks of climate change will be equivalent to losing at least 5% of the world's GDP, or $3.2 trillion, every year, now and forever, and that these damages could exceed 20% of GDP ($13 trillion) if more severe scenarios unfold.[36] Stern's figures have since been critiqued for being far too conservative and for underestimating the impacts of climate change.[37]

We are already seeing evidence of the mounting costs associated with climate change. As figure 14.2 indicates, global economic damages from natural catastrophes, many of them climate-related, have doubled over the past 10 years and totaled about $1 trillion over the past 15 years. Annual weather-related disasters have increased fourfold from 40 years ago, and insurance payouts increased elevenfold over the same period, rising by $10 billion per year for most of the past decade.[38] Munich RE, an insurance company, estimated that in 2011, 820 separate events caused $380 billion in damages, making it the "most expensive natural catastrophe year ever in national economic terms."[39] Hurricane Sandy, which flooded parts of New Jersey and New York in October 2012, caused up to $50 billion in damages in the state of New York alone.[40]

Unfortunately, as the adverse effects of climate change amplify, the distribution of costs will be far from equitable. Developing countries (many of which are least responsible for GHG emissions) will be most at risk. Indonesia, the Philippines, Thailand, and Vietnam are expected to lose 6.7% of their combined GDP by 2100, more than twice the global average, if the impact of climate change follows Intergovernmental Panel on Climate Change forecasts.[41] China and India, among others, could exhaust between 1% and 12% of their annual GDP in coping with climate refugees, changing disease vectors, and failing crops.[42] One wide-ranging survey of climate impacts in the Asia Pacific from the US Agency for International Development (USAID) predicted, among other things:

- Accelerated river bank erosion, saltwater intrusion, crop losses, and floods in Bangladesh that will displace at least 8 million people and destroy up to 5 million hectares of crops.
- More frequent and intense droughts in Sri Lanka, crippling tea yields, reducing national foreign exchange, and lowering incomes for low-wage workers.
- Higher sea levels inundating *half* of the agricultural lands on the Mekong Delta, causing food insecurity throughout Cambodia, Laos, and Vietnam.

- Increased ocean flooding and storm surges inundating 130,000 hectares of farmland in the Philippines, affecting the livelihoods of 2 million people.
- Intensified flooding in Thailand, placing more than 5 million people at risk and causing $39 billion to $1.1 trillion in economic damages by 2050.[43]

The USAID study concluded that the Asia Pacific will be subject to more land degradation, forced migration, erosion of prosperity, and economic disruption from sea-level rise than any other part of the planet. Between 1980 and 2009, the Asia Pacific suffered 85% of the deaths and 38% of the global economic losses arising from natural disasters.[44]

The effects of climate change, a process largely driven by energy production, will not only be expensive and unfair; according to those articulating the case for impending crisis, they could end our entire way of life. As columnist David Roberts concludes:

> If you aren't alarmed about climate, you aren't paying attention . . . Ponder the fact that some scenarios show us going up to 6 degrees by the end of the century, a level of devastation we have not studied and barely know how to conceive . . . that somewhere along the line, though we don't know exactly where, enough self-reinforcing feedback loops will be running to make climate change unstoppable and irreversible for centuries to come. That would mean handing our grandchildren and their grandchildren not only a burned, chaotic, denuded world, but a world that is inexorably more inhospitable with every passing decade. Take all that in, sit with it for a while, and then tell me what it could mean to be an "alarmist" in this context. What level of alarm is adequate?[45]

Increases in temperature could approach 6°C. Such temperature trends quickly become extreme when one considers that the increases could even cross what has been termed the "afterlife" threshold—where the impact on humanity is so great that it halts the long-term advancement of human civilization.[46]

Climate change, of course, is not the only externality. Millions of individuals are involuntarily resettled due to energy projects. Each year, about 4 million people are displaced by activities relating to hydroelectricity construction or operation; 80 million have been displaced in the past 50 years by the construction of 300 large dams.[47] Another wide-ranging international survey estimated that in India, at least 2.6 million people were displaced due to mining (for energy and other resources) from 1950 to 2009, and individual mines in Brazil, Ghana, Indonesia, and South Africa displaced (in each case) between 15,000 and 37,000 people from their homes.[48] In India, thousands of villagers in the region of Singrauli have seen their homes "uprooted and resettled" for the

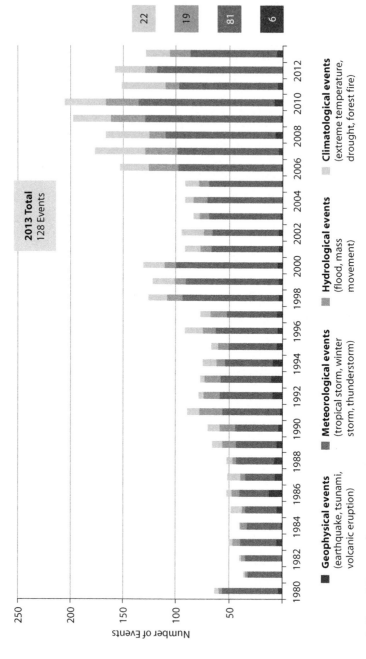

A. Natural Catastrophes

Number of Events

2013 Total
128 Events

Geophysical events
(earthquake, tsunami, volcanic eruption)

Meteorological events
(tropical storm, winter storm, thunderstorm)

Hydrological events
(flood, mass movement)

Climatological events
(extreme temperature, drought, forest fire)

22
19
81
6

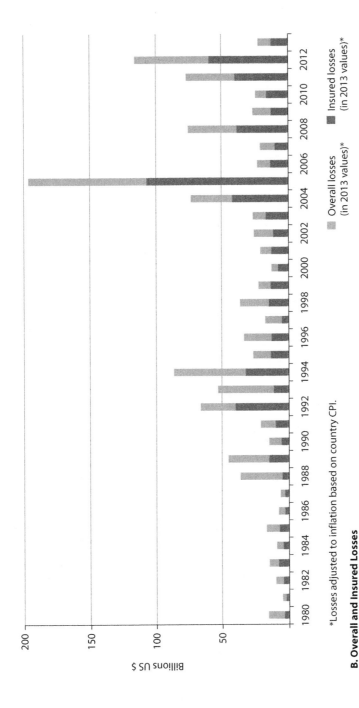

*Losses adjusted to inflation based on country CPI.

■ Overall losses (in 2013 values)*   ■ Insured losses (in 2013 values)*

B. Overall and Insured Losses

Figure 14.2. Natural Catastrophes and Overall Losses and Insured Losses Worldwide, 1980–2013 *Source:* Munich RE, *2013 Natural Catastrophe Year in Review* (Berlin: Munich RE, 2014). *Note:* CPI = consumer price index.

establishment of coal mines; those who protested or resisted were met with "police brutality and violence."[49] Even climate change adaptation projects being implemented under Reducing Emissions from Deforestation and Degradation have proceeded, at times, without proper consent and inflicted damages on local communities. An independent assessment from the Center for People and Forests warned that some governments have seized indigenous lands in order to capitalize on forest carbon revenues and utilized exploitable contracts that convince communities to accept terms that sign away land-use rights and set unfair liability standards, ultimately decreasing the production of local food and deepening poverty.[50] In short, for millions of people, energy projects do not bring hope; they bring despair.

There is evidence that institutions, procedures, and processes can also result in the unequal distribution of energy costs and benefits, eroding the vitality of communities that must bear the brunt of the negative externalities from energy production. Dirty infrastructure can create "national sacrifice zones" that condemn poorer communities to suffer disproportionately.[51] In Eastern Europe, the Roma have been displaced from so many countries and cities that they are forced to migrate to settlements akin to "environmental time bombs." Roma communities in the Czech Republic and Slovakia, for example, reside in housing located above abandoned mines, prone to flooding and susceptible to methane gas poisoning. Others live in abandoned factory sites or mines where children are exposed to toxins and suffer long-term effects on their health.[52] One study concluded that siting strategies involving nuclear power and waste sites, mines, and other energy facilities overwhelmingly select "peripheral communities" that are distant from vibrant (and often wealthier) urban communities.[53] The researchers explained that noxious energy facilities invariably migrate to communities that lack the political, social, and economic strength to oppose them, especially indigenous peoples and tribes, often at the extreme social and geographical periphery of society. The trend holds especially true for nuclear waste repositories.[54] As a case in point, the Japanese government has historically exhibited a tried and proven strategy of targeting communities in decline as prospective sites for nuclear power plants—communities that are then persuaded to accept these plants in return for promises of annual payments to villagers and enhanced public spending.[55]

Common Ground: Gains Have Been Made
at the Expense of Hidden Costs

What can we make of such divergent views concerning global energy security? On the one hand, things are looking up: annually, we are finding more and more energy to power our needs. On the other hand, our grandchildren will be playing in a burned, chaotic, denuded world. This debate reveals the complex, double-sided nature of the energy system; to borrow from Roman mythology, it has a Janus face.[56]

Without a doubt, some elements of the global energy system have improved markedly since the supply-side oil crises of the 1970s. The number of oil-exporting nations has increased, and the market control that OPEC nations once enjoyed has been severely undermined by diversification of supply-side options.[57] New reserves of oil, gas, coal, and uranium are being tapped weekly. Investments are pouring into renewable energy technologies, with a sixfold increase between 2004 and 2012.[58] These energy systems and technologies have proven to be socially advantageous, enabling human beings to pursue other aspects of life. Immense human achievements have been made possible through modern energy services, including a longer life expectancy through improved health care and the ability to land on the moon and send probes to Mars. A former CEO of the Electric Power Research Institute went so far as to declare that "energy is the elemental force upon which all civilizations are built, and technology provides the means to harness energy."[59]

Nonetheless, nothing in life comes without a cost. Decision makers simultaneously reap the benefits of technologies that accomplish much work for the human race and sow the costs required to maintain these benefits. Conventional energy systems are drawing down finite natural resources at unprecedented rates. And the sheer scale of energy production is producing externalities that are causing extreme environmental problems and unprecedented health costs. Such visceral environmental damages provoked the *Economist* to conclude that "using energy in today's ways leads to more environmental damage than any other peaceful human activity."[60]

Simply put, the progress that the world has enjoyed in terms of energy security over the past few decades has come with hidden costs that we are just now beginning to quantify. Still, there is a way to synthesize these views. As box 14.1 summarizes, we are consuming more energy and improving the lives of more people, but our economic gains have also engendered massive external costs that, when fully accounted for, offset many of those gains. Our ability to harness energy resources and access modern energy services has become more secure, but associated social and environmental impacts are becoming more pronounced. If we continue to use energy in the way we have used it thus far, with disregard for its broader consequences—especially in relation to climate change—this could entirely negate many of the energy security improvements we have made.

Countries around the world are already confronting dangerous changes in climate patterns, power outages, and deteriorating public health. All these ill-effects are consequences of conventional energy technologies constructed along a road that will soon come to a dead end. Mitigating these damages by

BOX 14.1

Global Energy Security and the Common Ground

Are we nearing a global energy crisis?

ONE SIDE: We continue to produce more energy globally and develop new technologies and policy mechanisms.

OTHER SIDE: The erosion of reliability, resource availability, affordability, and social and environmental quality places us on the brink of a global crisis.

COMMON GROUND: We can acknowledge the benefits energy gives us but recognize and mitigate broader social and environmental costs.

changing technologies does not "cost" us more. It simply ensures that all producers and consumers of energy are held accountable for the energy decisions they make. Ultimately, the question may not be "how can we improve global energy security?" or "how can we create a more stable climate?" but rather—given how many people still hold to the belief that the path we have traveled to this point will continue to serve us well—"how can we save ourselves?"

NOTES

1. National Academy of Engineering, *A Century of Innovation: Twenty Engineering Achievements That Transformed Our Lives* (Washington, DC: National Academies Press, 2003).

2. Daniel Yergin, "America's New Energy Security," *Wall Street Journal*, December 12, 2011.

3. Vaclav Smil, "Energy in the Twentieth Century: Resources, Conversions, Costs, Uses, and Consequences," *Annual Review of Energy* 25 (2000): 21–51.

4. B. K. Sovacool, R. Sidortsov, and B. Jones, *Energy Security, Equality, and Justice* (London: Routledge, 2013).

5. International Energy Agency, *Key World Energy Statistics 2013* (Paris: OECD, 2013).

6. Daniel Simmons, Testimony before the Committee on Energy and Commerce, U.S. House of Representatives, Hearing on HR 4255, the Accountability in Grants Act of 2012, September 11, 2012.

7. Ibid.

8. White House, "Develop and Secure America's Energy Resources" (2013), available at www.whitehouse.gov/energy/securing-american-energy.

9. Comparing data for 1973, from US Energy Information Administration (EIA), *Annual Energy Review*, September 2012, tables 1.2 and 10.1, and for 2013 (through May), from EIA, *Monthly Energy Review*, August 2013, tables 1.1 and 10.1.

10. Jimmy Carter, Address to the Nation, April 18, 1977.

11. Renewable Energy Policy Network for the 21st Century (REN21), *Renewables 2013 Global Status Report: Renewable Energy* (Paris: REN21, 2013).

12. ExxonMobil, *The Outlook for Energy: A View to 2040* (Irving, TX: ExxonMobil Corporation, 2013).

13. Ibid., 3–4.

14. World Economic Forum, "Our Mission," available at www.weforum.org/our-mission.

15. World Economic Forum, *The Global Energy Architecture Performance Index Report 2013*, prepared with Accenture (Davos, Switzerland: World Economic Forum, December 2012).

16. Ibid., 1.

17. A. Cherp et al., "Energy and Security," in *Global Energy Assessment*, ed. T. B. Johansson et al. (Cambridge: Cambridge University Press, 2012), 325–383.

18. B. K. Sovacool and M. A. Brown, "Competing Dimensions of Energy Security: An International Review," *Annual Review of Environment and Resources* 35 (November 2010): 77–108; Marilyn A. Brown et al., "Forty Years of Energy Security Trends: A Comparative Assessment of 22 Industrialized Countries," *Energy Research and Social Science* 4 (December 2014): 64–77.

19. B. K. Sovacool et al., "Evaluating Energy Security Performance from 1990 to 2010 for Eighteen Countries," *Energy* 36, no. 10 (October 2011): 5846–5853.

20. M. A. Brown and B. K. Sovacool, *Climate Change and Global Energy Security: Technology and Policy Options* (Cambridge, MA: MIT Press, 2011), fig. 1.1.

21. International Atomic Energy Agency (IAEA), "Top Ten Uranium Producing Countries in 2010," in *Uranium 2011: Resources, Production, and Demand* (Vienna: IAEA, 2011).

22. Brown and Sovacool, *Climate Change*.

23. Luc Gagnon, "Civilization and Energy Payback," *Energy Policy* 36 (2008): 3317–3322.

24. United Nations Economic and Social Commission for Asia and the Pacific (UNESCAP), *Energy Security and Sustainable Development in Asia and the Pacific*, ST/ESCAP/2494 (Geneva: UNESCAP, April 2008), 185.

25. Jeffrey Chow, Raymond J. Kopp, and Paul R. Portney, "Energy Resources and Global Development," *Science* 302, no. 5650 (November 2003): 1528–1531.

26. Elisabeth Rosenthal, "Ahead of the Pack on Cleaner Power," *International Herald Tribune*, September 30, 2010, VI.

27. Gallagher quoted in B. K. Sovacool, *The Dirty Energy Dilemma: What's Blocking Clean Power in the United States* (Westport, CT: Praeger, 2008), 29.

28. Philip Fairey, "The Unruly Power Grid," *IEEE Spectrum* 41, no. 8 (2004): 22–27.

29. White House, *Economic Benefits of Increasing Electric Grid Resilience to Weather Outages* (Washington, DC: White House, August 2013).

30. United Nations Development Programme (UNDP), *Energy for Sustainable Development* (New York: UNDP, August 15).

31. "85 Billion Needed for Stable Power; Supply Won't Improve Till December," *Daily Trust/All Africa Global Media*, June 25, 2008.

32. World Bank, *Project Paper Proposed Additional Financing Credit in the Amount of SDR $49.6 Million and a Proposed Additional Financing Grant in the Amount of SDR $10.5 Million to Nepal for the Power Development Project*, Report No. 48516-NP (Washington, DC: World Bank Group, May 18, 2009).

33. Harris Gardiner and Vikas Bajaj, "As Power Is Restored in India, the 'Blame Game' over Blackouts Heats Up," *New York Times*, August 1, 2012.

34. Brown and Sovacool, *Climate Change.*

35. Eileen Claussen and Janet Peace, "Energy Myth Twelve—Climate Policy Will Bankrupt the U.S. Economy," in *Energy and American Society*, ed. B. K. Sovacool and M. A. Brown (New York: Springer Publishing, 2007).

36. Nicholas Stern, *The Stern Review: Report on the Economics of Climate Change* (London: Cabinet Office, Her Majesty's Treasury, 2006).

37. See Partha Dasgupta, "Comments on the Stern Review's Economics of Climate Change" (University of Cambridge, November 11, 2006; revised December 12, 2006); Hal R. Varian, "Recalculating the Costs of Global Climate Change," *New York Times*, December 14, 2006.

38. B. Sudhakara Reddy and Gaudenz B. Assenza, "The Great Climate Debate," *Energy Policy* 37 (2009): 2997–3008.

39. Munich RE, *Natural Catastrophes 2011: Analyses, Assessments, Positions* (Berlin: Munich RE, 2012).

40. M. W. Walsh and N. Schwartz, "Estimate of Economic Losses Now up to $50 Billion," *New York Times*, November 1, 2012.

41. Asian Development Bank (ADB), *The Economics of Climate Change in Southeast Asia: A Regional Review* (Manila, Philippines: ADB, April 2009).

42. Economics of Climate Adaptation Working Group, *Shaping Climate-Resilient Development: A Framework for Decision-Making* (New York: Climate Works Foundation, 2009); Arief Anshory Yusuf and Herminia A. Francisco, *Climate Change Vulnerability Mapping for Southeast Asia* (Singapore: International Development Research Centre, January 2009); Royal United Services Institute, *Socioeconomic and Security Implications of Climate Change in China* (Washington, DC: CNA, November 4, 2009); CNA, "Climate Change, State Resilience, and Global Security" (CNA Conference Center, Alexandria, VA, November 4, 2009).

43. US Agency for International Development (USAID), *Asia Pacific Regional Climate Change Adaptation Assessment: Final Report Findings and Recommendations* (Washington, DC: USAID, April 2010), based on data from D. Anthoff et al., "Global and Regional Exposure to Large Rises in Sea-Level: A Sensitivity Analysis" (Working Paper 96, Tyndall Centre for Climate Change Research, Norwich, UK, 2006).

44. UNESCAP, *Low Carbon Green Growth Roadmap for Asia and the Pacific* (Bangkok: UNESCAP, 2012).

45. David Roberts, "If You Aren't Alarmed about Climate, You Aren't Paying Attention," *Grist*, January 10, 2013, available at http://grist.org/climate-energy/climate-alarmism-the -idea-is-surreal.

46. Andy Haines et al., "Health Risks of Climate Change: Act Now or Pay Later," *Lancet* 384 (September 20, 2014): 1073–1074.

47. World Commission on Dams, *The Report of the World Commission on Dams* (London: Earthscan, 2001).

48. Theodore E. Downing, *Avoiding New Poverty: Mining-Induced Displacement and Resettlement* (London: International Institute for Environment and Development, April 2002).

49. David Hunter, "Using the World Bank Inspection Panel to Defend the Interests of Project-Affected People," *Chicago Journal of International Law* 4 (2003): 201–211.

50. Patrick Anderson, *Free, Prior, and Informed Consent: Principles and Approaches for Policy and Project Development* (Bangkok: RECOFTC and GIZ, February 2011).

51. Robert D. Bullard, *Unequal Protection: Environmental Justice and Communities of Color* (San Francisco: Sierra Club Books, 1994).

52. Tamara Steger, *Making the Case for Environmental Justice in Central and Eastern Europe* (Budapest: CEU Center for Environmental Law and Policy, March 2007).

53. A. Blowers and P. Leroy, "Power, Politics and Environmental Inequality: A Theoretical and Empirical Analysis of the Process of 'Peripheralisation,'" *Environmental Politics* 3, no. 2 (Summer 1994): 197–228.

54. C. Michael Rasmussen, "Getting Access to Billions of Dollars and Having a Nuclear Waste Backyard," *Journal of Land Resources and Environmental Law* 18 (1998): 335–367.

55. D. P. Aldrich, *Site Fights: Divisive Facilities and Civil Society in Japan and the West* (Ithaca, NY: Cornell University Press, 2008).

56. In Roman mythology, Janus was a god who presided over change. Statues of Janus usually depict the god with two faces, one looking to the present and one looking to the past.

57. K. M. Campbell and J. Price, eds., *The Global Politics of Energy* (Washington, DC: Aspen Institute, 2008).

58. REN21, *Renewables 2013*.

59. Kurt Yeager, *Electricity and the Human Prospect: Meeting the Challenges of the 21st Century* (Palo Alto, CA: Electric Power Research Institute, 2004), 3.

60. "A Power for Good, a Power for Ill," *Economist*, August 31, 1991.

Can Energy Transitions Be Expedited?

In 1832, the American philosopher and poet Ralph Waldo Emerson, grieving the death of his wife and distraught over the hurdles confronting abolition of the practice of slavery, decided to seek solace through a sojourn in Europe.[1] On Christmas Day, he boarded the *Jasper*, a two-masted sailing ship carrying cargo for Malta. While in Europe, he visited Rome and became acquainted with John Stuart Mill, went to Switzerland to tour Voltaire's home in Ferney, and communed with William Wordsworth and Thomas Carlyle in England. When ready to return to the United States in October 1833 (less than a year later), Emerson was dismayed to find that no sailing ships were available for the Atlantic crossing. Instead, he had to embark on the steamship *Liberty*. Emerson ruminated on the energy transition that had occurred within the 11 months he had spent in Europe. On his outward journey, he had crossed the Atlantic on a vessel that was wind-powered, was controlled by craftsmen practicing ancient arts upon the open ocean, and was recyclable. He returned in a vessel that spewed oil into the water and smoke into the sky, was controlled by workers shoveling coal into boilers in the dark, and would eventually become a steel rust bucket.[2]

Are transitions in technologies that supply, convert, or use energy really capable of occurring so quickly? According to one view, energy transitions take an incredibly long time to occur. As the geographer Vaclav Smil writes, "All energy transitions have one thing in common: They are prolonged affairs that take decades to accomplish, and the greater the scale of prevailing uses and conversions, the longer the substitutions will take."[3] According to this perspective, fast transitions are anomalies, limited to countries with very small populations or unique contextual circumstances unlikely to be replicated elsewhere. Certainly, in this perspective, a global energy transition that is capable of extracting humanity from the perils of climate change is a tenuous proposition.

The opposing view suggests that there are many examples of energy transitions (on various scales) that have occurred expeditiously—some over a few years, others over a decade or so, and many within a single generation. On smaller scales, the widespread adoption of cookstoves, air conditioners, and flex-fuel vehicles provides excellent examples of rapid consumer uptake. On a national scale, substantial shifts to oil in Kuwait, natural gas in the Netherlands, and nuclear power in France occurred within roughly one or two decades. We'll present 10 such concrete examples of energy transitions that, in aggregate, affected almost a billion people and needed only 1 to 16 years to unfold.

One Side: Energy Transitions Are Long, Protracted Affairs

There is considerable evidence in support of the argument that energy transitions—defined as the time that elapses between the introduction of a new fuel or technology and its rise to 25% of national market share—take a significant amount of time to materialize.[4] The Global Energy Assessment, a major international, interdisciplinary effort in 2012 to better understand energy systems, noted that "transformations in energy systems" are "long-term change processes," occurring over the scale of decades or even centuries.[5] As two Stanford University scientists summarized, "It appears that there is no quick fix; energy system transitions are intrinsically slow."[6] Support for this argument comes from (1) important insights regarding "lock-in" and "path dependency," (2) historical experience, and (3) the numerous obstacles to enacting simultaneous changes on a global scale.

Path Dependency and Lock-in Make Future Transitions Difficult

Energy systems are largely path-dependent—large sums of financial, social, and political capital are "sunk" into them, creating strategic paths that resist attempts to alter course.[7] To compound the challenge, path dependencies give rise to institutional legacies that protect the status quo through political regulations, tax codes, and industry standards that favor incumbent technology.[8] Coalitions comprising industry stakeholders, financial institutions, and even educational institutions coalesce to insulate incumbent technologies from external threat.[9] The result is that conventional energy systems are like visits from in-laws: they tend to wear out their welcome. Transitions occur over long periods of time in "a messy, conflictual, and highly disjointed process."[10]

Scholars looking at energy transitions have argued that to counteract such resistance, truly "transformative change" requires alterations at every level of the energy system, in a nearly simultaneous manner.[11] That is, one must alter technologies, political will, legal regulations, market dynamics, and social attitudes and values, making a transition a highly confrontational process.[12] Indeed, the idea that energy transitions take an insufferable amount of time and effort is embedded in no less than four major academic theories (table 15.1), spanning the disciplines of science and technology studies, environmental science, environmental sociology, and political ecology. Sociotechnical transitions scholars emphasize the need for "transition management" to counteract the "momentum" of existing systems; ecological modernists highlight the

Table 15.1. Four Key Conceptual Approaches to Understanding Energy Transitions

Category	Sociotechnical transitions	Ecological modernization theory	Environmental sociology (social practice theory)	Political ecology
Related academic disciplines	Science and technology studies, evolutionary economics, structuration theory	Environmental science, environmental sociology	Sociology, anthropology, cultural theory	Human geography, ecology, political geography
Primary focus	Development or introduction of new technologies	Environmental regulation, reform, and governance	Everyday routines and practices	Conflict over natural resources
Barriers to change	Momentum, path dependency, carbon lock-in, multilevel perspective, transition management	Energy transitions, environmental reform, risk society, social movements	Changing practices, habits, socialization, normalization	Contestation, enclosure and exclusion, accumulation by dispossession, global production networks, neoliberalism
Units of analysis	Technologies, sociotechnical systems	Sectors, industries, institutions	Everyday practices or discourses	Ecological change, local communities, institutions
Selected authors	Arie Rip, Frank Geels, Johan Schot, Frans Berkhout, René Kemp, Wim A. Smit, Thomas Hughes	Ulrich Beck, Maarten Hajer, A. P. J. Mol, F. H. Buttel	Elizabeth Shove, Gordon Walker, Loren Lutzenhiser, Harold Wilhite	David Harvey, Michael Watts, Paul Robbins, James McCarthy, Gavin Bridge

Source: Modified from Mattijs Smits, *Southeast Asian Energy Transitions: Between Modernity and Sustainability* (Surrey, UK: Ashgate, 2015).

lengthy process of regulatory reform; environmental sociologists underscore how altering everyday routines and practices can take generations; political ecologists emphasize how neoliberal ideology has firmly entrenched capitalism in our social and political spaces so that alternatives are rarely imagined, let alone implemented.

This is why, in its forecasts about the future, the US Energy Information Administration predicts that in 2040, three-quarters of energy in the United States will still come from oil, coal, and natural gas.[13] The International Energy Agency (IEA) similarly projects that in 2035, under its "current policies" scenario, 80% of total primary energy supply worldwide will come from "traditional" fossil fuels.[14]

Past Experience Suggests That Major Transitions Take Decades to Centuries
Previous energy transitions have taken more time than it takes many steel bridges to rust away. Crude oil took half a century to transition from its introductory stages in the 1860s to capturing 10% of the global market in the 1910s. Then, it took 30 years more to achieve a 25% market share. Natural gas firms took 70 years to advance from 1% to 20% market share. It took the coal industry 103 years to account for just 5% of total energy consumed and an additional 26 years to reach a 25% market share.[15] The nuclear power industry took 38 years to achieve a 20% share, which occurred in 1995—and it has stagnated at that level ever since. As Smil points out, "It's taken between 50 and 70 years for a resource to reach a large penetration. When you look at the money, the infrastructure, the regulation, [and] the technologies, it takes many decades for any fuel source to make a large impact."[16]

In assessing promising new technologies, it merits noting that there is often a considerable time lag before any promising technology gathers market diffusion momentum. Smil notes that steam engines were designed in the 1770s but didn't enter widespread use until the 1800s, and the gasoline-powered internal combustion engine, first deployed by Benz, Maybach, and Daimler in the mid-1880s, only reached widespread adoption in the United States in the 1920s. Diffusion occurred even later in Europe and Japan for the automobile.[17] As Smil deduces from these precedents, "Energy transitions have been, and will continue to be, inherently prolonged affairs, particularly so in large nations whose high levels of per capita energy use and whose massive and expensive infrastructures make it impossible to greatly accelerate their progress even if we were to resort to some highly effective interventions."[18] This is why he calls energy systems "a slow-maturing resource . . . they grow up so . . . slowly."[19] As he remarks, "It is impossible to displace [the world's fossil fuel–based energy] supersystem in a decade or two—or five, for that matter. Replacing it with an equally extensive and reliable alternative based on renewable energy flows is a task that will require decades of expensive commitment. It is the work of generations of engineers."[20]

The argument that energy transitions are long-drawn-out affairs gains further support from energy analysts Peter Lund and Roger Fouquet. Lund found that market penetration of new energy systems or technologies can take as long 70 years.[21] Short "take-over times" of less than 25 years have been limited to a few end-use technologies such as water heaters and refrigerators. They are not common for major infrastructural systems such as those involving electricity or transport. A second study by Lund explored "how fast new energy technologies could be introduced on a large scale." It concluded that the earliest that wind could provide more than 25% of world electricity and solar could provide 15% would be 2050—40 years from the date of his study.[22] As Lund noted in his earlier study, "The inertia of energy systems against changes is large . . . among others [i.e., other reasons] because of the long investment cycles of energy infrastructures or production plants."[23] He later concluded that the "rate of adoption of these new [renewable energy] technologies would not exceed that of oil or nuclear in the past."[24] Fouquet studied various transitions involving both energy fuels and energy services from 1500 to 1920. He found that, on average, each transition was characterized by an innovation phase that exceeded 100 years, followed by a diffusion phase that approached 50 years.[25] Given that virtually all prominent scientific studies related to climate change note the need to significantly reduce greenhouse gas emissions by 2050, one can understand the alarm that these critiques engender.

Analysts Need to Focus on the Big Picture

Proponents of this view also argue that when one looks at the big picture, the challenge of facilitating a global energy transition is exacerbated by the sheer scale of the problem and beset by concerns that solutions will give rise to new problems. For starters, the scale and requisite timing of the transition are not static. The scale of the problem increases in direct proportion to the scale of energy demand. The IEA projects that total primary energy demand will increase by 33% by 2035 under its "new policies" scenario. This equates to an annual growth rate in energy demand of 1.2%. In other words, as each year unfolds, the challenge of facilitating a transition gets 1.2% harder. It also means that annual demand will double in less than 60 years.

There are past examples of how market growth can stymie progress—emergent technologies can grow in an absolute sense but still fail to make market headway. Hydroelectricity in the United States was a low-cost source of energy in the 1950s and 1960s. Accordingly, it tripled in capacity from 1949 to 1964. During this time, however, demand for electricity grew at a faster rate, so the hydropower industry's overall national share dropped from 32% to 16%.[26]

Similarly, from 2000 to 2010, global annual investment in solar photovoltaic (PV) increased sixteenfold, investment in wind energy quadrupled, and investment in solar heating tripled. This sounds impressive, yet the overall contribution of solar (heating and PV) and wind to total final energy consumption increased from less than 0.1% to slightly less than 1%,[27] almost inconsequential when viewed in the bigger picture.

The final part of the big picture that casts doubt on the ability to facilitate a speedy global transition away from carbon-intensive energy technologies stems from the potential negative repercussions of such a transition, should it actually be achieved. Construction of a single 1,000 MW nuclear reactor (the standard size) uses upwards of 179,000 tons of concrete, 36,000 tons of steel, and 729 tons of copper, among other items.[28] Many renewable sources of energy are even more material-intensive in various ways: hydroelectric dams, on a per gigawatt basis, need three times as much concrete as a nuclear reactor; wind turbines require 20,000 tons of fiberglass per installed gigawatt. Other harder-to-find critical materials—including lithium, cobalt, and indium—are essential to the batteries for electric vehicles and to the manufacturing of solar PV panels. Any technological transition will sire a new set of ecological (and economic) concerns because the scale of the change will require massive amounts of resources, whether steel for wind power systems, silicone for solar PV systems, or any other critical resource that supports alternative energy technologies.[29]

The Other Side: Energy Transitions Can Happen Quickly

Contrary to the arguments of those bemoaning the slow pace of energy transitions is the alternative view that, under certain conditions, sizable energy transformations can occur rather speedily. Those tiresome distant cousins or in-laws mentioned earlier can be sent on their way at breakneck speed, if there is a will. For instance, in the United States, most homes used whale oil lamps in 1850, but by 1859, about 85% of households had switched to coal-derived synthetic fuels, largely because the price of whale oil rose with the decline in whale populations.[30] In recent times, the penetration level of geothermal energy in Iceland grew from less than 1% of electricity supply and about 43% of space heating in 1970 to 26% of electricity and 90% of space heating in 2010.[31]

There are numerous modern-day examples of significant market transitions. California is on track to meet 33% of its total electricity demand from renewable sources by 2020, having surpassed 27.8% in 2014.[32] Germany's Energiewende (energy transition) has seen renewable energy contributions to the electricity mix grow from miniscule amounts in 1990 to 19.9% in 2012, making renewables the second-largest source of electricity on the grid, behind brown coal and ahead of nuclear power, natural gas, and hard coal.[33] India has seen its use of solar energy grow tenfold in just 18 months, jumping from less than 100 MW in 2013 to more than 1,000 MW in 2014.[34]

Japan is perhaps the best modern-day example of what can be accomplished. That nation, which is not tied to any other nation's electricity grid, dropped its nuclear power output from 292 TWh in 2010 (when it met 28.7% of national supply) to zero overnight as a result of the Fukushima disaster on March 11, 2011.[35] To compensate for the loss, from July 2012 to March 2014, Japan added a whopping 68,642 MW of new solar, wind, hydro, bioenergy, and geothermal electricity supply from more than 1.1 million distributed facilities.[36]

Some scholars have called these achievements short-term transitions,[37] or fast energy turnarounds.[38] Those who advocate the viability of these short-term transitions hold that (1) we have seen numerous fast turnarounds in terms of energy end use, (2) plentiful examples of national-scale transitions populate the historical record, and (3) we can learn from these trends so that favorable future energy transitions can be expedited. We examine here 10 clear-cut examples of fast energy turnarounds that provide food for thought in terms of whether these successes are replicable elsewhere.

History Shows Speedy Transitions in Energy End-Use Devices

At least five transitions in energy systems have occurred with remarkable speed: lighting in Sweden, cookstoves in China, liquefied petroleum gas stoves in Indonesia, ethanol vehicles in Brazil, and air conditioning in the United States.

Sweden was able to phase in a near universal shift to energy-efficient lighting in commercial buildings within about nine years. In 1991, Swedish Energy Authorities arranged for the procurement of high-frequency electronic ballasts for lights in office buildings, commercial enterprises, schools, and hospitals, which saved up to 70% compared with ordinary ballasts.[39] The authorities employed a multipronged IKEA-like approach, which entailed standardization, quality assurance, direct procurement, stakeholder involvement, and product demonstrations to support diffusion.[40] Due to these concerted efforts, the technology reached the commercial viability stage as early as 1996 and was able to leverage the momentum established by the government to increase adoption from about 10% in 1996 to almost 70% by 2000, as shown in figure 15.1. In essence, this meant that between 1991 and 2000, 2.3 million Swedish workers witnessed changes to the lighting in their offices.

The Chinese Ministry of Agriculture and Bureau of Environmental Protection and Energy (BEPE) oversaw an even more impressive project, the National Stove Improvement Program, from 1983 to 1998.[41] The BEPE adopted a "self-building, self-managing, self-using" policy that centered on rural people inventing, distributing, and maintaining energy-efficient cookstoves. It set up

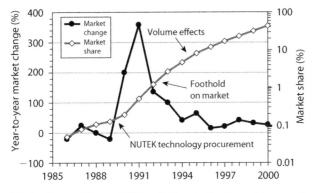

Figure 15.1. Market Share of Energy-Efficient Ballasts in Sweden, 1986–2000. *Source:* Peter Lund, "Effectiveness of Policy Measures in Transforming the Energy System," *Energy Policy* 35 (2007): 627–639.

pilot programs in hundreds of rural provinces. From its inception until 1998, the program was responsible for the installation of 185 million improved cook-stoves and facilitated the penetration of improved stoves from less than 1% of the rural Chinese market in 1982 to more than 80% by 1998—affecting over half a billion people (table 15.2). The number of cookstoves installed in China in 1994, during the height of the program, amounted to 90% of all improved stoves installed globally. As a consequence of this program, Chinese energy use per capita declined in rural areas at an annual rate of savings of 5.6% between 1983 and 1990.

Indonesia effected an energy transition of sorts through a large rural household program facilitating conversion from kerosene stoves to liquefied petroleum gas (LPG) stoves to help improve indoor air quality. Under leadership from Vice President Jusuf Kalla, the Indonesian "LPG Megaproject" offered households a free "initial package" consisting of a 3-kilogram LPG cylinder, an initial free gas fill, a one-burner stove, a hose, and a regulator. The government, in tandem, lowered kerosene subsidies (increasing its price) and constructed new refrigerated LPG terminals to act as national distribution hubs. Amazingly, in just three years—from 2007 to 2009—the number of LPG stoves nationwide jumped from a mere 3 million to 43.3 million, meaning that they penetrated almost two-thirds of Indonesia's 65 million households (affecting about 216 million people). Six of the nation's 34 provinces, including the Special Capital region of Jakarta, were declared "closed and dry," meaning that the program reached all of its targets and all kerosene subsidies were withdrawn.[42]

Brazil boasts one of the fastest substantive energy transitions on record. It launched its Proálcool program in November 1975 to boost ethanol production

Table 15.2. Households Adopting Improved Stoves in China, 1983–1998

Year(s)	NSIP households (million)	Households under provincial programs (million)	Total households/ year (million)	Total people/ year (million)
1983	2.6	4.0	6.6	21.1
1984	11	9.7	20.7	66.2
1985	8.4	9.5	17.9	57.3
1986	9.9	8.5	18.4	58.9
1987	8.9	9.1	18	57.6
1988	10	7.5	17.5	56
1989	4.5	5	9.5	30.4
1990	3.6	7.8	11.4	36.5
1991–98	7.8	57.2	65	208
Total	66.7	118.3	185	592

Source: M. A. Brown and B. K. Sovacool. "China's National Improved Stove Program, 1983–1998," in *Climate Change and Global Energy Security: Technology and Policy Options* (Cambridge, MA: MIT Press, 2011), 292–301.

Note: NSIP = National Stove Improvement Program.

and substitute ethanol for petroleum in conventional vehicles. By 1981, just six years later, 90% of all new vehicles sold in Brazil could run on ethanol—an impressive feat. A more recent development, connected in part to the Proálcool program, is also noteworthy. In 2003, the Brazilian government started incentivizing flex-fuel vehicles (FFVs) through reduced tax rates and fuel taxes. These FFVs were capable of running on any blend of ethanol from zero to 100%, giving drivers the option of choosing between various blends of gasoline and ethanol, depending on price and convenience. In 2004, when FFVs first appeared on the market, they accounted for 17% of new car sales. By 2009, market diffusion had reached 90% , as shown in figure 15.2—meaning that about 2 million FFVs were purchased during the first five years of the program.[43]

Air conditioning in the United States is a final example, albeit in an opposite direction when it comes to energy conservation. In 1947, mass-produced, low-cost window air conditioners entered the market, enabling many people to enjoy air conditioning without the need to buy a new home (for centralized air conditioning) or completely overhaul their existing heating systems.[44] That year, only 43,000 units were sold. Yet, by 1953, the number had jumped to one million, as air conditioners were embraced by builders eager to mass-produce affordable yet desirable modern homes.[45] Consequently, according to the US Census in 1960, more than 12% of people (occupying 6.5 million housing units) owned an air conditioner. By 1963, ownership had risen to 25%, and by 1970, ownership had reached 35.8%. This amounted to 24.2 million homes and more than 50 million people.[46] By 2009, 87% of single-family homes possessed air conditioning.[47] Annually, the United States now consumes more electricity for

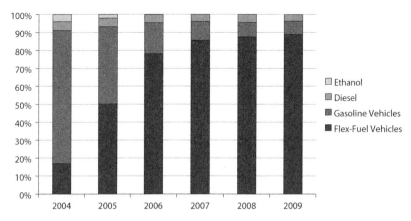

Figure 15.2. Flex-Fuel Vehicle Sales as Percentage of Overall Car Sales in Brazil, 2004–2009. *Source:* M. A. Brown and B. K. Sovacool. "Brazil's Proalcohol Program and Promotion of Flex-Fuel Vehicles," in *Climate Change and Global Energy Security: Technology and Policy Options* (Cambridge, MA: MIT Press, 2011), 260–274.

air conditioning than the entire continent of Africa consumes for all electricity uses.[48] In fact, the United States currently uses more energy (about 337 billion kWh) for air conditioning than all other countries' air conditioning usage combined.[49] Although the example of air conditioning diffusion in the United States is not a fuel mix transition, the case study strongly suggests that consumers, when properly incentivized, will quickly embrace technological change.

Fast Transitions Have Occurred in National Energy Supply

Those who contend that fast transitions are possible also point to five other energy transitions that occurred on a larger scale—at the national level—and entailed sweeping changes in energy mixes. These transitions relate to crude oil in Kuwait, natural gas in the Netherlands, nuclear electricity in France, combined heat and power in Denmark, and coal retirements in Ontario, Canada.

Substantial investments in oil catalyzed an almost complete shift in Kuwait's national energy profile over about nine years. These investments catapulted oil from constituting a negligible amount of total national energy supply in 1946 to 25% in 1947, and ultimately exceeding 90% of total primary energy use in 1950.[50] The impact of this transition is staggering when one considers that in 1945, the Kuwaiti oil industry was characterized by five-gallon barrels transported to customers by camels, donkeys, or wooden pushcarts. By 1949, the industry was characterized by huge volumes of oil transported by motorized trucks, tankers, and pipelines to filling stations all over the nation.[51]

By 1965, Kuwait had become the world's fourth-largest producer of oil (behind the United States, USSR, and Venezuela, and ahead of Saudi Arabia).[52] Even Vaclav Smil, who is dubious about the viability of facilitating an expeditious global energy transition, concedes that "in energy terms, Kuwait thus moved from a pre-modern society dependent on imports of wood, charcoal, and kerosene to an oil superpower in a single generation."[53] As was the case with the adoption of air conditioning in the United States, the incentives to embrace oil engendered an astonishingly quick pace of technological diffusion.

In 1959, the massive Groningen natural gas field was discovered in the Netherlands, fostering what would become a rapid transition away from oil and coal to natural gas.[54] That year, coal supplied about 55% of the Dutch primary energy supply; crude oil contributed 43% and natural gas amounted to less than 2% of the national energy mix. By December 1965, one year after gas deliveries from Groningen began, there were signs that a shift was underway. The role of natural gas expanded that year to 5% of the Netherland's primary energy. To facilitate a broader transition, the government decided in December 1965 to abandon all coal mining in Limburg province within a decade. This was an act that would do away with some 75,000 mining-related jobs, affecting more than 200,000 people. What made this transition politically possible was that the government strategically implemented countermeasures to alleviate transitional job losses. It provided subsidies for creating new industries, relocated government industries from the capital to regions of the country hardest hit by the mine closures, launched retraining programs for miners, and offered shares in Groningen to Staatsmijnen (the state mining company). Six years later, in 1971, natural gas accounted for 50% of the national energy mix. The lesson to be learned from this Dutch transition is that planning for transitional losses goes a long way to attenuating some of the opposition to change.

The French transition to nuclear power was also swift. Following the oil crisis in 1974, Prime Minister Pierre Messmer launched a large nuclear power expansion program intended to generate all of France's electricity from nuclear reactors and displace the republic's heavy dependence on imported oil. As the maxim went at the time, "No coal, no oil, no gas, no choice."[55] The Messmer Plan proposed the construction of 80 nuclear power plants by 1985 and 170 plants by 2000. Work commenced on three plants—Tricastin, Gravelines, and Dampierre—immediately following announcement of the plan, and France ended up constructing 56 reactors between 1974 and 1989. As a result, nuclear power grew from 4% of national electricity supply in 1970 to 10% in 1978 and almost 40% by 1982. As energy and technology scientist Arnulf Grubler noted, "The reasons for this success lay in a unique institutional setting allowing centralized decision-making, regulatory stability, dedicated efforts for standardized reactor designs and a powerful nationalized utility, EDF, whose substantial in-house engineering resources enabled it to act as principal and agent of reactor construction simultaneously."[56] Again, the French nuclear power case study demonstrates that where there is a will, there is a way.

Though Denmark is perhaps more famous for its transition to wind energy, a far more accelerated transition that occurred in the 1970s and 1980s involved different technologies. This transition entailed a shift away from oil-fired electricity to other fossil fuels and combined heat and power (CHP) plants. From 1955 to 1974, almost all heating in Denmark was provided by fuel oil, which meant that the oil crises of the 1970s had a particularly painful impact on the country's economy.[57] In response, the Danish Energy Policy of 1976 articulated the short-term goal of reducing oil dependence, building a "diversified supply system," and meeting two-thirds of total heat consumption with "collective heat supply" by 2002. Specifically, it sought to reduce oil dependence from 90% to 20%, an ambitious goal that involved the conversion of 800,000 individual oil boilers to natural gas and coal. In a mere five years—from 1976 to 1981—Danish electricity production changed from oil dominance to 95% coal-based. The government's support for CHP was strengthened by the 1979 Heat Supply Act, which was designed to "promote the best national economic use of energy for heated buildings and supply them with hot water and to reduce the country's dependence [on oil]."[58] As a result, CHP production increased from trivial amounts in 1970 to supply 61% of national electricity and 77% of the country's district heating in 2010.[59]

Our final example of a rapid transition is relevant because, rather than transitioning toward a specific technology, it involves transitioning away from an undesirable technology. In 2003, the government of Ontario announced plans to retire all coal-fired electricity generation by 2007—something it accomplished, albeit a few years behind schedule. Ontario's oldest coal plant—the 1,140 MW Lakeview facility—was closed in April 2005, followed by sequential closures of plants in Thunder Bay (306 MW), Atikokan (211 MW), Lambton (1,972 MW), and Nanticoke (3,945 MW) between 2007 and 2014. Coal generation thus declined from 25% of provincial supply in 2003 to 15% in 2008, 3% in 2011, and 0% in 2014. To achieve this transition, Ontario invested more than $21 billion in cleaner sources of energy, including wind, hydroelectricity, solar, and nuclear power, as well as $11 billion in transmission and distribution upgrades.[60] Ontario is on track to see renewable sources of electricity grow to 46% of supply by 2025.[61] The reason the Ontario shift is relevant is that it was designed to transition out of coal—a common objective in support of any strategy for mitigating global climate change.

Future Energy Transitions Can Be Expedited

The final argument that supports the viability of an expeditious transition is that although some transitions might have taken a great deal of time historically, policymakers can learn from best practices to ensure that contemporary or future energy transitions can be accelerated. Future energy transitions, because they can draw on synergistic advances in multiple domains at once—cutting across materials, computing, combustion, gasification, nanotechnology, biological and genetic engineering, 3D printing, and the industrial Internet—can truly be accelerated in ways that past transitions (generally) could not. As Harvard energy scientist Kathleen M. Araujo argues, "Countries can, in fact, alter their energy balance in a significant way—stressing low carbon energy sources—in much less time than many decision-makers might imagine. Critical substitution shifts within [Brazil, France, Denmark, and Iceland] were accomplished often in less than 15 years. Moreover, these transitions were effectuated even amidst circumstances at times involving highly complex energy technologies."[62]

Former vice president Al Gore encapsulated this optimism about change in 2008 when he exhorted, "Today I challenge our nation to commit to producing 100 percent of our electricity from renewable energy and truly clean carbon-free sources within 10 years."[63] Gore went on to say that a complete change in energy production was "achievable, affordable and transformative" within the course of one decade. Innovations in both technology and policy design can accelerate technological change and achieve an energy transition in ways not possible even just a few decades ago.

Common Ground: Energy Transitions Are Path-Dependent and Cumulative

How can these two almost incommensurable views be reconciled? In keeping with the dichotomy presented in this chapter, when a transition does occur, some might say, "My, that was unexpected," while others might say, "Gosh, what took it so long?" The synthesis presented in box 15.1 holds that change need not be entirely revolutionary or substitutive; transitions can, instead, be evolutionary and cumulative. The motorized automobile–inspired (at least partially) transitions to oil in Kuwait and to FFVs in Brazil were an amalgamation of earlier inventions fused together: the internal combustion engine, the wheel, the casting of steel, electric lights, tires, the assembly line, and so on. The CHP, biomass, wind, and solar technology behind the transitions in Denmark and Ontario benefited from advances in the fossil fuel chain, including combined cycle turbines, batteries, and compressed air energy storage. Thus, a transition often appears, not as an exponential line on a graph, but as a punctuated equilibrium that dips and rises. Sometimes, transitions are not completed because the new fuel sources do not fulfill all the needs met by the old fuel sources. Older sources of energy—such as muscle power, animal power, wood power, and steam power—still remain in use throughout the world today. They have not been entirely replaced by fossil, nuclear, and modern renewable energy.[64] One analyst at MIT recently commented that "we'll use renewable energy more as technology makes it cheaper, but we're likely to keep using more of the other sources of energy, too."[65]

This common ground gives rise to three concluding insights. First, transitions can be like boxing upsets that seem to happen with a punch that comes out of nowhere; however, the speed of the transition masks all of the background activities that made the transition possible. The general public sees just the final punch thrown by the underdog, not the years of training that got the boxer into the ring. To Ralph Waldo Emerson, the change from sailing ships to steam ships might have seemed frightfully quick (less than a year) because his journey to Europe occurred during the pinnacle of the transition, but the change was almost a century in the making. Similarly, the American transition to oil, according to Smil, took about 80 years to reach a 25% share,[66] yet during the most accelerated phase of that transition—from 1900 to 1925—oil's share grew from 2.4% to 24%, giving an impression that it was quite sudden.[67]

How one chooses the starting point of a transition also affects the perception of speed. Consider air conditioning. Whether one takes the time of first conception (Nikola Tesla developed electric motors that made the invention of oscillating fans possible in 1885), the first invention (Willis Carrier invented the first modern system in 1902), or the first successful commercial application (Henry Galson developed an affordable mass-produced system in 1947) greatly alters the perceived speed of market penetration.[68] Depending on how one looks at it, Brazil's transition to flex-fuel vehicles took 1 year (from the start of the national program to large-scale diffusion), more than 20 years (from the invention of FFVs in 1980), almost 30 years (from the start of the national ethanol program), or more than eight decades (from the invention of a Brazilian engine capable of using ethanol in the 1920s).

In the case of national transitions, we see similar ambiguity. Kuwait's transition to oil can be seen as beginning in 1934, with the first concession given to the Kuwait Oil Company; or in 1937, when the first exploratory wells were drilled in the Burgan field; or in 1946, when commercial production began (the starting point we take); or even in 1949, when the first refinery was established. Similarly, the French nuclear power program could be viewed as commencing in 1942, with the first chain reaction in the Manhattan Project; or in 1945, with formation of the Commissariat à l'Énergie Atomique; or in 1948,

---

BOX 15.1

Energy Transitions and the Common Ground

Can energy transitions be expedited?

ONE SIDE: On the global scale, transitions are long, protracted affairs, made difficult by path dependency and carbon lock-in.

OTHER SIDE: Rapid transitions in end-use or national supply are possible, made all the more likely by future learning and innovation.

COMMON GROUND: We must understand that energy transitions are path-dependent and cumulative rather than fully revolutionary or substitutive.

---

when France's first research reactor was commissioned; or in 1974, with the launch of the Messmer Plan.

The difficulty in defining when a transition begins is connected to a second insight: what might seem to be an abrupt transition can actually be a bundle of more discrete advances that accumulate over time. As O'Connor concludes, "Big transitions are the sum of many small ones. Looking at overall energy consumption will miss the small-scale changes that are the foundation of the transitions."[69] For example, the ascent of oil at the beginning of the twentieth century can be interpreted as a series of progressive advances, including:

· The switch from animal power to internal combustion engines for private vehicles and the social rejection of electric vehicles.
· The conversion of steam engines to diesel in marine vessels and trains.
· The shift from candles and kerosene for lighting to oil-based lamps.
· The adaptation of coal boilers to oil boilers for the generation of electric power.
· The shift from wood-burning fireplaces and coal stoves to oil and gas furnaces in homes.

Similarly, the air conditioning transition in the United States was the result of progressive innovations in air circulation equipment, heat exchangers, heat pumps, halocarbon refrigerants, customization, mass production, and marketing.[70] Often these "minor transitions," when they occur in a concerted manner, enable the "major transitions" that are so easily identifiable.

Third, energy transitions are complex and irreducible to a single cause, factor, or blueprint. They can be influenced by endogenous factors within a country, such as aggressive planning from stakeholders in Denmark, Indonesia, or Ontario, accelerated by political will and stakeholder involvement. Or they can be affected by exogenous factors outside a country, such as military conflict (think of the World Wars spawning the French nuclear program), a major energy accident (Chernobyl, Fukushima), or a global crisis (the oil shocks of the 1970s, the collapse of communism in the early 1990s). Other transitions, such as the adoption of air conditioning or the ascendance of oil in Kuwait, were almost entirely market-driven.

The implication is that energy transitions are not predicated on a magic formula. The United Kingdom, for instance, had the same access to natural gas as the Netherlands, yet it was unable to cultivate the same type of change-over.[71] The experience of tiny, affluent countries such as Denmark and Kuwait may be relevant for countries in a similar class (such as Belgium, Brunei, or Qatar) but less so for nations like India or Nigeria. Moreover, the sociocultural or political conditions behind transitions in Brazil and China, which at the time were a military dictatorship and a communist regime, respectively, are incompatible with policymaking in modern democracies in Europe and North America. Furthermore, history seems to suggest that past transitions—including many of the case studies examined in this chapter—are based on discoveries of new, significant, and affordable forms of energy or technology, but in the future, it may be scarcity rather than abundance that influences decisions.[72] In short, some households elect to bid tiresome distant cousins or in-laws a quick adieu, while other households grin and bear it—and suffer the consequences.

NOTES

1. Robert D. Richardson Jr., *Emerson: The Mind on Fire* (Berkeley: University of California Press, 1994).

2. William McDonough and Michael Braungart, *Cradle to Cradle: Remaking the Way We Make Things* (New York: Farrar, Straus and Giroux, 2002), 128.

3. Vaclav Smil, *Energy Myths and Realities: Bringing Science to the Energy Policy Debate* (Washington, DC: Rowman and Littlefield, 2010), 140–141.

4. Ibid., 136–141.

5. Global Energy Assessment, *Global Energy Assessment—Toward a Sustainable Future* (Cambridge: Cambridge University Press; Laxenburg, Austria: International Institute for Applied Systems Analysis, 2012).

6. N. P. Myhrvold and K. Caldeira, "Greenhouse Gases, Climate Change and the Transition from Coal to Low-Carbon Electricity," *Environmental Research Letters* 7 (2012): 014019.

326 Energy Security and Energy Transitions

7. Janelle Knox-Hayes, "Negotiating Climate Legislation: Policy Path Dependence and Coalition Stabilization," *Regulation and Governance* 6, no. 4 (December 2012): 545–567.

8. M. A. Brown et al., *Carbon Lock-in: Barriers to the Deployment of Climate Change Mitigation Technologies*, ORNL/TM-2007/124 (Oak Ridge, TN: Oak Ridge National Laboratory, November 2007).

9. A. Goldthau and B. K. Sovacool, "The Uniqueness of the Energy Security, Justice, and Governance Problem," *Energy Policy* 41 (February 2012): 232–240.

10. James Meadowcroft, "What about the Politics? Sustainable Development, Transition Management, and Long Term Energy Transitions," *Policy Sciences* 42, no. 4 (November 2009): 323.

11. G. C. Unruh, "Understanding Carbon Lock-in," *Energy Policy* 28 (2000): 817–830.

12. F. W. Geels and J. W. Schot, "Typology of Sociotechnical Transition Pathways," *Research Policy* 36 (2007): 399–417; J. W. Schot and F. W. Geels, "Strategic Niche Management and Sustainable Innovation Journeys: Theory, Findings, Research Agenda, and Policy," *Technology Analysis and Strategic Management* 20, no. 5 (2008): 537–554.

13. US Energy Information Administration (EIA), *Annual Energy Outlook* (Washington, DC: US Department of Energy, 2013).

14. Based on table 2.1 in International Energy Agency (IEA), *World Energy Outlook* (Paris: OECD, 2012), 51.

15. Vaclav Smil, "A Skeptic Looks at Alternative Energy," *IEEE Spectrum*, June 28, 2012.

16. Smil quoted in Stephen Lacey, "Why the Energy Transition Is Longer Than We Admit," *Renewable Energy World*, April 22, 2010, available at http://blog.renewableenergyworld.com/ugc/blogs/2010/04/why-the-energy-transition-is-longer-than-we-admit.html.

17. Vaclav Smil, *Energy Transitions: History, Requirements, Prospects* (Santa Barbara, CA: Praeger, 2010).

18. Ibid., 150.

19. Smil, "Skeptic Looks at Alternative Energy."

20. Ibid.

21. Peter Lund, "Market Penetration Rates of New Energy Technologies," *Energy Policy* 34 (2006): 3317–3326.

22. Peter Lund, "Exploring Past Energy Changes and Their Implications for the Pace of Penetration of New Energy Technologies," *Energy* 35 (2010): 647–656.

23. Lund, "Market Penetration Rates," 3320.

24. Peter Lund, "Fast Market Penetration of Energy Technologies in Retrospect with Application to Clean Energy Futures," *Applied Energy* 87 (2010): 3582.

25. Roger Fouquet, "The Slow Search for Solutions: Lessons from Historical Energy Transitions by Sector and Service," *Energy Policy* 38, no. 11 (November 2010): 6586–6596.

26. Peter A. O'Connor, "Energy Transitions," *Pardee Papers*, no. 12 (November 2010): 10–11.

27. Data from IEA, *World Energy Outlook 2000* (Paris: OECD, 2000), and IEA, *World Energy Outlook 2011: Fuel Shares in World Total Final Consumption* (Paris: OECD, 2011).

28. B. K. Sovacool, "Exploring the Hypothetical Limits to a Nuclear and Renewable Electricity Future," *International Journal of Energy Research* 34 (November 2010): 1183–1194.

29. Nayantara D. Hensel, "An Economic and National Security Perspective on Critical Resources in the Energy Sector," in *New Security Frontiers: Critical Energy and the Resource Challenge*, ed. Sai Felicia Krishna-Hensel (London: Ashgate, 2012), 113–138. See also US Department of Energy (DOE), *Critical Materials Strategy* (Washington, DC: DOE, December 2011).

30. Amory Lovins, "A Farewell to Fossil Fuels: Answering the Energy Challenge," *Foreign Affairs*, March/April 2012, 135.

31. Kathleen M. Araujo, "Energy at the Frontier: Low Carbon Energy System Transitions and Innovation in Four Prime Mover Countries" (PhD dissertation, MIT, February 2013).

32. California Energy Commission, "Total System Power for 2012: Changes from 2011" (2012), available at http://energyalmanac.ca.gov/electricity/total_system_power.html.

33. Amory B. Lovins, "Separating Fact from Fiction in Accounts of Germany's Renewables Revolution" (Rocky Mountain Institute, August 15, 2013), available at http://blog.rmi .org/separating_fact_from_fiction_in_accounts_of_germanys_renewables_revolution.

34. Debjoy Sengupta, "Indian Solar Installations Are Forecast to Be Approximately 1,000 MW," *Economic Times of India*, May 25, 2014.

35. "Before and After Fukushima," *Asia Pacific Consulting Newsletter*, January 2014.

36. Japan Renewable Energy Foundation, Ministry of Energy, Economy, and Industry, and Japan Agency of Natural Resources and Energy, *Renewable Energy Installations under FiT*, as of March 2014 (Tokyo: Japan Renewable Energy Foundation, Ministry of Energy, Economy, and Industry, and Japan Agency of Natural Resources and Energy, June 17, 2014).

37. Jan Rotmans, René Kemp, and Marjolein van Asselt, "More Evolution Than Revolution: Transition Management in Public Policy," *Foresight* 3, no. 1 (2001): 15–31.

38. Staffan Jacobsson and Volkmar Lauber, "The Politics and Policy of Energy System Transformation—Explaining the German Diffusion of Renewable Energy Technology," *Energy Policy* 34, no. 3 (February 2006): 256–276.

39. Peter Lund, "Effectiveness of Policy Measures in Transforming the Energy System," *Energy Policy* 35 (2007): 627–639.

40. Alan Ottossen and Staffan Stillesjo, *Procurement and Demonstration of Lighting Technologies for the Efficient Use of Electricity* (Stockholm: Royal Institute of Technology, Sweden, and NUTEK, 1996).

41. Kirk R. Smith et al., "One Hundred Million Improved Cookstoves in China: How Was It Done?" *World Development* 21, no. 6 (1993): 941–961. See also M. A. Brown and B. K. Sovacool, "China's National Improved Stove Program, 1983–1998," in *Climate Change and Global Energy Security: Technology and Policy Options* (Cambridge, MA: MIT Press, 2011), 292–301.

42. Hanung Budya and Muhammad Yasir Arofat, "Providing Cleaner Energy Access in Indonesia through the Megaproject of Kerosene Conversion to LPG," *Energy Policy* 39 (2011): 7575–7586.

43. M. A. Brown and B. K. Sovacool, "Brazil's Proalcohol Program and Promotion of Flex-Fuel Vehicles," in *Climate Change*, 260–274.

44. National Academy of Engineering, *Air Conditioning and Refrigeration Timeline* (Washington, DC: National Academies Press, 2013).

45. Rebecca A. Rosen, "Keepin' It Cool: How the Air Conditioner Made Modern America," *Atlantic*, July 14, 2011.

46. US Bureau of the Census, *U.S. Census of Population and Housing 1960* (Washington, DC: Bureau of the Census, 1960), vol. 1; US Bureau of the Census, *U.S. Census of Housing 1970* (Washington, DC: Bureau of the Census, 1970), vol. 1.

47. EIA, "Air Conditioning in Nearly 100 Million U.S. Homes" (August 19, 2011), available at www.eia.gov/consumption/residential/reports/2009/air-conditioning.cfm.

48. Stan Cox, "Climate Risks Heat up as World Switches on to Air Conditioning," *Guardian* (London), July 10, 2012.

49. Michael Sivak, "Will AC Put a Chill on the Global Energy Supply?" *American Scientist*, September/October 2013.

50. Data compiled from Kuwait Ministry of Planning, Statistics and Census Sector, *Statistical Review, 1950–1970* (Kuwait City: Kuwait Ministry of Planning, 1988).

51. Petroleum Media and Public Relations Department at Oil Ministry, *History of Petrol Stations in the State of Kuwait* (Kuwait City: Kuwait Oil Ministry, 2005).

52. Smil, *Energy Transitions*.

53. Ibid., 89.

54. Ibid., 50–56.

55. Quoted in Araujo, "Energy at the Frontier."

56. Arnulf Grubler, "The Costs of the French Nuclear Scale-up: A Case of Negative Learning by Doing," *Energy Policy* 38 (2010): 5186.

57. B. K. Sovacool, "Energy Policymaking in Denmark: Implications for Global Energy Security and Sustainability," *Energy Policy* 61 (October 2013): 829–831.

58. Danish Government, Heat Supply Act 1957, available at www.iea.org/policiesand measures/pams/denmark/name-21778-en.php.

59. As an aside, national planners managed a *third* transition, away from coal, in the 1990s. The Danish parliament passed the "coal stop," functionally outlawing the construction of new coal-fired power stations, with exceptions for only two 450 MW plants.

60. Ministry of Energy, Canada, *Achieving Balance: Ontario's Long-Term Energy Plan* (Toronto: Ministry of Energy, 2013).

61. Ibid.

62. Araujo, "Energy at the Frontier," 24.

63. Andrew Revkin, "The Annotated Gore Energy Speech," *New York Times*, July 17, 2008.

64. This point is articulately made by David Edgerton in *The Shock of the Old: Technology and Global History Since 1900* (Oxford: Oxford University Press, 2007), and by Martin Melosi in "Energy Transitions in Historical Perspective," in *The Energy Reader*, ed. Laura Nader (London: Wiley Blackwell, 2010), 45–60.

65. Kevin Bullis, "How Energy Consumption Has Changed Since 1776," *MIT Technology Review*, July 3, 2013, available at www.technologyreview.com/view/516786/how-energy-consumption-has-changed-since-1776.

66. Smil, *Energy Transitions*, 9.

67. Joseph A. Pratt, "The Ascent of Oil: The Transition from Coal to Oil in Early Twentieth-Century America," in *Energy Transitions: Long-Term Perspectives*, ed. Lewis J. Perelman, August W. Giebelhaus, and Michael D. Yokel (Boulder, CO: Westview Press, 1981), 9–34.

68. Will Oremus, "A History of Air Conditioning," *Slate Magazine*, July 15, 2013.

69. O'Connor, "Energy Transitions."

70. National Academy of Engineering, *Air Conditioning and Refrigeration.*

71. Smil, *Energy Transitions.*

72. On this point see Jörg Friedrichs, *The Future Is Not What It Used to Be: Climate Change and Energy Scarcity* (Cambridge, MA: MIT Press, 2013). See also Pratt, "Ascent of Oil."

Values and Truth, Fact and Fiction in Global Energy Policy

In 1975, after the first oil shock and energy crisis of the decade had subsided, the US National Academies of Science launched an ambitious enterprise: it created the Committee on Nuclear and Alternative Energy Systems (CONAES), with the goal of providing a "detailed analysis of all aspects of the nation's energy situation."[1] To ensure broad stakeholder representation, an interdisciplinary committee was created. It enlisted members from universities, government laboratories, oil companies, instrument manufacturers, electric utilities, banks, and law firms. It solicited participation from engineers, physicists, geophysicists, economists, sociologists, ecologists, a physician, a banker, and a public interest attorney. The idea was that within two to three years, this committee should be able to produce a 150- to 200-page report presenting recommendations, based on a consensus of experts, as to which technologies the government should support and which it should not (meshing nicely with our conclusion in chapter 3).

Unfortunately, the result was a series of meetings that could have been hosted by Jerry Springer. As Philip Handler, president of the National Academy of Sciences, noted in 1978, "That first meeting of CONAES was remarkable; the tension seemed almost physical; profound suspicion was evident; first names were rarely used; the polarization of views concerning nuclear energy was explicit. Four years later, that polarization persists, and many of the same positions are still regularly defended."[2]

By 1981, contention and disagreement had swelled the report to 718 pages. Despite repeated rounds of external review, the experts were not able to reach any type of common ground. Eventually, the powers that be gave up and released a report with chapters split into competing "sides," encumbered by a sea of caveats. To get members to sign off on the publication, an appendix was included in which committee members could offer their "personal comments when they wished to clarify or take exception to statements in the text."[3]

The process of preparing the report demonstrates the difficulties faced by any group trying to arrive at agreement on energy issues. "It simply can't be done," CONAES committee chairman Harvey Brooks concluded, "at least not within any group that honestly represents the spectrum of defensible views in today's academic, intellectual, and industrial community."[4] If some of the brightest and best experts were unable to agree on the role of energy technology for just one country, should we be surprised that, globally, the problem of disharmony and disagreement is only amplified?

In this final chapter, we attempt to explain the contention manifest in our 15 questions. We argue that at least six causes of contention underpin the conflicting frames on energy issues: competing interests, rapid changes in technology or data, uncertainty, marginalization of certain stakeholders, competing values, and flat-out hubris. With the hope that readers might want to become part of the solution, we also offer some maxims for avoiding or minimizing contention, which center on understanding the sources of your own frame and the frames of others.

Causes of Contention

As we mentioned in the introduction to this book, a distinguishing feature of the volume is that it focuses on subjective frames—differing conceptions of reality, or worldviews—rather than on objective facts. Indeed, there are no less than eight competing energy frames permeating opposing positions on our 15 questions, as listed in table 16.1—and this list is probably far from exhaustive. Each of these frames influences how energy is conceptualized, what variables of analysis are important, how energy resources are valued, and indeed, what merits attention as an energy problem. One of the most popular frames, that of the "technological optimist," holds that we can fix practically any problem with technology. This cognitive frame of mind weaves through our chapters on energy efficiency, peak resources, shale gas, renewables, electric vehicles, biofuel, geoengineering, clean coal, and nuclear power. The technological optimist mindset embraces the notion that we can keep living as we do, as long as we keep innovating. This worldview is contested by notions that it is individual behavior or consumer demand that must be changed (the "conscientious consumer" frame) or that we should prioritize protection of the environment above and beyond the delivery of energy services (the "environmental preservationist" frame). Although all frames are not in conflict at all times, many harbor the potential for conflict. When it comes to complex issues like energy, with its numerous systemic influences, there is bound to be something that people with different frames can disagree over.

Viewing energy governance as a political venue populated by stakeholders with competing frames suggests that advocates of a particular energy system should recast their arguments based on whom they are addressing. For example, nuclear power can be opposed not only on national security grounds (weapons proliferation) and economic grounds (cost overruns and liability from accidents) but out of concerns about environmental ethics (damage from uranium

Table 16.1. Eight Competing Energy Frames

Icon	Frame	Explanation	Chapter(s) in which evident	Key proponents	Central value for energy resources	Focus of concerns
LED lightbulb	Technological optimists	Energy is merely a property of heat, motion, and electrical potential. We can design various technologies to provide it and to repair whatever damage is done.	2 (energy efficiency) 4 (peak resources) 5 (shale gas) 6 (renewables) 7 (electric vehicles) 8 (biofuel) 10 (geoengineering) 11 (clean coal) 12 (nuclear power)	Physicists, scientists, engineers, politicians	Efficiency	Inefficiency and entropy; environmental restrictions on expanded supply
Dollar bill	Free-market libertarians	Energy is a commodity, or collection of commodities, such as electricity, coal, oil, and natural gas. It is best managed by the free market.	2 (energy efficiency) 3 (government intervention) 4 (peak resources) 11 (clean coal) 13 (energy independence)	Economists, financiers	Price	Cartels and inefficient economic behavior; energy problems arising not as the result of imminent depletion of domestic or foreign reserves but from government policy errors exacerbated by the cartel-like actions of oil-producing nations

(continued)

Table 16.1. (continued)

Icon	Frame	Explanation	Chapter(s) in which evident	Key proponents	Central value for energy resources	Focus of concerns
Flag planted on a hilltop	Defenders of national security	Energy supply is a strategic resource that must be defended militarily.	3 (government intervention) 7 (electric vehicles) 8 (biofuel) 12 (nuclear power) 13 (energy independence) 15 (energy transitions)	Security experts, defense analysts, political scientists	Energy access; geopolitical stability	Uneven geographical concentration of energy resources; political instability of producing and consuming countries; declining availability of fuel substitutes
Giving a leg-up	Energy philanthropists	Energy services are a fundamental human right.	9 (climate change mitigation and adaptation) 12 (nuclear power) 14 (global crisis)	Nongovernmental organizations, aid groups, economic development theorists	Equity; empowerment	Indoor air pollution; inequality; energy poverty
Polar bear	Environmental preservationists	Energy production and distribution can be an environmental bane.	1 (pollution) 2 (energy efficiency) 9 (climate change mitigation and adaptation) 11 (clean coal) 14 (global crisis)	Environmentalists, consumer and public interest organizations, affluent households	Environmental footprint	Overconsumption of energy; externalities; rapid depletion of natural resources and ecosystem services

Symbol	Perspective	Statement	Chapters	Proponents	Values	Concern
Scales	Justice advocates	Energy decisions must respect free, prior, informed consent and be equitable in their distribution of costs and benefits.	3 (government intervention) 9 (climate change mitigation and adaptation) 12 (nuclear power)	Lawyers, ethicists, philosophers	Equity, transparency	Unfair or inequitable energy planning; forced relocation of communities living near energy infrastructure
Pyramid	Neo-Marxists	The global energy system exploits class inequality.	9 (climate change mitigation and adaptation) 14 (global crisis) 15 (energy transitions)	Activists, socialists, unions, labor economists, political ecologists	Access, especially by class	Concentration of wealth; unfettered growth and expansion at the expense of communities and the environment; centralization and consolidation
Open book	Conscientious consumers	We consume energy to affirm or even realize our social values and lifestyles.	1 (pollution) 2 (energy efficiency)	Anthropologists, psychologists, sociologists, behavioral economists, corporate sustainability managers	Convenience, cleanliness, price	Energy illiteracy; incompatible or unsustainable values

Table 16.2. Six Causes of Contention in Energy Deliberations

Cause of contention	Explanation	Academic disciplines supporting this claim
Competing interests	Energy is big business and no one wants to lose when the loss amounts to one's livelihood.	Political economy, political science, economics, geography
Complexity and change	Stakeholders base their support on data and technology projections that are contentious and change rapidly.	Engineering, industrial processes, innovation studies, energy policy
Risk and uncertainty	Differing interpretations of hazards and their implications can convince people to make poor decisions.	Risk management, project management, social psychology
Undemocratic exclusion and injustice	Energy systems can exclude or marginalize people from the decision-making or licensing process.	Social justice, contemporary ethics, legal studies, policy analysis
Values and ideology	Distinct systems of values and beliefs can lead to competition over what should be prioritized.	Political science, sociology, anthropology, cultural studies
Energy evangelism	Energy is such a heated topic that the outcome can become a matter of religious or political faith—downgrading or ignoring opposing information.	Sociology of expectation, group psychology, communication studies

mines) or justice (exclusion from the decision-making or licensing process). Similarly, when trying to convince someone about the merits of energy efficiency, it can be sold as a boon for the environment (the least-cost way to save emissions), a leg up for national security (an effective way for lessening energy dependence), or an enabler of local employment (providing more jobs per unit of energy saved/delivered than alternatives). Indeed, this explains why multiple worldviews were represented in the discourse presented in each chapter.

Our list of frames in table 16.1 also implies that energy discussions of any real depth will ultimately sire disagreement rather than consensus. To further unpack this statement, we elaborate here on what we see as six likely sources of contention, drawn from a mosaic of academic research (table 16.2).

Competing Interests

Energy is big business, meaning that there is so much at stake that it can become a battleground for competing interests. In a typical year, almost one in seven dollars in your pocket eventually finds its way to the energy sector. Direct energy expenditures in 2012 amounted to $1.42 trillion in the United States, or about $4,560 per capita (10.4% of gross domestic product).[5] International purchases of oil and gas amount to roughly $1.2 trillion per year (meaning that two-thirds of all oil and gas is traded internationally), in addition to another $1 trillion in annual revenues from the extractive industries sector, to which coal is the largest contributor. No less than 200 billion barrels of crude oil, worth some $20 trillion, are traded as stocks or futures each year. These staggering amounts say nothing about investments "sunk" into energy infrastructure over the past century,[6] which could add another $30 to $50 trillion to the equation.[7]

Entrenched interests are everywhere. One estimate places the global workforce at roughly three billion workers, with 21% of that workforce engaged in industrial activities directly connected to energy extraction, production, and consumption. This figure—630 million workers—excludes those employed in energy-intensive sectors such as agriculture or building construction.[8] Economic interests committed to energy production begin at the extraction phase, in exploring and drilling for oil and natural gas, mining coal and uranium, cultivating biomass, building dams, and harvesting wind and solar energy for power production. At the manufacturing stage—refining oil, processing natural gas, cleaning coal, pelletizing and refining biomass, and transporting energy commodities—a prodigious number of activities take place, and each activity is supported by embedded investment, contractual commitments to workers, and political ties.[9] There are also thousands of energy-related companies at later stages of energy conversion and use. For instance, the United States has more electric utilities—inclusive of large investor-owned utilities, rural electricity cooperatives, government power providers, and smaller distribution and transmission utilities—than it has Burger King restaurants.[10] As in the upstream situation, each of these downstream enterprises has entrenched investments that engender opposition to change. So eliciting change is not just about convincing behemoths such as the ExxonMobils of the world to embrace wind power. The challenge is less akin to turning a supertanker and more akin to trying to align a cluster of marbles atop a table on a sailboat in rough seas.

Entrenched interests of this type suggest that contention and power struggles are inevitable and unavoidable features of energy decision making; they can perhaps be managed but are never eliminated. Energy (and climate policy) is not just "a tradeoff between the present and the future, but also a tradeoff between winners and losers at any given time."[11] As such, analysts need to be more open to the probability that a given energy pathway will distribute benefits inequitably and more willing to accept that there will always be relative losers. Controversy can be particularly sharp when the effort to reduce environmental or health risks may jeopardize other socially valued objectives such as employment and economic growth.[12]

Complexity and Change

Rapid changes in technological capability, resource availability, and prices wreak havoc on data analysis. Unfortunately, data in the energy sector turn over much faster than peoples' convictions—meaning that many people continue to cling to positions supported by obsolete information.

Consider how economic changes in conventional energy alone can alter analyses. Figure 16.1 depicts the market prices of all four major energy fuels from 2004 to 2011. During this period, oil prices oscillated from a low of $37 to a high of $96 per barrel; natural gas prices from a low of $11 to a high of $15.60 per million cubic feet (Mcf); coal from a low of $42 to a high of $98 per ton; and uranium from a low of $15 to a high of $88 per pound.

Energy prices are increasingly subject to capricious change in response to a variety of influences. The price of oil, for instance, jumped dramatically during the oil shocks of the 1970s, during both Gulf Wars, and during the "Arab Spring" of 2011, among other events. The price of natural gas at the Henry Hub trading point in New York skyrocketed from $6.20 per million BTUs (MMBtu) in 1998 to $14.50 in 2001, then dropped precipitously for almost a year, only to rebound again.[13] Hurricane Katrina caused similar price spikes for both oil and gas when the storm disrupted natural gas refining and reprocessing infrastructure in the Southeastern United States. Transportation bottlenecks and demand surges in major developing countries such as India and China have been partly to blame for coal price increases. Other influences such as constricted rail service, flooding, hurricanes affecting barge routes, mine closures, and restrictions on mountain top removal also increasingly influence energy market dynamics. Even the predominant fuel for nuclear power plants, uranium, has exhibited considerable volatility. The cost of uranium jumped from $7.25 per pound in 2001 to $47.25 in 2006, an increase of more than 600%. With price swings of this magnitude, discussions about comparative economic value take on bipolar characteristics. Someone debating the economic merits of natural gas over coal in 2006 would have been in a far weaker position than in a debate over the same issue in 2008. What a difference two years can make in the energy sector.

Complexity and rapid change are eroding humans' ability to rationally manage energy systems. Work on improving cooperation for solving social problems suggests that some variables are key to success.[14] These variables include the availability of high-quality, accurate information and predictable changes in technology and institutions. Conversely, complexity and change are corrosive for effective governance and cooperative efforts because change gives rise to arguments over whether the emergent trend is sustainable, subject to regression, or just temporary.

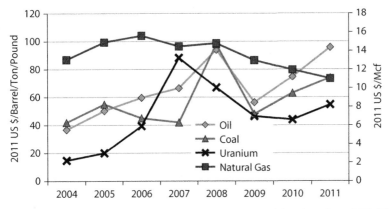

Figure 16.1. Yearly Average Energy Prices, 2004–2011 *Sources:* Coal data from CAPP/NYMEX Coal Futures Settlement Prices of Central Appalachian; oil data from the Cushing, OK, WTI Spot Price FOB; natural gas data from the Henry Hub Gulf Coast Natural Gas Spot Price; uranium data from the Nuexco exchange spot price. *Note: Left axis:* uranium—weighted average price, uranium spot contracts, dollars per pound U_3O_8 equivalent; oil—US Crude Oil First Purchase Price, dollars per barrel; coal—dollars per ton, Central Appalachian bituminous. *Right axis:* natural gas—Annual Residential US Natural Gas Price, dollars per million cubic feet (Mcf).

The complex nature of many technological markets also hampers predictions about the future. Here are a few humorous but revealing examples of just how off-base experts can be:

- Octave Chanute, American aviation pioneer, 1904: "Airplanes will eventually be fast, they will be used in sport, but they are not to be thought of as commercial carriers. To say nothing of the danger, the sizes must remain small and the passengers few, because the weight will, for the same design, increase as the cube of the dimensions, while the supporting surfaces will only increase as the square."[15]
- Clark Woodward, admiral in the US Navy, 1939: "As far as sinking a ship with a bomb is concerned, you just can't do it."[16]
- Thomas Watson, future CEO of IBM, 1943: "I think there is a world market for maybe five computers."[17]
- Dekka Recording, 1962, after rejecting the Beatles: "We don't like their sound, and guitar music is on the way out."[18]
- Lawrence Rocks and Richard Runyon, energy analysts, 1972: "[China] will never be an economic super-state because of her low energy resources."[19]
- Margaret Thatcher, future prime minister of the United Kingdom, 1974: "It will be years—not in my time—before a woman will become Prime Minister."[20]

All of these erroneous forecasts were based on historical extrapolation of trend data that still proved to be wrong. In other words, these forecasters had no empirical evidence to support an alternative perspective, and so they clung to their beliefs in the face of disconcerting evidence—a ripe cause of contention that we see in many of our chapters.

Risk and Uncertainty

It is normally assumed that knowledge reduces uncertainty and so improves decision making. But as many of our chapters show, aleatory uncertainty (that which we cannot know) in complex systems (wicked problems) leads to a situation in which stakeholders with more knowledge (albeit still incomplete) become further entrenched in their positions and make worse decisions. The best example of this concerns support for nuclear power. Compared with the average insurance company underwriter, nuclear engineers have far more technological knowledge about nuclear plant operations. Yet, despite this advanced knowledge, these engineers are far more likely to discount aleatory risk to support nuclear power. Indeed, we know this to be true because no Japanese power plants have been underwritten by insurance policies. After Fukushima, it turns out that the individuals who were less "well-informed" made the better decisions.[21] Similarly, contention can arise over whether consequences are viewed as reversible or irreversible. When effects are irreversible (such as the alteration of a river for a hydroelectric dam or the extinction of a species), they may be even more difficult to resolve.

Conflicts over the feasibility, viability, or desirability of a given technology often boil down to different interpretations of "systemic risk"—hazards that are complex, uncertain, and ambiguous and have the potential to reverberate throughout political, social, and economic dimensions.[22] Consider two competing frames—free-market libertarian and environmental preservationist—and their disagreement over a technological option such as carbon capture and storage (CCS). Those embracing free-market libertarianism perceive extreme risk in abandoning the status quo (dependence on fossil fuel). To them, the riskiest course of action is to transition away from conventional energy because this will erode corporate profits and alter control over the global energy system. At risk, for them, are millions of jobs and trillions of dollars of infrastructure. Their conclusion is to support CCS. Conversely, the environmental preservationists perceive the greatest risk to lie in continuing to support a technology that continues to devastate the environment. To them, doing nothing births future crises, and technical fixes represent unknown risks that can never be fully predicted or controlled. They argue that applying technology on a global environmental scale is suicidal—it should be avoided as a matter of self-preservation and precaution. Their stance is to oppose CCS.

Nuclear power offers a good example of how even shared frames can be in conflict. Justice advocates might point to the risks of having no energy at all—societies in the developing world left, literally, in the dark, with the extreme injustices and health effects of energy poverty. To these individuals, nuclear power is a necessary evil, a tool that can expand access to modern electricity networks and minimize the amount of life-endangering soot and smoke choked down by mothers and young children. Other justice advocates might oppose nuclear energy due to the risks involved in its fuel cycle (such as contamination of indigenous community lands by uranium tailings), its connection to weapons of mass destruction, or the massive consequences of a serious accident (typically affecting underprivileged and/or minority populations the most). Both groups agree that the risks are huge, but they view them in opposite ways with opposing courses of recommended action.

Undemocratic Exclusion and Injustice

Another source of contention is exclusion and injustice in matters of energy: marginalizing people or excluding them entirely from the decision-making process. This can involve involuntary resettlement, lack of consent for an energy project, marginalization of communities living near energy infrastructure, or exclusion from input into the policymaking process. Chapter 5 (shale gas), chapter 6 (renewable electricity), and chapter 12 (nuclear power) all present evidence of how such exclusion can germinate into actual opposition to a particular energy system. More egregious are the cases presented in chapter 14 (global energy security) of people being forcibly resettled due to mining, energy infrastructure, and climate change projects around the world—people who ended up losing their homes, their livelihoods, and in some situations their dignity. A global study warns that the impact of such displacement often extends beyond loss of land to include joblessness, homelessness, marginalization, food insecurity, increased health risks, social disarticulation, and the loss of civil and human rights.[23]

These actions provoke conflict because they are seen as unjust—as violating time-honored notions of due process and justice. Procedural justice refers to equity in the process of allocating costs and benefits (i.e., transparency and access), whereas distributive justice refers to equity in the final allocation of costs and benefits (who gets what and who is charged what). As geographers John Farrington and Conor Farrington put it, "A just society is one that inter alia grants the opportunity of participation in society to all of its members, and a society will certainly be unjust if it does not grant this opportunity to all of its members. Thus, a just society is inter alia a socially inclusive one, and a society is unjust if it is a socially exclusive one."[24] Such procedures are an instrumental part of preserving basic fundamental liberties, and sociologist Claire Haggett suggests that without due process, public support for any given technological system will decline precipitously. As she writes, "While fiscal regulations and subsidies, technical efficiency and political deliberations all affect the deployment of renewables, the stark fact remains that all of this matters little if there is no public support for a development."[25] Conflict will probably be lessened if principles of procedural and distributive justice are followed.[26] If injustice of either type is evident, the possibility of stakeholder opposition will be far greater.

Values and Ideology

Sometimes, a conflict over a given energy option is not only about risk assessment—that is, the hazards—but also about values. Unlike a preference, a value is a nonnegotiable principle that causes a stakeholder to prioritize one thing over another: values provide "a standard for assessing our behavior and that of others."[27] As coalition expert Paul Sabatier explains, core values rarely change over short time spans, and when they become a factor in an energy decision, the importance of all other variables tends to be deemphasized.[28] Therefore, competing values, if present in an energy decision, can result in heated conflict.

An astute reader might ask at this time, how can nuclear power (or any other technology) be value-laden? What, after all, is the core value that could underpin support for nuclear power? Some nuclear power supporters base their support on a concern for the environment. James Hansen, for example, argues that climate change is by far the most serious threat humankind faces and that alternative sources of noncarbon energy are insufficient to expedite the energy transition needed to avert the worst perils attributed to climate change.[29] Therefore, one could argue that for Hansen and colleagues, the underlying value is an environmental ethic: environment first. Yet, for this "value" to underpin support for nuclear power, a non-value-laden assumption is necessary: alternative sources of noncarbon energy are insufficient. In this example, as in many of the other value-laden debates discussed in this book, a value has become attached to a given solution, not because it is inherently attached, but because the proponent of the solution linked that solution ineluctably with that particular value. This is an important distinction: energy debates are not about values; they are about ideologies masquerading as values. Ideologies are different; they are not always healthy, nor are they always premised on real values.

In many of our chapters, we saw contentious perspectives sustained by a broad array of ideologies. Consider the chapters on climate change (looking at mitigation and adaptation, geoengineering, and clean coal). The pathways dif-

Table 16.3. Comparative Analysis of Mitigation, Clean Coal, Geoengineering, and Adaptation

Category	Mitigation	Clean coal	Geoengineering	Adaptation
Timing	Costs now, benefits delayed	Costs now, benefits later	Costs now, benefits very soon	Costs whenever, benefits may be relatively soon after
Temporal incidence	Costs now, benefits to later generations	Costs now, benefits now and to later generations	Benefits mostly to the generation bearing the costs	Benefits mostly to the generation bearing the costs
Geographical incidence	Local costs, global benefits	Local costs, local and global benefits	Local or broader costs, global benefits	Local costs, often relatively local benefits
Sectoral incidence	Focus on emissions from energy consumption	Focus on energy-intensive industries (cement, iron, and steel) and power plants	Only a few options are likely to garner political support	Highly heterogeneous
Relation to uncertainty	Must act early despite greater uncertainty	Must act early despite greater uncertainty	May act later after reducing uncertainty	May act later after reducing uncertainty
Governance issues	Dominated by national goals and international negotiations	Dominated by traditional energy companies and those with significant storage capacity	International oversight needed because of possible actions of rogue nations and individuals acting on their own	Dominated by state and local agencies, but need for coordination is great

fer in terms of timing, temporal and geographical incidence, relation to uncertainty, and governance (table 16.3). These differences can become polarized—people who support putting off dealing with the problem are irresponsible; those who suggest prompt action are hasty—and when this happens, groups become ideologically divided. One cannot support the alternative view because it would mean spurning a value of great importance, regardless of whether the ideology in question is really supported or justified by the value.

The debate over the advisability of geoengineering or CCS is predicated on ideologies pertaining to human progress and technological confidence. One view attaches a high degree of confidence to technological ingenuity, based on historical achievements. Through technology, we are living longer, healthier, and more literate, entertained, luxurious, and fulfilling lives. As E. O Wilson caricatured, "Genius and effort have transformed the environment to the benefit of human life. We have turned a wild and inhospitable world into a garden.

Human dominance is Earth's destiny. The harmful perturbations we have caused can be moderated and reversed as we go along."[30] Geoengineering and clean coal exemplify this ideology of technological ingenuity overcoming a (sometimes) hostile environment. We are, in this perspective, a planet of engineers who are progressively tasked with fine-tuning an engine that we understand well enough to accommodate any contingency.

A competing ideology argues that the earth's environmental endowments and ecological systems have evolved over billions of years to form a complicated adaptive system that is locked together by numerous pliable yet ultimately fragile connections. As Wilson, again, put it, "The biosphere creates our special world anew every day, every minute, and holds it in a unique, shimmering physical disequilibrium. On that disequilibrium the human species is in total thrall. When we alter the biosphere in any direction, we move the environment away from the delicate dance of biology."[31] Interfering with this system, which we neither control nor comprehend, threatens our own existence. Geoengineering and clean coal, in this view, ignorantly and recklessly interfere with supple biological and terrestrial systems that have cradled and nursed humanity through the eons.[32] We are, in this perspective, a planet of untrained tinkerers who are progressively tasked with fine-tuning an engine that runs on a technology with which we are unfamiliar, using a set of tools that might or might not be sufficient for the job.

Such competing ideologies breed conflict precisely because the thing becomes a symbol for the ideology. To argue that we should embrace simpler technological options is tantamount to saying that we have lost control of our destinies—even though opponents are not necessarily making such an argument. That is the inherent problem with ideological conflict: the thing under contention becomes a proxy for the ideology itself. It does not matter whether there are better alternatives to support the ideology or whether the ideology is even based on verifiable facts. When a thing becomes an ideology, it becomes *the* ideology.

Energy Evangelism

As alluded to earlier, sometimes contention arises not from rational thought or a clash of ideologies but from misplaced expectations. Put another way, hope can affect how one favors a certain energy system over another, meaning that actors become converts to a particular symbolic vision. Visions for the future are key elements in the process of technological development and acceptance.[33] Two science and technology specialists, writing about "sociotechnical imaginaries" percolating into nuclear research in South Korea and the United States, point out that national "imaginations can penetrate the very designs and practices of scientific research and technological development."[34] To this day, proponents of nuclear power are still chasing the vision put forth in the 1950s of nuclear energy becoming "too cheap to meter."[35] The continuing controversy over nuclear energy is as much about a series of serious mismatches between expectations and experience as it is about, say, the cost of reactors or the risk of accidents.[36]

Such technological visions and/or rhetorical fantasies can sow contention in at least three ways. First, they become exclusionary and self-replicating, convincing those who do not share these visions to leave a project or disciplinary field entirely.[37] Second, they convince sponsors to underestimate costs and overestimate benefits. In their comparative survey of nuclear power programs in several countries, John Byrne and Steve Hoffman noted that nuclear power has been and continues to be evaluated in the "future tense," that is, in terms of what it will bring rather than what it has already wrought. In short, advocates are guilty of "sweeping away current concerns for future gains."[38] Third, technological fantasy can breed contention by convincing powerful stakeholders to endorse a technology with almost religious fervor, pinning it to utopian narratives about how grandly society will be changed once a particular technology is adopted. For another example of this phenomenon, one need only harken back to the US Republican presidential campaign of 2008, with Sarah Palin leading a room full of adults in an exalted chant of "Drill, baby, drill." More recently, the Tea Party has created a historical alliance with the Sierra Club, known as the Green Tea Coalition, that has been fighting with evangelical zeal for increasing the use of solar power.[39]

Developing deep attachment to technologies in which one has a financial, reputational, political, or vocational stake is not new. One study found this "utopian" and "religious" theme was present in historical deliberations about steam engines, automobiles, hydroelectricity, and nuclear energy.[40] Even experts have been shown to suffer from varying degrees of "trained incapacity,"[41] "selective remembrance,"[42] and "occupational psychosis,"[43] related terms that describe how people prepare to see the world in certain ways, while simultaneously developing a bias that blinds them to other perspectives. This is extremely corrosive to deliberative discourse, particularly in rapidly evolving technology markets. As one study noted, "Public discourse suffers because our society has mechanisms only for resolving conflicting interests, not conflicting views of reality."[44]

Six Maxims for Readers

In reflecting on the 15 questions in this book and the causes for contention discussed above, we wanted to offer some parting guidance to help improve analytical skills in energy governance and decision making. The following six maxims or solutions can help bring far better perspective and understanding to the analysis of energy problems. For readers who are members of the general public, this will make you better citizens; for policymakers, this will make you better practitioners:

1. Know the players: To reveal competing interests, understand where the power lies and how it manifests itself in energy decisions.
2. Inform yourself: To counter the rapidity of change, keep up-to-date and educate yourself about energy technologies and issues.
3. Be prudent about risk: To manage risk and uncertainty, attempt to make energy decisions that are based on clear ethical principles and are well-informed by science.
4. Seek diversity and inclusivity: To avoid undemocratic exclusion and opposition by special interest groups, remember that energy decisions must meet the needs of a broad spectrum of citizens and stakeholders.
5. Practice self-reflection: To understand underlying ideologies, strive to become aware of your own ideological frames that might prohibit a balanced analysis.
6. Embrace technological agnosticism: To avoid energy evangelism, look beyond a given energy technology to the services it provides, and recognize that many systems can deliver the same solution.

Know the Players

To address competing interests, our first maxim is to seek to know the players: make the interests behind an energy system transparent, acknowledge trade-offs, and expect push-back. Readers can start by making an attempt to understand the undercurrents in support of a given energy system. In short, continually ask, "Energy *for whom?*" or "*Who* benefits from this frame?"

Understanding the relationship between power and technological dominance is important on three levels. First, it reminds us that the existing energy regime—with its gas stations, oil refineries, electricity substations, transmission lines, extensive natural gas pipelines, coal mines, and varying types of generating and consuming technology—was and is by no means inevitable. The success of incumbent technologies is the product of coercion, competition, and politicking. Since the current system was created and entrenched by people, it can also be changed by people, but to do so requires competitive engagement with powerful foes.

Second, clarifying why certain stakeholder groups support certain energy technologies allows us to study and analyze the enabling factors that create winners and losers. The implication of this is that a technology can acquire market appeal in two ways: by possessing superior technology or by possessing stakeholder appeal. The Danish wind power industry is a case in point. Initially, the Danish government's strategy was to encourage large manufacturing concerns to lead a wave of wind power development that was predicated on economies of scale. When it became apparent that larger firms were not interested in this market niche and that support for wind power came largely from farmers and farming cooperatives, the government altered its policy to encourage cooperative investment.[45] The success of this is now evident when viewing the vistas in virtually every rural area throughout Denmark.

Third, revealing competing interests highlights the fact that competition will always exist among certain energy options, meaning that we should expect push-back because there will inevitably be losers with any change. Satisfying everybody or every energy objective is an elusive aim. As evidence of this, one study investigated five distinct strategic approaches designed to lessen a country's dependence on imported fuels, to provide energy services at the cheapest price possible, to enable universal access to electricity grids, to mitigate greenhouse gas emissions, and to foster energy systems that can operate under conditions of water stress and scarcity.[46] The authors concluded that each of the five strategies was, more often than not, in conflict with the others. A group that supports climate change mitigation might advocate a ramped-up presence for nuclear power, whereas a group supporting water security might seek to phase out nuclear power. No single strategy optimized all energy security criteria.

In sum, for most stakeholders in the energy sector, energy policy is a zero sum game, where change means that someone gains at someone else's expense.[47] Although some of the conflicts that arise when trying to bolster energy security could be attenuated through better strategic planning, there is no silver bullet when it comes to optimizing energy security—or pleasing all interests. Conflict and power relations are inescapable in the global energy system.

Inform Yourself

To counter rapid changes in energy technologies, prices, resources, and so on, we urge readers to stay informed. Critical to this challenge is to ensure that the sources of your knowledge are diverse so as to avoid becoming biased by the media or others. We also urge policymakers and planners to support public education outreach programs. Thomas Jefferson is attributed with the saying that "a democratic society depends upon an informed and educated citizenry," but for education to occur, people have to be informed "even against their will."[48]

With that said, information and education programs must be carefully tailored to suit the audience. Information is less likely to be used if accessing or interpreting it requires the assistance of an expert. When stakeholder response is an objective of an education campaign, change directed at behavior perceived to be directly under the individual's control, involves few barriers or adjustments, and includes built-in incentives (or lacks disincentives) tends to be the easiest to initiate.[49] Psychologists Renee J. Bator and Robert B. Cialdini, for example, found that public information campaigns can accomplish their goals if they (1) recognize saturation and realize that their message must compete with thousands of others, (2) set achievable goals that emphasize moderate and easy changes in behavior, and (3) target specific audiences and thoroughly understand the demographics, lifestyles, values, and habits of each audience.[50] When structured this way, public information campaigns have changed norms and shifted social attitudes. This is exemplified by specific programs for mitigating household hazardous waste disposal and littering, which reduced these undesirable behaviors by 10% to 20%.[51]

Unfortunately, delivering information to stakeholders in the proper manner is just part of the battle. As the old adage suggests, you can lead a horse to water but you can't make it drink. Recent studies in psychology show that many consumers don't *want* to be better informed about problematic issues such as climate change; instead, they seek to deny that the problem exists so as to assuage feelings of guilt and shame.[52] The implication is that individuals will work to avoid feelings of responsibility for energy insecurity and climate change; some will even cultivate optimistic biases, downgrading any negative information they receive and counterbalancing it with almost irrational exuberance.[53]

Lamentably, formal education is often counterproductive in terms of instilling environmental awareness. Some research suggests that the educational system, far from producing independent thinkers who want to change the world, more often than not serves to entrench the types of material consumption that are responsible for many of our environmental woes.[54] The educational system, according to one education historian, is about creating "masses of industrious workers, loyal subjects, and faithful church members," socializing them into the modern economy.[55] It is therefore unrealistic and perhaps even counterproductive to hope that "education" will solve energy problems, if it indoctrinates us into the global capitalist system underlying many of the energy problems identified in this book. Education is a good start, to be sure, but it remains an imperfect solution. This is why our other maxims are also needed.

Be Prudent about Risk

To learn to discern risk and uncertainty, another maxim must be pursued: strive to be comprehensive in your search for information and look for the hidden linkages. At its core, emergent technology can be considered a response to some earlier flawed technology. Therefore, all new technologies will inevitably possess weaknesses that an analyst must try to identify. A truly prudent energy strategy is one that is comprehensively informed, interdisciplinarily aware, and ethical.

Just as the technological options in the energy sector are diverse, so, too, are the criteria for judging the acceptability of a given technology. The following types of questions can be raised whenever one considers the desirability of a particular energy technology or pathway:

1. Does it harm the environment?
2. Does it degrade the social structure of local communities?
3. Does it damage traditional culture?
4. Does it benefit local economies and utilize local resources?
5. Does it provide education or local participation?
6. Does it promote efforts aimed at conservation and efficiency?
7. Does it foster the well-being of future generations?

While the importance of such questions may appear obvious, most assessments of energy technology continue to ignore the entire range of possible impacts that a given energy system can have on society.

Further complicating evaluation is that some technological decisions serve certain social and environmental goals while directly undermining others. For instance, the deployment of a large nuclear power plant in a small rural community could greatly benefit a select few in the local economy and might even be of value in stimulating industrial growth, but it would also put the community at risk for the sake of electricity that will largely be exported to remote power markets. Similarly, building a large dam may help displace a polluting coal plant (thus improving the environment) but, in the process, destroy aquatic habitats and force widespread relocation of homes and businesses.

Risk profiles not only vary with technology but also change over time. Some technologies, notably energy efficiency and small-scale renewables, produce more easily managed risks. As energy sustainability specialist Mark Diesendorf pointed out, when a nuclear power plant explodes, it is a global disaster; when a solar plant explodes, it is otherwise just another sunny day.[56] For many technologies, risk profiles depend critically on how the technologies are designed and operated. Our chapters have touched on a host of best-practice principles for risky energy systems such as shale gas, coal, and nuclear power. Many of the adverse effects of shale gas are attenuated when waste and fracking fluid discharges are properly monitored, when methane leaks are accounted for, and when siting in environmentally or geologically sensitive areas is avoided. Improvements in occupational safety at coal mines, integrated closure programs and remediation activities after mine closures, and pollution controls at power plants can make coal cleaner. Fully accounting for decommissioning and accident liability costs, enriching uranium with renewable electricity, and following state-of-the-art safety procedures can improve the risk profile of nuclear reactors. Risks for these energy systems can never be eliminated, but there are techniques to better manage them.

Seek Diversity and Inclusion

To minimize exclusion, our fourth maxim suggests that diverse viewpoints and public needs must be comprehensively woven into energy policy decisions. This helps appease competing factions and reduces the costs associated with stakeholder dissent and opposition. Inclusion of input from diverse actors spread across many disciplines, social classes, cultures, and geographical locations also enhances feedback, reduces groupthink, and improves decision making. Public policy analyst Harvey Brooks noted that scientific disputes have always been value-laden, and no practical way of disentangling social interests from technical issues exists.[57] Brooks concluded that policy issues could be resolved only by bringing experts and generalists from the public together so that the values and preferences of the masses were heard. He suggested that this strategy leverages two types of expertise: specialists provide expertise from their fields, while generalists provide expertise on the preferences of society. "Only continual confrontation between generalists and experts," Brooks concluded, "can synthesize the values of society and the facts of nature into a policy decision that is both politically legitimate and consistent with the current state of technical knowledge."[58]

In a just society, citizens have a right to knowledge and information, a right to participation, a right to guarantees of informed consent, and a right to life or protection from danger.[59] These rights need to be exercised, however, because to adequately address many of the hazards in modern society—dangerous chemicals and wastes, nuclear power, genetically engineered organisms—the public must be engaged in the policymaking process.

One useful tool for fostering diversity, inclusion, and justice is critical stakeholder analysis, a technique for identifying actors connected to a particular project or energy system. Critical stakeholder analysis can jumpstart dialogue and facilitate discussions among previously disconnected actors, making this process an important component of democratic decision making. It can also reveal power asymmetries among stakeholders. The process of identifying stakeholder interests can promote a common understanding of key agendas and help incentivize collaboration. By making the power relations of stakeholders more visible, critical stakeholder analysis can improve social responsibility and result in acceptable change.[60]

As such, we encourage active participation by all parties in energy discussions so that the energy technology preferences selected for integration into society better match interests and values. Moreover, we must all remain aware that decisions made today affect not only the lives of all who currently tread this planet but the lives of all who come after us. We have an obligation to balance our interests with theirs.

Practice Self-reflection

Our fifth maxim encourages enhanced self-reflection: we all must become more aware of our hidden values and ideological frames and the weaknesses of the assumptions sustaining them. By understanding why we embrace the energy perspectives we do, we can begin to understand how we prioritize issues and, accordingly, how this differs from the way that others prioritize things.

Part of this process involves realizing that even monkeys fall from trees. As proof, psychologist Philip Tetlock studied 284 "experts" who made their living commenting or offering advice on political, social, and economic trends.[61] At the end of the study, after these experts had made 82,361 forecasts, Tetlock found that the specialists were not significantly more reliable than nonspecialists in predicting events. Louis Menand adds that the experts surveyed by Tetlock performed worse than they would have if they had simply assigned an equal probability to the occurrence of different outcomes. "Human beings who spend their lives studying the state of the world," Menand concludes, "are poorer forecasters than dart-throwing monkeys, who would have distributed their picks evenly."[62]

Why are even smart people so prone to making these mistakes? Experts fall in love with their hunches, and they really hate to be wrong. Most people, including experts, tend to dismiss new information that does not fit with what they already believe. Experts use a double standard: they are tough in assessing the validity of information that undercuts their worldview, but lax in scrutinizing information that supports their worldview.

The problems that stem from blinkered or biased perspectives on an issue can be reduced considerably by nurturing a habit of skepticism about one's own knowledge. Sociologist Steve Woolgar refers to this as benign introspection.[63] Sociologist Michael Lynch adds that enhancing self-awareness can include training oneself to recognize the philosophical roots and historical context of one's views—or what we have called frames. By becoming more self-aware, we become more conscious of personal biases and learn to reflect critically on the wellsprings of our own personal values.[64]

Embrace Technological Agnosticism

To stop energy evangelists from establishing a cult of ill-informed followers, our final suggestion is to encourage technological agnosticism by focusing on energy services rather than energy systems. We tend to forget that energy provision is a means to an end, not an end in itself. Energy is useful only insofar as it performs tasks that serve human needs. We do not consume electricity or oil for the fun of it; rather, we consume it to provide thermal comfort, cooked food, hot water, television shows, recorded music, and a host of other services. We don't absolutely need to drill, mine, leech, extract, and deplete natural resources at breakneck speed to achieve this, but we do need a way to provide humanity with the service of energizing our lifestyles.

Such a statement, while obvious to many energy analysts, has somewhat profound implications.[65] Practicing technological agnosticism reorients the direction of energy policy interventions. Proper policy no longer centers on securing barrels of oil or tons of coal as an end in itself, but focuses on optimizing human mobility and comfort. This might include the promotion of walking, cycling, and running paths to enhance mobility rather than focusing only on refineries or roads for cars. Technological agnosticism is centered in the notion that many technologies can provide the same energy service. Indeed, at the current scale of energy demand, there is no single technology that can satisfy all energy needs without creating other problems; instead, a portfolio of options is the only viable approach. As Oxford climate change policy researchers Prins and Rayner suggest, agnosticism implies that rather than a silver bullet, the solution lies in silver buckshot.[66]

Engines and Mirrors, Values and Truth, Fact and Fiction

Overall, this book has attempted to educate and inform readers by pulling back the curtain to reveal some of the precepts underlying contention in energy policy. Asymmetric knowledge, vested interests, and competing ideologies produce fertile ground for deep-rooted disagreement. In the energy world, the only truth is that there is no such thing as a single, overarching perspective when it comes to our 15 questions. Nonetheless, we offer readers some parting thoughts.

Energy systems can be thought of as both engines and mirrors of society. They support goods and services that propel economic development. As economist E. F. Schumacher put it, energy "is not just another commodity, but the precondition of all commodities, a basic factor equal with air, water, and earth."[67] Yet energy systems also reflect human values and provide insight into the leanings of power. This means that purely technical solutions to our energy problems will continue to be contested as long as technologies create winners and losers and the values and ideologies underpinning these technologies remain hidden. Controversy and dissent are basic elements of energy policy.

Given that key questions about energy have less to do with facts and more to do with assumptions and frames, more information or better data will seldom fully resolve conflict. Most people will dismiss data that challenge their worldviews, even if the information is reliable. Provision of information is not always an efficacious mechanism for altering frames; one merely needs to look to politics across the world, where attempts to challenge ideologies can very well do the opposite and further strengthen incumbents. This is particularly true when there are entrenched investments—a prevalent characteristic of the energy sector. Consequently, we can expect energy to remain a contentious topic even after our 15 questions have faded from relevance. So, rather than attempting to marshal "facts" whenever one encounters seemingly illogical support for a particular technology, a better strategy is to endeavor to define vested interests and ideologies and to ask what is at stake and who benefits. This will not yield a solution, but it will explain the seemingly irrational.

As we have seen, energy decisions are not determined by objective fact but by contextual truth, supplemented by a dose of invented, soothing fiction. In many ways, understanding why some ineffective energy technologies move forward while other promising technologies fail to make it to the premier leagues requires advanced detective skills. We need to follow the money, look for clues that reveal entrenched ideologies, and try to strip away the fiction so we can better understand how context shapes truth.

Rather than rejecting opposing viewpoints as ravings from the insane, we should view contention as a pool of clues that can help us better understand competing motives. Rather than treating knowledge as static, we should view it as perpetually evolving as resources are exhausted, prices change, values alter, and technologies mature. Most of all, we should sharpen our ability to hold two competing sides or theses in our head and not only continue to function but manage to synthesize these sides into a higher, more progressive common ground. If we do this, we might be able to shed light on the seemingly paradoxical positions that others take on important energy issues like those examined in this book. Then, as detectives, when we are presented with a specific energy reality, we might be able to understand how we got here, and the means and motives that will drive change in the future.

NOTES

1. National Research Council (NRC), *Energy in Transition 1985–2010: Final Report of the Committee on Nuclear and Alternative Energy Systems* (Washington, DC: National Academies Press, 1982)

2. Handler quoted in ibid., vi.

3. NRC, *Energy in Transition*, viii.

4. Harvey Brooks, "Energy: A Summary of the CONAES Report," *Bulletin of the Atomic Scientists*, February 1980, 23.

5. US Energy Information Administration (EIA), *Annual Energy Outlook 2014*, DOE/EIA-0383 (Washington, DC: US Department of Energy, 2014).

6. Bradford L. Barham and Oliver T. Coomes, "Sunk Costs, Resource Extractive Industries, and Development Outcomes," *Nature, Raw Materials, and Political Economy* 10 (2005): 159–186.

7. Paul McLeary, "Estimating Sunk Investment in U.S. Energy Infrastructure," *Energy Business Blog*, May 10, 2004.

8. B. Kohler, "Sustainability and Just Transition in the Energy Industries," in *Sparking a Worldwide Energy Revolution: Social Struggles in the Transition to a Post-Petrol World*, ed. K. Abramsky (Oakland, CA: AK Press, 2010), 569–576.

9. Marilyn A. Brown and Gyungwon Kim, "Energy and Manufacturing: Technology and Policy Transformations and Challenges," in *Handbook of Manufacturing in the World Economy*, ed. Jennifer Clark, John R. Bryson, and Vida Vanchan (Cheltenham, UK: Edward Elgar, 2015), 121–146.

10. B. K. Sovacool, *Dirty Energy Dilemma* (Westport, CT: Praeger, 2008).

11. J. B. Ruhl, "The Political Economy of Climate Change Winners," *Minnesota Law Review* 97 (2012): 208.

12. Cass R. Sunstein, *Laws of Fear: Beyond the Precautionary Principle* (New York: Cambridge University Press, 2005).

13. EIA, *Annual Energy Outlook 2006* (Washington, DC: US Department of Energy, 2006).

14. Elinor Ostrom et al., "Revisiting the Commons: Local Lessons, Global Challenges," *Science* 284 (April 9, 1999): 278–282; Thomas Dietz, Elinor Ostrom, and Paul Stern, "The Struggle to Govern the Commons," *Science* 302, no. 5652 (2003): 1907–1912.

15. Chanute quoted in Laura Nader, *The Energy Reader* (New York: Wiley Blackwell, 2010), 549.

16. Woodward quoted in ibid., 10.

17. Watson quotation from Things People Said, "Bad Predictions," available at www.rinkworks.com/said/predictions.shtml.

18. Dekka quotation from ibid.

19. Lawrence Rocks and Richard Runyon, *The Energy Crisis: The Imminent Crisis of Our Oil, Gas, Coal, and Atomic Energy Resources and Solutions to Resolve It* (New York: Crown Publishers, 1972), 141.

20. Thatcher quotation from Things People Said, "Bad Predictions."

21. Scott V. Valentine, *Wind Power and Politics* (Oxford: Oxford University Press, 2014).

22. A. Klinke and R. Ortwin, "Systemic Risks as Challenge for Policy Making in Risk Governance," *Forum: Qualitative Social Research* 7, no. 1 (January 2006): art. 33, p. 3. See also Roman Sidortsov, "Reinventing Rules for Environmental Risk Governance in the Energy Sector," *Energy Research and Social Science* 1 (March 2014): 171–182.

23. Theodore E. Downing, *Avoiding New Poverty: Mining-Induced Displacement and Resettlement* (London: International Institute for Environment and Development, April 2002).

24. John Farrington and Conor Farrington, "Rural Accessibility, Social Inclusion and Social Justice: Towards Conceptualization," *Journal of Transport Geography* 13 (2005): 11.

25. Claire Haggett, "Public Engagement in Planning for Renewable Energy," in *Planning for Climate Change: Strategies for Mitigation and Adaptation for Spatial Planners*, ed. Simin Davoudi, Jenny Crawford, and Abid Mehmood (London: Earthscan, 2009), 300.

26. Susan Cozzens, "Distributive Justice in Science and Technology Policy," *Science and Public Policy* 34, no. 2 (March 2007): 85–94.

27. Dale Jamieson, "Ethics, Public Policy, and Global Warming," in *Climate Ethics: Essential Readings*, ed. Stephen M. Gardiner et al. (Oxford: Oxford University Press, 2010), 80.

28. Paul A. Sabatier, "An Advocacy Coalition Framework of Policy Change and the Role of Policy-Oriented Learning Therein," *Policy Sciences* 21, no. 2–3 (1988): 129–168.

29. James Hansen, *Storms of My Grandchildren* (London: Bloomsbury Press, 2009).

30. E. O. Wilson, *The Future of Life* (New York: Random House, 2002), 144.

31. Ibid., 146.

32. Aldo Leopold, *Sand County Almanac* (Oxford: Oxford University Press, 1947).

33. Mads Borup et al., "The Sociology of Expectations in Science and Technology," *Technology Analysis and Strategic Management* 18, no. 3–4 (2006): 285–298.

34. Sheila Jasanoff and Sang-Hyun Kim, "Containing the Atom: Sociotechnical Imaginaries and Nuclear Power in the United States and South Korea," *Minerva* 47, no. 2 (2009): 119–146.

35. B. K. Sovacool, *Contesting the Future of Nuclear Power* (Singapore: World Scientific Publishing, 2011).

36. F. Diaz-Maurin and Z. Kovacic, "The Unresolved Controversy over Nuclear Power: A New Approach from Complexity Theory," *Global Environmental Change* 31 (2015): 207–216.

37. Harro Van Lente, "Navigating Foresight in a Sea of Expectations: Lessons from the Sociology of Expectations," *Technology Analysis and Strategic Management* 24, no. 8 (2012): 769–782.

38. John Byrne and Steven M. Hoffman, "The Ideology of Progress and the Globalization of Nuclear Power," in *Governing the Atom: The Politics of Risk*, ed. John Byrne and Steven M. Hoffman (New Brunswick, NJ: Transaction Publishers, 1996), 43.

39. Chris Martin, "Tea Party's Green Faction Fights for Solar in Red States," *Bloomberg Business*, November 12, 2013, available at www.bloomberg.com/news/2013-11-12/tea-party-s-green-faction-fights-for-solar-in-red-states.html.

40. B. K. Sovacool and B. Brossmann, "The Rhetorical Fantasy of Energy Transitions: Implications for Energy Policy and Analysis," *Technology Analysis and Strategic Management* 26, no. 7 (September 2014): 837–854.

41. Thorstein Veblen, *The Instinct of Workmanship and the State of the Industrial Arts* (New York: Macmillan, 1914).

42. B. K. Sovacool and M. V. Ramana, "Back to the Future: Small Modular Reactors, Nuclear Fantasies, and Symbolic Convergence," *Science, Technology, and Human Values* 40, no. 1 (January 2015): 96–125.

43. Kenneth Burke, *Permanence and Change: An Anatomy of Purpose* (Berkeley: University of California Press, 1984), 7.

44. Amory Lovins, "Soft Energy Technologies," *Annual Review of Energy* 3 (1978): 507.

45. S. V. Valentine, *Wind Power Politics and Policy* (Oxford: Oxford University Press, 2014).

46. B. K. Sovacool and H. Saunders, "Competing Policy Packages and the Complexity of Energy Security," *Energy* 67 (April 2014): 641–651.

47. B. K. Sovacool and C. J. Cooper, *The Governance of Energy Megaprojects: Politics, Hubris, and Energy Security* (Cheltenham, UK: Edward Elgar, 2013).

48. Sovacool, *Dirty Energy*, 213.

49. Amanda R. Carrico et al., "Costly Myths: An Analysis of Idling Beliefs and Behavior in Personal Motor Vehicles," *Energy Policy* 37, no. 8 (2009): 2881–2888.

50. Renee J. Bator and Robert B. Cialdini, "The Application of Persuasion Theory to the Development of Effective Proenvironmental Public Service Announcements," *Journal of Social Issues* 56, no. 3 (2000): 527–541.

51. Michael P. Vandenbergh, "From Smokestack to SUV: The Individual as Regulated Entity in the New Era of Environmental Law," *Vanderbilt Law Review* 57 (2004): 515–610.

52. Ezra M. Markowitz and Azim F. Shariff, "Climate Change and Moral Judgment," *Nature Climate Change* 2 (March 2012): 243–247.

53. Per Espen Stoknes, "Rethinking Climate Communications and the 'Psychological Climate Paradox,'" *Energy Research and Social Science* 1 (March 2014): 161–170.

54. David W. Orr, *Earth in Mind: On Education, Environment, and the Human Prospect* (Washington, DC: Island Press, 2004).

55. Henry J. Perkinson, *The Imperfect Panacea: American Faith in Education* (New York: New York University Press, 1995).

56. M. Diesendorf (paper presented at Barriers to Energy Transition Conference, City University of Hong Kong, Hong Kong, May 18, 2014).

57. Harvey Brooks, "The Resolution of Technically Intensive Public Policy Disputes," *Science, Technology, and Human Values* 9, no. 1 (Winter 1984): 39–50.

58. Ibid., 48.

59. Phillip J. Frankenfeld, "Technological Citizenship: A Normative Framework for Risk Studies," *Science, Technology, and Human Values* 17, no. 4 (1992): 459–484.

60. P. De Leon and D. M. Varda, "Toward a Theory of Collaborative Policy Networks: Identifying Structural Tendencies," *Policy Studies Journal* 37, no. 1 (2009): 59–74; J. S. Dryzek and A. Tucker, "Deliberative Innovation to Different Effect: Consensus Conferences in Denmark, France and the United States," *Public Administration Review* 68, no. 5 (2008): 864–876.

61. Philip Tetlock, *Expert Political Judgment: How Good Is It? Can We Know?* (Princeton: Princeton University Press, 2005).

62. L. Menand, "Everybody's an Expert," *New Yorker*, December 5, 2005, 33.

63. Steve Woolgar, "Reflexivity Is the Ethnographer of the Text," in *Knowledge and Reflexivity: New Frontiers in the Sociology of Scientific Knowledge*, ed. Steve Woolgar (London: Sage, 1988), 1–13.

64. Michael Lynch, "Against Reflexivity as an Academic Virtue and Source of Privileged Knowledge," *Theory, Culture, and Society* 17, no. 3 (2000): 363–375.

65. B. K. Sovacool, "Security of Energy Services and Uses within Urban Households," *Current Opinion in Environmental Sustainability* 3, no. 4 (September 2011): 218–224. See also B. K. Sovacool, "Conceptualizing Urban Household Energy Use: Climbing the 'Energy Services Ladder,'" *Energy Policy* 39, no. 3 (March 2011): 1659–1668.

66. G. Prins and S. Rayner, "The Wrong Trousers: Radically Rethinking Climate Policy" (Joint Discussion Paper of the James Martin Institute for Science and Civilization, University of Oxford, and the MacKinder Centre for the Study of Long-Wave Events, London School of Economics, 2007).

67. E. F. Schumacher, *Schumacher on Energy: Speeches and Writings of E. F. Schumacher*, ed. G. Kirk (London: Cape, 1977), 1–2.

Page numbers in *italics* refer to figures and tables.

ABOUT THE AUTHORS

BENJAMIN K. SOVACOOL is director of the Center for Energy Technologies and a professor of business and social sciences in the Department of Business and Technology Development at Aarhus University in Denmark. He is also a professor of energy policy at the Science Policy Research Unit (SPRU), School of Business, Management, and Economics, University of Sussex, United Kingdom.

MARILYN A. BROWN is an endowed professor in the School of Public Policy at the Georgia Institute of Technology, where she created and leads the Climate and Energy Policy Laboratory. She is a presidential appointee to the Board of Directors of the Tennessee Valley Authority and a co-recipient of the 2007 Nobel Peace Prize for co-authorship of the report on Mitigation of Climate Change.

SCOTT V. VALENTINE is an associate professor in both the School of Energy and Environment and the Department of Public Policy at the City University of Hong Kong. He was formerly an associate professor and founding associate director of the International Masters of Public Policy Program at the Graduate School of Public Policy, University of Tokyo.

Printed in Great Britain
by Amazon